*Engineering War and Peace in
Modern Japan, 1868–1964*

JOHNS HOPKINS STUDIES IN THE HISTORY OF TECHNOLOGY

Merritt Roe Smith, *Series Editor*

Engineering War and Peace in Modern Japan, 1868–1964

TAKASHI NISHIYAMA

Johns Hopkins University Press

Baltimore

Johns Hopkins University Press
2715 North Charles Street
Baltimore, Maryland 21218-4363
www.press.jhu.edu

Library of Congress Cataloging-in-Publication Data
Nishiyama, Takashi, 1969–
Engineering war and peace in modern Japan, 1868–1964 /
Takashi Nishiyama.
       pages   cm
Includes bibliographical references and index.
ISBN-13: 978-1-4214-1266-5 (hardcover : alk. paper)
ISBN-10: 1-4214-1266-7 (hardcover : alk. paper)
ISBN-13: 978-1-4214-1267-2 (electronic)
ISBN-10: 1-4214-1267-5 (electronic)
1. Military engineering—Japan—History—19th century. 2. Military engineering—Japan—
History—20th century. 3. World War, 1939–1945—Engineering and construction. 4. World
War, 1939–1945—Japan. 5. Railroad engineering—Japan—History—20th century.
6. Railroads—Design and construction—Japan—History—20th century. I. Title.
UG105.N57 2014
358´.22095209041—dc23   2013022923

A catalog record for this book is available from the British Library.

*Special discounts are available for bulk purchases of this book. For more information,
please contact Special Sales at 410-516-6936 or specialsales@press.jhu.edu.*

To my parents, Kiyoshi and Keiko Nishiyama

Beat your plowshares into swords, and your pruninghooks into spears: let the weak say, I am strong.

—*Joel 3:10*

And he shall judge among the nations, and shall rebuke many people: and they shall beat their swords into plowshares, and their spears into pruninghooks: nation shall not lift up sword against nation, neither shall they learn war any more.

—*Isaiah 2:4*

CONTENTS

This book is the culmination of an intellectual odyssey that lasted over a decade. It had its genesis in my many conversations with my mentor and colleague in the United States, James Bartholomew. He has shaped my lifelong exploration in the field of history of technology in Japan. He read and commented on all parts of my project as it took shape, showing himself to be patient, kind, and generous at all times. He was also always critical without being harsh in his meticulous feedback, which I cherished at every stage. Of all of the chapters, chapter 1 in particular shows his gentle and detailed guidance in its approach, framework, and focus—those familiar with his scholarly work may well recognize him in these pages. All that I owe to him is beyond words.

My utmost appreciation also goes to my unfailing mentor in Japan, Hashimoto Takehiko. He kindly hosted my two-year stay at the Research Center for the Advanced Science and Technology, Tokyo University, where he and his students at the time opened my eyes and made me rethink about the history and historiography of technology in Japan. Without his generosity and mentorship, little of my research and writing in Japan would have been possible; without his unreserved support even after my stay at Tokyo University, my transformation into a historian would have been inconceivable. For more than a decade, both Jim and Hashimoto-sensei have tirelessly supported my efforts with their time, understanding, and advice, as well as with much encouragement. They steered my research morally, academically, and internationally across geographical borders—all of which I am eternally grateful for. Both are my sensei in scholarship and in life.

Other colleagues of mine kindly gave their time to read earlier versions of these chapters and to offer valuable feedback along the way. Philip Brown commented on previous drafts of various chapters. His suggestions were always concrete, practical, and constructive, which was a blessing in the completion of this book and in other projects. Like Jim and Hashimoto-sensei, Phil coached me in many aspects of my professional preparation. John F. Guilmartin fostered, defined, and refined my interest and thinking in the field of history of military technology. A former pilot in the United States Air Force and a veteran with impressive accolades, Joe always cheered me up with his experience, wit, and advice. Chapters 2 and 3, which are about military aviation, resulted in part from my countless conversations with him. Mizuno Hiromi and especially Walter Grunden offered their thoughtful and helpful feedback on parts of my drafts. Daqing Yang generously shared his expertise with me, which shows in chapter 5. I would also like to thank Robert Kargon and Laurence Principe for kindly inviting me to speak at Johns Hopkins University, where I received useful suggestions, particularly from Stuart Leslie, which I incorporated into chapters 4, 6, and 7. I am also grateful to the following colleagues who inspired me with their experiences, research, and insights: Emily Anderson, Yōko Brown, Hyungsub Choi, Maureen Donovan, Shigehisa Kuriyama, Christopher Reed, Franziska Seraphim, Otsubo Sumiko, Michiko Takeuchi, and Sharon Traweek. At the State University of New York, Brockport, I was privileged to have wonderful colleagues especially in the world-history caucus. Alison Parker offered wholehearted support for my leave of absence, and James Spiller provided important suggestions on scholarship and teaching.

My postdoctoral research on the MIT campus was indispensable in my intellectual journey as I sought the meanings of technology and culture, and of war and peace. Rosalind Williams took me under her wing before, during, and after my stay at the Dibner Institute for the History of Science and Technology, opening the door for me to teach in the Science, Technology, and Society Program, from which I benefited immensely. I owe a debt of gratitude to David Kaiser and Merritt Roe Smith for sharing their thoughts on war, science, and technology during casual conversations. Richard Samuels helped me reformulate my research question about Japan and the time period after 1868, which he explored in his well-known monograph, *Rich Nation, Strong Army*. I would like to express my heartfelt thanks to John Dower who allowed me to audit his history courses on cultures of war. Every moment of my conversations with him in and outside of his office was stimulating in many ways. In good faith, he prompted me to question many assumptions I had wrongly held for quite some time, raising and then refining new questions that eventually formed the cores of chapters 2, 3, and 4, and the conclusion.

I obtained generous financial support from the Dibner family who hosted me and others like me at the Dibner Institute, where much of my research and (re)writing took shape beyond my doctoral dissertation. Director George Smith, a former engineer and a philosopher of science, pushed me to rethink the engineering profession, as well as the mechanics and meanings of physical phenomena that are important in my book, such as flutter vibrations. I spent especially precious time with the following colleagues who offered useful advice and moral support: Thomas Archibald, Bruno Belhoste, David Cahan, Claire Calcagno, Karine Chemla, Deborah Cramer, Dane Daniel, Gerald Fitzgerald, Olival Freire, Kristine Harper, Matthew Harpster, Arne Hessenbruch, Ben Marsden, David Pantalony, Peter Shulman, Seth Shulman, Conevery Valencius, Sara Wermiel, and Chen-Pang Yeang. My office neighbors—Karine, Dane, Chen-Pang, and Kristine—were a source of daily inspiration. I also enjoyed friendship and casual yet stimulating conversations with Wen Hua-Kuo, Chihyung Jeon, and Wen-Ching Sung. This book builds on insights I gained from interactions with each one of these invaluable colleagues. The cozy, ideal learning environment at the Dibner Institute was made possible through the caring assistance of the staff—Trudy Contoff, Rita Dempsey, and Bonnie Edwards—who often went beyond the call of duty.

My research and (re)writing were also generously sponsored by a National Science Foundation Doctoral Dissertation Fellowship (2003), an Aviation/Space Writers Award from the National Air & Space Museum (2004), a Twentieth-Century Japan Research Award from the University of Maryland (2006), and the Dr. Nuala McGann Drescher Affirmative Action Program (2010). I owe special gratitude to the D. Kim Foundation for its generosity and noble, unfaltering commitment to the advancement of the field of history science and technology in Asia. Its director Dong-Won Kim gave me much-needed encouragement and wisdom along the way, for which I am grateful.

This book represents my humble attempt and ongoing struggle to contribute transnationally to various academic communities in both the United States and Japan. My colleagues at Tokyo University and Tokyo Institute of Technology honed my thinking continuously throughout the project. The Hashimoto-ken Colloquium at the Research Center for Advanced Science and Technology was a chief source of novel ideas and contagious joy; for this project, it was a locus of stimulating discussion, constructive criticism, and intellectual challenges. Namely, I would like to thank Ito Kenji, Nakamura Masaki, Boumsoung Kim, Nakazawa Satoshi, Dan Plafcan, and Satō Yasushi. We often disagreed passionately in the classroom and then continued our discussion over dinner—and I miss that fruitful environment. Matsumoto Miwao offered feedback that reshaped some of my research questions about technology transfer. Nakayama Shigeru's encouragement made my day. At the Tokyo Institute of Technology, Yamazaki Masakatsu and Kaji Masanori in particular showered me with very useful suggestions on my research approach and on archival materials in Japan. I would also like to thank Mizusawa Hikari, Okada Taishi, Kihara Hidetoshi, Hinokawa Shizue, Sato Ken'ichi, Kato Shigeo, Setoguchi Akihisa, Tsukahara Togo, and Furukawa Yasu for introducing me to a network of historians of science, technology, and medicine in Japan. I was fortunate to befriend them and receive their support in numerous ways.

This book could not have been completed without various archival and non-archival materials written in six different languages. Famous state-sponsored projects—namely, the Shinkansen and, above all, wartime research by the Japanese military—proved more difficult subjects to study than originally anticipated. Many wartime documents in Japan were either confiscated by the Allies or destroyed before their arrival in the summer of 1945, or, worse yet, remained in the hands of military historians and amateur collectors alike across the country. The most useful sources for this study were private collections I accessed by joining or establishing personal networks with individuals working on military technology of wartime and railway technology of postwar Japan. I am greatly indebted to the following individuals who shared with me crucial primary sources that formed the basis of my research: Kawamura Yutaka, Koyama Tōru, Matsuura Yoshinari, and Tanasawa Naoko. In addition, Satō-san (Yasushi) and Kaji-san, along with Sven Saaler and Matthias Dorries, kindly provided me with other sources in Japanese, German, and French, respectively, that proved vital to my research.

A few chapters of this book originate in my previous journal article publications, two in English and two in Japanese: *Technology and Culture* (2007), *Comparative Technology Transfer and Society* (2003), *Kagaku, Gijutsu, Shakai* (Japanese Journal for Science, Technology and Society, 2004), and *Kagakushi Kenkyū* (Journal of History of Science, 2011). I would like to thank the editors of these journals, especially Bruce Seely and John Stedmeier, for their advice which sharpened my arguments. I am also grateful to the anonymous reviewers of my journal publications and of this book.

Last but not least, my most personal thanks are reserved for my family: my wife Huili Lin and my son Seiji for their loving support. I dedicate this book to my parents who, from Japan, demonstrated their unfailing faith in me all along: my mother Keiko Nishiyama, and a former aeronautical engineer and my father, Kiyoshi Nishiyama, who first introduced me to the world of aviation.

I have followed the Japanese convention in which the family name precedes the given name, except in the bibliography. I have used the modified Hepburn system of romanization. Well-known names such as Tokyo and Osaka are not italicized and thus lack macrons. Names in Chinese follow the pinyin system.

Engineering War and Peace in
Modern Japan, 1868–1964

Technology and Culture, War and Peace

On March 28, 2000, Japan's sole public broadcaster, NHK (Nippon Hōsō Kyōkai), introduced the enormously popular documentary program *Project X: Challengers* and televised it for the next five years. Most of the 191 episodes were "success stories" of multiple guests, mostly males. The program lionized the ordinary citizens who had contributed to Japan's double-digit economic growth in the 1960s (the decade of an "economic miracle") in contrast to the 1990s (its sluggish economy signaling an "economic debacle"). In the episode that aired on May 9, 2000, three former military engineers in their eighties recalled their wartime and postwar years. The one-hour program dramatized their roles and nationalism in the development of the Shinkansen, Japan's high-speed rail service that has remained technologically successful since its commercial introduction in 1964. By this account, the skills and values that the engineers had developed during World War II seemed perfectly suited for the postwar national rail project. Technological progress in this view had been steady and linear—almost preordained. This episode, among others, sold well in book format and was eventually translated into English, Spanish, Russian, and Arabic.[1]

The popularity of *Project X* encapsulated important issues surrounding technology, culture, war, and peace. First, it showed how a nation could portray its engineers as "victors" after defeat. Engineers became the means for the nation to explain how military failure nonetheless could give rise to postwar success. Victors are said to write history, and in this case, ordinary citizens of the defeated nation rewrote history, presenting themselves as "victors" in achieving technological triumph after war. The program also posed vexing questions about the relative credit due to war and peace in the narrative of Japan's twentieth-century technological transformation.

Clearly, two world wars and then the Cold War carried vast implications for technological development around the world. From 1868 to 1945, Japanese technological progress depended considerably on almost incessant armed conflicts outside its borders. The Sino-Japanese War (1894–1895), the Russo-Japanese War (1904–1905), World War I (1914–1918), and World War II in Asia and the Pacific (1937–1945) prodded Japan to strengthen its technological muscles. Each war elevated the importance of military science and technology.

In the process, the nationalistic slogan "Rich Nation, Strong Army" proved sound and persuasive. Technological transformation in preparing for war, militarism, and industrialism, each reinforced the other under the imperative of national security. In addition to the millions of ordinary citizens on the home front, the newly formed modern armed forces and their supporting engineers at home all focused their efforts on countering the rise of Western imperial power in East Asia. The combination of militant ideology, trained conscripts, and innovative technology proved highly effective in building imperialism in Japan (and elsewhere). The Ministry of Education and armed forces played key roles in designing an engineering workforce for war. By the 1930s, three key variables— (1) engineering training in institutions of higher education, (2) civilian and military research and development facilities, and (3) civilian firms in war-related industries —worked in tandem. Technology was for war, and, reciprocally, war favored the technologically adept. On December 7, 1941, with strong optimism, or a fatally unrealistic, fascism-tainted vision, Japanese leadership steered the country into war with the United States. At the time Japan, with a well-educated engineering workforce, stood as the most industrialized and wealthiest nation in Asia.

Unconditional surrender followed. It was the first time for the country to face defeat. American strategic bombing demolished Japanese wartime industry, reducing urban centers to rubble and ashes. The cities of Hiroshima and Nagasaki may partially have survived atomic attack, but the bombing left indelible scars on the psyches of the citizens. Tokyo, the nation's capital and the hub of railway networks, had come under the Allies' assault 122 times during the war, and on August 15, 1945, when the fighting ended, parts of the city seemed beyond repair. The war devastated the spirit of millions of ordinary Japanese people.

Yet in just two years during the Allied occupation (1945–52), the new Ministry of Education heralded defeat as paradoxically constructive. In a short, illustrated textbook, *The Story of the New Constitution* (1947), which it issued to benefit seventh graders across the country, the ministry described the meanings, roles, and responsibilities of "a New Japan." Page 18 carried a representation of an ideal, peace-oriented society. In the graphic, the phrase "Renouncing War" runs across the center of a black pot, into which an unseen hand drops wartime products—military aircraft, bombs, and army tanks. The smoke of damage and defeat billows from the magical black pot. From the bottom flow the shining symbols of modern, peacetime technology—a commercial ship, a commuter train, a truck, a tower, and a building. By 1952, the vanquished nation turned defeat into revival and then, by the end of the 1960s, staged a decades-long journey of remarkable reconstruction that continues to shape the lives of the Japanese in the twenty-first century.

War and the means of war link inextricably to technological development. The case of the United States in the twentieth century, during which a succession of foreign conflicts fueled scientific and technological advances, such as aircraft, infantry weapons, and submarine and antisubmarine warfare, vividly illustrates the point. The Manhattan Project (besides owing to the war effort) makes clear the reverse—how scientific and technological breakthroughs can immediately influence military decisions. Overall, however, the impact of war on technology seems more elusive and difficult to explicate than the impact of technology on war. The lasting influence of war on tangible, material forms of technology tend to be more observable than war's effects on extrinsic, amorphous concepts attached to objects, such as ideas and values.[2] War has far-reaching effects. But what about the end of, or absence of, war as in lack of direct military engagement? A continuing focus on the history of technology in victor nations of the West—the United States and France, for example, in World War II—may obscure important factors in the technological success of a regime. Here, one encounters the case of Japan, a country that achieved great success in technology and science before and during war but, even more remarkably, after the experience of total war and unconditional surrender.

Academic inquiry into something unique to a nation requires caution. Historians have situated technological change in nationally comparative frameworks, focusing on their systems, practices, and values, calling attention to national styles of innovation. Usually, national differences in doing things matter more than the similarities therein. Cultural interpretations, which have long dominated popular perceptions in Japan and the West, ascribe Japanese methods to the country's unique historical legacy.[3] Was there anything quintessentially Japanese that drove the nation's science and technology at times of war and peace? Did Japanese scientists and engineers choose to do what they did *because* they were Japanese? Such questions have little value because they treat national culture as monolithic, static, and ahistorical; worse yet, they perpetuate stereotypes. The more meaningful issues are as follows: Was there anything distinct about the

From *The Story of the New Constitution*, issued in 1947 by the Ministry of Education.

technology and culture that arose—not on account of nationality, but because of experiences of total war and total defeat? What cultures of war and defeat were useful for (re)building the nation and technology? What elements of continuity and discontinuity can we observe in trans-war history of technology despite, and because of, war and defeat?

While downplaying Japan's cultural exceptionalism, the following study draws upon cases from the West and non-West to highlight the country's strategy for technology transfer and diffusion in wartime and peacetime—a strategy that suited distinctly to the geopolitics and geographical attributes of an archipelago nation in Asia that shares no land borders with other countries. This cultural analysis of technology in trans-war society shows that military defeat altered the technological landscape far more than had the country's earlier victories in war. From 1868 to 1964, Japan built modern research and development, embraced war and defeat, and militarized and demilitarized its technology and culture—and the process embodied some ironic, unintended consequences from the war that likely carried more transformative power than the intended ones. Japanese technology resulted from a value-laden, internally conflicting, and contingent process, which adapted constructively to the end of destruction abroad and at home. The malleable, accommodating, and innovative nonweapons technology that reflected the experience of defeat—as embodied in the Shinkansen high-speed rail service—can be viewed as technology of defeat.

Relying on documents previously unavailable to both Western and Japanese scholars, this study views Japan's scientists and engineers as a lens through which we can examine technology, culture, war, and peace. What follows is a close cultural study of the engineers' changing connections to their laboratories, research institutions, and local/regional surroundings, as well as economic and political worlds within and outside of Japan at times of war and peace. This book focuses on the state-sponsored engineering projects of Japan and highlights research and development efforts by (former) military engineers in such projects, because the state policies of militarization and demilitarization from 1868 to 1964 most directly affected their communities. They were the architects of the national projects. Before 1945, they built technology for war. Defeat dissolved (at least temporarily) the entire military structure, thereby derailing their careers in the new peace-oriented society. These engineers retained no viable ideology or vision of wartime. Yet for them wartime technology and defeat were sources of power whereby they actively and constructively molded Japan for peace.[4]

These engineers formed a sociological generation as much as a biological one; they played socioculturally prescribed roles that were particular to the time under study. My story will knit together two different, mutually reinforcing strands of history: the personal life histories of engineers and their collective, social history. In Japan as around the world, engineering communities were not divorced from the rest of the society and the international landscape. Personal growth and communal change are inseparable, remain relative to each other, and help define each other, especially through historical experiences of crisis.[5] How did total war and defeat shape engineering communities as well as their views toward technology and nation?

In their profession, engineers have sets of guiding values that support the goals of their institutions and communities. By any political, economic, or sociological measure, engineers provide a critically important workforce in industrial societies. Engineers solve mundane, technical problems. They often disagree rightly, rationally, continuously, and bitterly with intended and unintended consequences. Such tensions are often deeply rooted in engineering culture; it is particular to a laboratory and institution and is crystallized in tools, knowledge, and engineering communities.[6] Technological transformation embodies a series of cultural tensions among engineers, their research laboratories, and their institutions at times of war and peace. By highlighting these tensions, we can see how and why some engineering cultures prevailed over others and with what consequences in the national and international settings. A cultural analysis of engineering communities can reveal that a series of tensions were compatible with the unifying values that underlay Japan's creative adaptability to war and peace.

So, how did Japan build its technology and engineering communities for war before 1945? Military historians as well as historians of science and technology often have failed to weave military history more tightly into the fabric of social history.[7] Rarely can technology and the military thrive in a society that neglects the importance of higher education in engineering. We begin with an overview of the history of engineering education at institutions of higher learning because examining the engineers' education, research activities, and cultures may reveal how and why the Japanese communities gained their technological edge before 1941, lost it between 1941 and 1945, but ultimately regained it.

Designing Engineering Education for War, 1868–1942

Ōki Takatō was an idealist with determination. On February 12, 1873, five years after the Meiji Restoration, this first governor of Tokyo (1868–69) and Minister of Education (1871–73) submitted his proposal for modernizing Japan's education system. "The growth of the talents of a civilized people," he wrote, "is imperative for the wealth, power, security, and well-being of a nation." Thus, it is "necessary to build schools and establish educational methods" to achieve these goals. The education system that he had envisioned consisted of "seven or eight educational divisions" all formed on "the basis of population and land area"—and "each region shall have a university."[1] None of his ideas had been tested before, but institutional arrangements proved immensely important in shaping the contours of modern Japan. What he could not have foreseen at that time was the pace and direction of the education system buildup, especially in the field of engineering, amid a series of external wars that Japan engaged overseas.

Japan's national landscape changed rapidly from 1868 to 1942, from peace to war and then back again. Meanwhile, the nation pursued its ambition to create an empire abroad. After 1868, the newly formed Meiji regime initially established modern engineering programs first in the capital, Tokyo; engineering programs subsequently spread through imperial and private universities as well as technical schools across the country. The growth of modern engineering studies—especially aeronautics—marked four successive phases across the decades: (1) 1895–1897, (2) 1905–1911, (3) 1918–1924, and (4) 1938–1942. It was no coincidence that these four phases corresponded to the Sino-Japanese War (1894–1895), the Russo-Japanese War (1904–1905), World War I (1917–1919), and World War II in Asia and the Pacific (1937–1945). Each of these four international conflicts functioned as a catalyst, providing the central and often local governments of Japan with both the rationale and the financial resources to build more educational institutions across the country. The external wars confirmed the necessity of engineers to a greater degree. The case should not be overstated, and multiple contingent factors, notably local oppositions and financial constraints, delayed or often halted the process of expanding engineering education in various regions. Overall, however, the four armed conflicts in succession reduced the financial, legal, and political constraints on engineering education, thereby empowering the modern nation for war.

Building the Infrastructure for Engineering Education, 1860s–1890s

The national government played a crucial role in implementing forward-looking, costly engineering programs across the country, starting from the capital. After the Meiji Restoration, the city of Edo was transformed into Tokyo, the hub of new industrial Japan. The location, from where the Tokugawa shogunate had ruled from 1603 to 1868, gained new importance for the modern goals adopted during nation building. As the modern eastern capital, Tokyo stood in contrast to Kyoto, the traditional, conservative, longtime capital of an earlier time.[2]

The transformation of Tokyo from the 1860s did not entirely neglect the experiences of Edo. The Meiji government's approach to the modern urbanization of the city remained rooted in existing patterns of landownership rather than top-down public planning, private enterprise, or individual concerns.[3] The new regime, however, drove the large-scale technological transformation nationwide as its chief sponsor. Most of the funds for technological development before 1868 had come from private and domain sources, but the Meiji leadership altered this pattern, channeling the requisite capital gleaned from national taxation into engineering projects across the country. One of them was establishing institutions of higher education—most notably, imperial universities. Without the state government directing the process, the development of Japan's educational infrastructure—and, thus, modernization—would have been slow, rough, and difficult at best. By 1930, the functional meanings of modern capital became more science and technology oriented as Tokyo formed the center of engineering education and, later, research and development, for a massive military buildup.

The single-most important government agency in Japan's early industrial growth was the Ministry of Public Works. Established in Tokyo (1870), it managed a wide variety of engineering projects that involved mining, steel production, lighting, railroad construction, and telegraphic communication. Foreign knowledge and experiences were indigenized and infused into entirely new realms or realms formally dominated by Japanese craftsmen. For this purpose, the Meiji regime employed more than 3,000 foreigners from the West, mostly in the first 20 years. The foreign engineers were needed in the country, as it was a latecomer to industrialization, especially in the areas of transportation and communication. Inviting engineers from abroad was costly. A foreign worker's salary was about three to ten times as much as that of a high-ranking Japanese government official.[4]

The government projects to build a modern nation needed not only foreign engineers but also homegrown engineers and educational establishments. The Ministry of Education, created in 1871, took charge of nurturing future engineers at institutions of higher learning, such as the Imperial College of Engineering (Kōbudaigakkō). Around this time, Japan and other latecomers to industrialization espoused foreign models of institutionalized education in engineering, setting up so-called institutes of technology, roughly akin to the Massachusetts Institute of Technology in the United States.[5] Staffed by British scientists and engineers and established in 1873, Japan's first institution for engineering education was both innovative and inventive. The 26-year-old Scottish principal, Henry Dyer, adopted its original model in part from Scotland with some inspiration from a top government-funded, polytechnic establishment of its kind, Eidgenossische Polytechnische Schule of Switzerland. Residing in dormitories, a total of 211 graduates actively absorbed foreign knowledge and culture around the clock. All instruction was in English, and the students had at least one English meal every day. They chose an engineering field from the six options—civil engineering, mechanical engineering, communication, architecture, metallurgy, and mining—and gained a moderately theoretical understanding of the technology at that time. After six years of study, the graduates worked at the Ministry of Public Works for seven years.[6]

The creation of Tokyo University in 1877 further cemented the merger between the central government and engineering education. Often referred to by its condensed name, Tōdai, it represented the nation's unfaltering commitment to the creation of its best engineers. The first major task of the university was to combine different engineering traditions that had existed previously. The theory-oriented programs at former Tokugawa institutions and the empirically driven education under the guidance of the Ministry of Public Works were merged in 1886 into the College of Engineering and administratively placed under the Ministry of Education.[7] Thus, in Japan, engineering formed a vital part of the university from the beginning. This marked a contrast to Germany, where the field was excluded from the universities because its practicality was deemed incompatible with the German ideal of personal cultivation.[8] In 1886, the Imperial University Ordinance laid the foundation of Japan's university system and codified the purpose of the imperial university. The first article stated that the institution was to serve the interests of the state, thereby integrating higher education into nation building and the pursuit of nationalism. The decree prepared the path for Ministers of Education to interfere in the academic autonomy at the university.[9]

Engineering programs spread, in part, due to the social prestige attached to the Faculty of Engineering at Tokyo University. Rarely the case in Western Europe, the field of engineering was respected in Japan because of its strong connection with the Japanese government and its origins in technologically advanced Western countries. At the time of creation, 10 out of 11 professors had studied abroad and were from the former samurai class of the former Tokugawa regime. The formal position of engineering in the academic system was thus higher in Japan than in Europe.[10] Networks of Tōdai graduates cemented the linkage between the government and emerging industries. From 1888 to 1897, graduates of the Departments of Civil and Mechanical Engineering worked for the Ministry of Home Affairs as well as for government and commercial agencies in railroad construction, which depended heavily on engineers from Germany, the United States, and above all, Great Britain. Graduates from the Department of Electrical Engineering worked at electric light companies and the Ministry of Communications. Those from the Department of Applied Chemistry moved on to the Ministry of Agriculture as well as to gas and cement companies.[11] Commonly referred to as "the institution of the highest learning," by an imperial decree, Tokyo University became the first institution of higher learning to earn the authoritative title, "imperial." The prestige of the university was codified and bestowed. The Tokyo University Faculty of Engineering stood at the pinnacle of a pyramidal, hierarchical system of engineering education—a development that signified the social legacy of Tokugawa Japan; by one account, as late as 1890, 86 percent of the engineering graduates were from the former samurai class.[12]

The university's ties with the central government and with industry provided a vital financial source for education and research. Some alumni helped to channel a large sum of money from the Ministry of Education to the Faculty of Engineering. Other alumni held important positions in private industries. By their generous endowments, the Faculty of Engineering was the wealthiest body of the university from 1878 through 1945. The institution as a whole was accorded roughly 11 million yen, and the Faculty of Engineering received the largest share of 31 percent, or 3.4 million yen. This amount was slightly smaller than the combined endowments to the Faculties of Medicine (11 percent), Science (7 percent), and Agriculture (17 percent).[13]

Graduates from Tokyo University strengthened many institutional connections that involved the military. The Department of Naval Engineering, for instance, produced phalanxes of competent engineers for the shipping industry, government agencies, and the navy. Among the 104 graduates from 1883 to 1903, 10 graduates moved on to the maritime transportation business, 14 to advanced engineering education, 20 to the Ministry of Communications, 26 to the ship-yard, and 34 (the largest group) to the navy.[14] At the strong request of the army and navy, in 1887, Tokyo University added special departments of arms technol-ogy and explosives. This initiative—unprecedented in Western countries—was a means to secure engineers for military arsenals.[15] Graduates from the Faculty of Engineering formed a crucial part of the human nexus and became a means to transform the society via government-led industrialization.

Expanding the Infrastructure for Engineering Education, 1890s–1930s

From the 1890s to the 1920s, expansion in engineering education took shape through the years of three successive wars abroad. Japan's attempts to expand its influence in Asia resulted in the Sino-Japanese War (1894–1895), the Russo-Japanese War (1904–1905), and World War I—all of which were accompanied by the further buildup of engineering education. These wars not only justified the importance of engineering education for the nation, but also linked education and the state more tightly than before. The central, and often, local governments gained both interests and resources to build more institutions of higher educa-tion for engineering beyond Tokyo, from Sapporo of Hokkaidō in the north to Fukuoka of Kyūshū in the south. To a degree, the Faculty of Engineering at Tokyo University served as a model for the expansion across the nation. Its curriculum, originally consisting of five fields of study—civil engineering, mechanical en-gineering, electrical engineering, mining/metallurgy, and applied chemistry—formed the basis of engineering education at all the other imperial universities that were subsequently established. The armed conflicts reduced the financial, legal, and political constraints on expenditures, thus, enabling the country to ex-pand engineering education in distinct stages.

The first phase of expansion (1895–1897) took shape from the time of the Sino-Japanese War (1894–1895), as exemplified in the creation of Kyoto University, Japan's second imperial university. A tentative plan for a university in Kyoto had been conceived by Minister of Education Mori Arinori, who had helped establish Tokyo University. Historical rivalry between east and west Japan—the kantō and kansai regions—enhanced a sense of competition favorable to the growth of sci-ence and technology, but a lack of funds had rendered a university establishment in Kyoto unfeasible until the Sino-Japanese War changed the political dynamic. During the fighting, the Japanese homeland remained safely untouched by the theaters of war on the Korean peninsula and in China. In the aftermath of the war, reparations from China and the ensuing postwar economic boom brought in new capital. Seeing the welcome opportunity and the increasing need for more engineers and technicians, the project to create Kyoto University progressed. A committee for establishing the university was officially formed in 1895, submitting to the Ministry of Education a report advising (1) that the new university be created in Kyoto, (2) that the size of the new institution be about two-thirds of that of Tokyo University, (3) and that the new school consist of four faculties, each with 200,000 yen for setup and maintenance costs. Subsequently, Kyoto University was inaugurated in 1897 with faculties in science, engineering, and medicine; it expanded by later adding law. Until 1924, the imperial universities in Tokyo and Kyoto were the only ones in Japan to offer instruction in law, medicine, engineering, literature, science, agriculture, and economics.[16]

For the central government, imperial universities equipped with science and engineering laboratories were costly to set up and even costlier to maintain. Largely following a French model, the Japanese institutions adopted the chair system: a full professor led his academic unit, which typically consisted of an associate professor, a few lecturers/assistants, and several graduate students.[17] In the 1890s, funds were still lacking, and as a solution, basic science and engineering were soon combined into a single faculty at Kyoto University.[18] This initiative blurred the organizational boundary between science and engineering, minimizing the number of programs available in Kyoto. The newly created institution was, thus, smaller than Tokyo University. It had 7 departments and 21 chairs in science and engineering; thus, it was about half the size of Tokyo University, with 14 departments and 38 chairs.[19] In 1898, the educational curricula in engineering expanded slightly with the addition of electrical engineering, mining/metallurgy, and chemistry for manufacturing.[20] Overall, the founding of Kyoto University in 1897 set an important precedent for the subsequent expansion of engineering education at institutions of higher learning.

Concurrently, institutionalized technical knowledge became available at technical schools in two urban centers of the nation: Osaka and Tokyo. Systematic training began to replace an apprenticeship that had honed the skills of workers.[21] This type of establishment was originally set up to produce foremen on the factory floor, or technicians equipped with practical, hands-on skills to support the engineers with university degrees. In 1896, the city of Osaka hosted one such school. In the city historically known as a major commercial hub of the nation with maritime transportation, the Osaka Technical School offered dyeing and weaving as craft skills and, later, shipbuilding as fields of study. Tokyo Technical

School had its origin in 1881 and like its Osaka counterpart, officially earned the title of technical school in 1901. Its curricula reflected traditional craftsmanship and offered dyeing, weaving, and ceramics as fields of study. These two schools offered formal programs that reflected local tastes, tradition, and needs of local light industries rather than heavy industries; the latter remained more tied to imperial universities with programs such as in mining civil engineering.[22]

The second phase of expansion (1905–1911) in engineering education followed the Russo-Japanese War of 1904–1905. An important development in this context was the Faculty of Engineering at Kyūshū University, the third imperial university in the country. The planning for its creation started around 1898. Indemnity from the Sino-Japanese War underlay the construction of the Yawata Ironworks, which became operational in 1901 and a major production center of steel. This local and national initiative was launched, in part, to reduce Japan's heavy dependence on foreign iron and steel, some of which was imported from Western Europe until the Russo-Japanese War disrupted the flow. The war elevated the strategic importance of Kyūshū. It stimulated heavy industries such as shipbuilding, steelmaking, mining, and electric power needed to establish military, transportation, and industrial infrastructure.[23] More initiatives unfolded in parliament. In January 1901, the House of Representatives and the House of Peers both passed a bill to expand the nation's higher education. Thirty members from the parliament submitted an opinion paper for the creation of more high schools and two new imperial universities, one in Kyūshū, another in Tōhoku. "Our country is lagging way behind strong Western nations in terms of higher education," claimed Representative Fuji Kinsaku. He continued, "We need to establish more universities" as this was "the nation's urgent business and obligation."[24]

Seeing the ripe opportunity, the prefectural assembly decided in 1906 to contribute financially to create the institution for engineering education in Kyūshū. Industrialists such as the Furukawas kept the construction project financially afloat.[25] Created in 1911, the Faculty of Engineering at Kyūshū University used the model of Tokyo University, offering programs in civil engineering, mechanical engineering, electrical engineering, applied chemistry, mining, and metallurgy.[26] Around that time, the postwar economic boom had spread engineering education across the nation, as seen through the establishment of technical schools in the cities of Nagoya (1905), Kumamoto (1906), Sendai (1906), and Yonezawa (1910). Except for Yonezawa, the educational curricula in these institutions focused on civil engineering, mechanical engineering, and mining/metallurgy. The Russo-Japanese War helped promote the key programs of engineering education in remote regions.

The next and the largest wave of expansion (1918–1924) in engineering education was a product of World War I. Behind the development was an ominous change that attested to the decisive role of technology in executing military operations. Advanced science and engineering manifested themselves as such deadly weapons as tanks, airplanes, submarines, and biochemical gas. Nations at war planned, organized, and mobilized their home fronts for the mass production of goods. Industrial capacity, backed by ample natural and other material resources, proved vitally important for waging this new kind of war—a war of attrition in a crude form. Correspondingly, the war fueled the demand for steel production, increasing domestic production and the export of war goods. Even in Japan, which played a minor role in the war, engineers with baccalaureate degrees were in high demand in the heavy industries. In postwar society, the engineering profession was considered a promising career choice. Referred to as "hotshots," the engineers with university degrees were half-jokingly offered daughters in marriage, a custom that reflected the fathers' ambition and faith in the engineers' future prospects.[27]

The war empowered engineering education at three previously established imperial universities. Most notably, Tokyo University began to offer a greater variety of subjects to study than before. It newly gained a total of 54 chairs from 1919 to 1923; the Faculty of Engineering obtained the most, that is, 19, followed by the Faculty of Science (11), Faculty of Literature (10), Faculty of Agriculture (7), Faculty of Law (4), and Faculty of Economics (3).[28] A new credit-hour system was introduced in 1921 to add more flexibility to engineering education. This practice obscured the traditionally rigid distinction between requirements and electives, thereby allowing students to freely take courses of their own choice and to fulfill credit hours toward graduation.[29] Kyoto University underwent a similar transformation. It divided the combined Faculty of Science and Engineering into two, and the Faculty of Engineering expanded the number of chairs twofold. In 1919, it added new subjects of study, such as architecture, strength of materials, and industrial chemistry, to the preexisting programs that consisted of civil engineering, mechanical engineering, electrical engineering, mining, and metallurgy.[30] Similarly, the Faculty of Engineering at Kyūshū University became the second university in the country to establish a Department of Shipbuilding.

Concurrently, engineering education expanded to the north. Hokkaidō University began as an agricultural institution. It originally offered engineering programs in agriculture, geology, and surveying, but its early attempt to expand the curricula ended prematurely when the Faculty of Engineering was abolished in 1896.[31] World War I renewed the effort, however. An Imperial Ordinance on University (1919) channeled 212,600 yen from parliament to the university to recreate the dissolved faculty.[32] In 1924, the new curricula consisted of civil engineering, mining, mechanical engineering, and electrical engineering.[33] Tōhoku University followed suit despite some financial constraints. The institution originally began with faculties in science and agriculture. It lacked engineering programs because neither local industrialists nor the Miyagi prefectural government were clear about what they wanted in the university until around 1910.[34] World War I added a new vision to the institution, and the Faculty of Engineering, created in 1919, began to offer programs in materials engineering, mechanical engineering, electrical engineering, and chemical engineering, once local residents agreed to bear much of the cost.[35] Engineering education spread beyond such imperial universities. The Imperial Ordinance of 1919 created 10 new high schools and 19 new technical schools for technical education across the country.[36]

Nevertheless, this expansion of engineering education across the nation did not continue into the mid-1920s. Historically, building engineering education at the national level was costly and would require sound political and economic justifications such as external war, mobilization, or an economic boom. None of these factors existed from 1926 to 1936. Neither local nor private initiatives were sufficient in evidence, especially after the crash of the New York Stock Exchange in October 1929. At Tokyo University, for instance, the Faculty of Engineering had a period with no chair appointments from 1924 to 1933; it gained only three new chairs from 1934 to 1938 (Table 1.1).[37] This relative hiatus in engineering education was observable across the nation. The only exception was in 1929, before the Great Depression, when two technical schools, one in Tokyo, the other in Osaka, gained the status of imperial university and became the Tokyo Institute of Technology and Osaka Institute of Technology, respectively. From 1925 to 1937, no engineering department or faculty was newly created across *all* imperial universities in the country.[38]

TABLE 1-1
Number of Chairs at Tokyo University from 1893 to 1945

	Economics	Agriculture	Literature	Law	Science	Medicine	Engineering	Total
1893–1918	13	34	29	29	29	33	39	206
1919–23	3	7	10	4	11	0	19	54
1924–28	1	1	2	1	1	2	0	8
1929–33	1	1	0	0	0	1	0	3
1934–38	1	0	2	0	0	2	3	8
1939–43	1	3	0	1	3	1	8	17
1944–45	0	3	0	0	3	2	15	23
Total	20	49	43	35	47	41	84	327

It was not that Japan lacked strong incentive to promote the engineering profession. The World Engineering Congress, held in Tokyo from October 29, 1929, showcased the high hopes of Japanese politicians and engineers among an audience representing 42 foreign countries. It was the first world congress devoted specifically to engineering as a field of study. There were more than 1,200 participants, with the largest foreign delegation from the United States, followed by Great Britain, China, France, Germany, Italy, and Sweden. The event provided a good opportunity for Japan to learn from others in twelve fields of engineering. And it was also a chance to raise Japanese industry's reputation overseas when the country was suffering from an unfavorable balance of payments. This "major historical event," as reported in a newspaper of the time, deserved "special mention in the history of Japanese engineering."[39]

To a degree, the stock market crash on Wall Street, which occurred on the same day that the congress opened, was responsible for the hiatus in the expansion of engineering education. The crisis produced many job seekers in the stagnant economy, including graduates from institutions of higher learning, especially those from liberal arts programs. During 1929–1930, the employment rate among graduates from the Tokyo University Faculties of Law and Economics—known as the prestigious production center of government bureaucrats—remained depressingly low at approximately 50 percent.[40] In the nation's industrial economy, light, textile industries were scarred more than heavy and chemical industries, which began to show signs of growth despite the weak economy. From 1930 to 1935, the industrial production in mining increased by 60 percent, and the production by chemical, metalworking, and machinery industries grew from 33.8 percent to 47.7 percent.[41] Relatively unharmed by the Great Depression, the civilian sector had continued to absorb new graduates from engineering programs across universities. In 1890, 131 recent graduates moved into civilian industries, 385 did so in 1900 followed by 846 in 1910, 3,230 in 1920, and 25,331 in 1934. This growth in the engineering market reflected the efforts of universities to meet socioeconomic needs. In 1890, the total number of engineers with university degrees was 314, but this number increased to 859 over the next 10 years. The number continued to grow from 1,921 in 1910 to 5,025 in 1920, and to 41,080 in 1934.[42]

The Faculty of Engineering at Tokyo University remained resilient through the Great Depression. Its graduates were in high demand from 1920 to 1934 and maintained an employment rate of roughly 90 percent. The overwhelming majority of them—1,090 of the 1,610 graduates during 1933–1937, or 68 percent—sought careers in private business, and 394 graduates, or 24 percent, moved into the public sector, including the military.[43] Even when the economy hit bottom, all of the students in the Department of Mechanical Engineering were offered jobs in various companies by the summer of their graduating year. Job offers for the class of 1937 in that department were four times more than the total number of graduates. This happy outcome sent one faculty member on a busy mission. He spent his time paying visits or writing notes of apology to various private companies for declining their offers of employment.[44]

In retrospect, much of the educational infrastructure in engineering was completed by the mid-1920s, before the Great Depression. Meiji Japan put the status of engineering on par with medicine and higher than that of science. The regime established the faculties of engineering and medicine at *all* the nationally funded universities in Japan—a development uncommon in Western Europe or in the United States, where the engineering profession was less esteemed than those of science, medicine, or law. Many Japanese political leaders from the 1870s to the 1920s seemed more interested in producing applied scientists—physicians, agricultural experts, and, above all, engineers—than they were in people with training in law. In 1896, for instance, Tokyo University had 127 chairs, comprising 18 in science, 20 in medicine, 20 in agriculture, 20 in literature, 22 in law, and 24 in engineering. This pattern of distribution continued into 1920, when the entire system expanded to 479 chairs. Literature occupied 11 percent of the total; science and agriculture, 13 percent each; law and economics, 16 percent; medicine, 22 percent; and engineering, the largest share, 25 percent.[45] Engineering education was dispersed, democratized, and institutionalized by the mid-1920s through the experiences of three successive wars abroad: the Sino-Japanese War, the Russo-Japanese War, and World War I.

Mobilizing Engineers for War, 1937–1942

The external war against China from July 1937 created an urgent need to expedite the production of engineers in society, thereby ending the relative stagnation in engineering education. The local skirmish between Japanese and Chinese troops over the Marco Polo Bridge in the south of Beijing that year proved to be the beginning of a protracted, full-scale, Second Sino-Japanese War. The increase in war demands meant good news for recent engineering graduates from institutions of higher learning. From 1930 to 1937, the employment rate of university students had already improved from 39.1 to 57.8 percent, and that of students at technical schools increased from 43.8 percent to 61.8 percent. The labor market welcomed expertise in science and engineering. Factories and companies sought 90,000 recruits in such fields in the spring of 1939, when there were 12,000 graduates from these programs across the country, meaning that there were 7.5 jobs awaiting each recent graduate.[46]

Soon after the war erupted, the first Konoye cabinet prepared a government structure to mobilize engineers for "national defense." Their supply and demand were choreographed thereafter as the Ministry of Education prescribed annual quotas for engineering students. Each institution of higher learning followed this instruction, nurturing and sending its fixed number of students to their new, designated places of employment.[47] Minister Kido Kōichi took on a new assignment apparently at the request of the army, to lead the Ministry of Public Welfare and to improve the health of the labor force for war. The ministry, created in 1938, began to manage the demand for engineering students based on a formal legal framework (gakkō sotsugyōsha shiyō seigenrei). This legislation in August 1938 was a way to reverse a sudden, severe shortage of technicians and engineers across the nation. By law, each civilian firm asked the ministry in writing for the desired number of new technicians and engineers and at which factories and offices. Information about the availability of engineering graduates flowed from the Ministry of Education, and accordingly, the Ministry of Public Welfare approved such requests with some modifications, thereby spreading the new workforce across various industries.[48] At least for a time, both the Ministries of Education and Public Welfare were placed under the leadership of Kido Kōichi. Careful coordination between the two ministries formed the structural basis for mobilizing engineers for war.

While the Law Regulating the Number of New Graduates favored war-related heavy industries, it prompted mixed responses in society. Prestigious, well-capitalized giants in the aircraft industry—such as Mitsubishi and Nakajima—saw little benefit from the quota system. In their thinking—rightly, as it turned out—they could attract and gain the best engineering students from the nation's top universities. However, small, less known firms such as Showa Aircraft Company and Manshū Aircraft Company welcomed the system because it worked in their favor. It guaranteed that each aircraft company, regardless of its reputation, would obtain a fixed number of new engineering graduates every year. Some students proved less lucky than others under the system. Each student could voice his preferred places of employment in his senior year, but his preference was not definitive; under the government's direction during wartime, he might be moved away from his desired firm against his wish.[49] This quota system was hardly an effective solution to the personnel shortage in the industry. In March 1941, on average, each civilian aircraft firm could obtain only 10 percent of the graduates it needed from institutions of higher learning.[50] The deficiency was even more acute in civilian-oriented industries that, according to a newspaper of the time, had "virtually no university graduates allotted from important fields such as mechanical and electrical engineering." Some companies created their own training schools, but without a university education, the workers' level of competence remained low.[51]

This shortage of highly educated engineers in the country stemmed from its institutional infrastructure, which proved inadequate, especially in the field of aircraft engineering. Tokyo University was the *only* institution in the entire country to offer the program until 1937. The Department of Aeronautics, originally created in 1919, expanded only slowly thereafter. At the outbreak of the Second Sino-Japanese War, two tracks were created at the department for more specialized studies in the airframes and engines of flying machines. After this expansion, in 1938, the quota of students in each track increased to 17 and 7, respectively, per year.[52]

The media of the time reported on the urgent need to produce more aeronautical engineers. Seventeen days after the 1937 Marco Polo Bridge incident, on July 24, the *Asahi Shinbun* (one of Japan's largest daily newspapers) wrote that given "the shortage of engineers in various fields . . . the expansion of engineering manpower is the pressing need of the hour."[53] The scarcity of aeronautical engineers in particular was a concern serious enough for this national newspaper to offer a bold proposal. In October 1937, the article "Proposing the Aeronautical Institute of Technology" pointed out the country's "dire need to devise a new plan for aircraft design technology" and to "strengthen Japan's air power with specialized education." The Aeronautical Research Institute at Tokyo University, according to this proposal, should be converted from the nation's only establishment for academic, basic research to an educational institution. It was "a matter of course for a Japan with air power to have a university specialized in aeronautics and to mass produce [aeronautical] engineers; in reality, it is becoming too late" for the country to do so.[54] This insightful judgment proved correct in retrospect.

Only after 1937 did other universities across the country start to act on the acute shortage of aeronautical engineers. For one, Tokyo Institute of Technology launched its own program in 1939, stating the following:

> The poor quality of engineering on [factory] sites is the most critical issue. The reason for this is none other than the small number of engineering students in the Department of Aeronautics at Tokyo University, which has produced aeronautical engineers of the highest-quality. The graduates from the program have failed to meet the [market] demand for aircraft designers and researchers, leaving the field

work to technicians with little qualification. . . . Without at least three years of specialized studies about aircraft, engineers at the production floor [in the industry]
should not be considered [technically] competent, or able to keep pace with the
times. The nation needs to act. Tokyo Institute of Technology is aimed mainly at
nurturing engineers for work on site in the aircraft industry.[55]

Similarly, other institutions of higher education—Kyūshū University (1937),
Osaka University (1938), Tōhoku University (1939), and Nagoya University (1940)
—began programs in the field of aeronautics. Once the Japanese-American War
broke out in December 1941, Kyoto University (1942) followed suit. This massive
expansion of education in the field quickly multiplied the number of aeronautical
engineers. In 1936, the number of graduates in aeronautics was only 8, all from
Tokyo University, but in 1944, it increased to 243 graduates from national and
private universities.[56]

This rapid expansion of the field signified a fourth wave (1939–1942) in engineering education. Institutions of higher learning responded to the call of national emergency for war voiced by Minister of Education Araki Sadao (1877–
1966). This ultranationalist army general was determined to press for change
on the basis of his background. After the Russo-Japanese War and with greater
intensity after World War I, he began to express concern about communism.
When appointed to head the Ministry of Education in June 1938, Araki—former
president of the War College and minister of war—decidedly moved to infuse nationalism and militarism into education at all levels. The most notable example
was his attempt to manage the appointments of professors and presidents at all
the imperial universities.[57] During his twenty-one-month service in two cabinets,
he was most successful in promoting education, research, and development in
science and technology, as embodied in a board of investigation for the advancement of the sciences (kagaku shinkō chōsakai). A group of academic scientists,
including Tanakadate Aikitsu from Tokyo University, had noted its importance
earlier. Once the board was formed in August 1938, Araki presided as the minister
of education, appointing forty-three members drawn widely from the army, navy,
academic societies, and various government ministries. The board discussed,
among many things, how to expand facilities for research and development, how
to strengthen education in science and engineering, and how to increase the
number of graduates from science and engineering programs. Many ideas were
put to use under Araki's strong leadership. For instance, in 1939, the Ministry of
Education set up a new subsidy of 3 million yen per year to research and develop-

ment in science. This signified the nation's enormous commitment to science and war. Until that year, an annual grant of a mere 73,000 yen had remained the only source of monetary incentive for research and development in the field.[58]

Meanwhile, Tokyo University was transformed into a mass production center of future engineers for war. The Ministry of Education under Araki raised the annual quota for the number of freshmen from roughly 330, which remained the norm after 1922, to 460 in 1939—a massive increase of 40 percent in one year alone.[59] Money was available to accommodate this rapid growth of the student body and the need for more facilities. In contrast to the programs in liberal arts and humanities with fewer financial resources, the Faculty of Engineering doubled in size, setting up its second Faculty of Engineering outside of Tokyo.[60] In April 1942, despite delays in construction and gas installation, ten departments were newly created and welcomed freshmen on the Chiba campus.[61] During the war, Tokyo University added 8 chairs to the Faculty of Engineering in 1939–1943 and 15 more in 1944–1945. This total increase of 23 exceeded the total number of new chairs, that is, 17, that were added to all the other faculties combined.

Other institutions of higher learning, private and public, newly created or strengthened their preexisting engineering programs. In 1939 alone, the Ministry of Education established seven new higher technical schools from northern to southern Japan: one in the city of Muroran in Hokkaidō; others in the cities of Morioka, Taga, Osaka, Ube, and Niihama in the main Honshū island; and one more in Kurume (Kyūshū). Engineering education expanded most prominently at imperial universities. In 1942, Nagoya University added the Faculty of Science and Engineering to its campus, offering five fields of study that were directly related to the nation's war efforts (mechanical engineering, electrical engineering, applied chemistry, metallurgy, and aeronautics).[62] From 1937 to 1945, Osaka University, Kyoto University, and Tokyo Institute of Technology each added four new engineering programs. Furthermore, Tōhoku University introduced three new ones. Kyūshū University and Hokkaidō University each introduced two. Waseda University, a private college founded in 1882, began to offer technical curricula from 1938 in five fields of study, all directly related to war: applied metallurgy (1938), telecommunications (1942), petroleum technology (1943), civil engineering (1943), and industrial management (1943). Some institutional creations were made possible by financial contributions from local industrialists. Nongovernment circles had historically understood the reciprocal relationship between industrial development on the one hand and science and technology on the other. Funded privately by the retired president of a paper company, Fujiwara Institute of Technology began offering mechanical and electrical engineering as well as applied chemistry from 1942. This institution was incorporated into Keiō University two years later, forming its Faculty of Engineering.[63]

Focusing on the quantity rather than the quality of engineers for war, the Ministry of Education presided over a rapid increase in the production of workforce at these institutions of higher learning. Newly created in 1939, the seven higher technical schools across the country turned out 1,192 graduates in 1941 and 2,112 graduates in 1944, doubling their output in three short years. Similar patterns showed in engineering programs at private and imperial universities, where 1,489 engineering students graduated in 1936 and twice that number (3,125) completed their education in 1944. Among the class of 1944 were 578 graduates from mechanical engineering, 386 from electrical engineering, 250 from mining and metallurgy, 240 from architecture, 236 from civil engineering, and 213 from applied chemistry. Meanwhile, the term of all the three-year academic programs at institutions of higher learning was shortened twice, first by three months in October 1941 and then by six months in the following year.[64] Determined to maintain high-quality education, especially Tokyo University faculty strongly opposed the idea and approached the Ministry of Education and Privy Council to recess the talks. The government, however, rammed the bill authorizing an abbreviated curriculum through the diet.[65]

In retrospect, the massive expansion of engineering education, especially after 1939, was an ad hoc, quick, and belated response to the war in progress. Japan failed to expand engineering education until around 1938. By that time, the nation was already heavily involved in the war with China and was badly prepared for the subsequent modern war of science and technology against the Allies. The massive expansion of institutional infrastructure from 1939 to 1942 masked the inherent problem of a time lag; freshmen entering in April 1939 became fully serviceable in wartime industries only after their three years of engineering education in the spring of 1942. The lag was more serious at Tokyo University. Its Second Faculty of Engineering, created in 1942, produced the first class of graduates in September 1944 and the second class in September 1945, after the war. Moreover, it took at least a few years on the job to turn the fresh university graduates into competent, experienced engineers. This failure in engineering education stemmed from the nation's lack of experience in fighting an external war that lasted three years or more.

A dire consequence of the problem was the acute shortage of experienced engineers and technicians in wartime industries, a phenomenon that was observable even before the Pearl Harbor attack in December 1941. Probably the most revealing example was in aircraft development, where a limited number of competent, senior aeronautical engineers bore responsibility. At the Nakajima Aircraft Company, Koyama Yasushi, a Tokyo University graduate, led his design teams in the development of a series of warplanes such as the Ki-43 and Ki-84 fighters, both known for wide deployment by the army air corps. As the Second Sino-Japanese War progressed, Koyama's professional responsibilities depleted his energy, causing insomnia. He also lost 20 kilograms. In August 1940, he was hospitalized, and then returned to work after a month of bed rest. Soon, fatigue sent him back to the hospital for medical care, where he learned on his sickbed about the December 1941 Pearl Harbor attack.[66]

A starker example could be found at the Mitsubishi Aircraft Company. One of its important assets was a navy design team led by Horikoshi Jirō, who had graduated from the Department of Aeronautics at Tokyo University and created the famously agile, formidable aircraft A6M Zero Fighter. At least initially, his team of roughly 30 members was filled with optimism, enthusiasm, and the vigor of youth. Their average age in 1938 was merely 24 years.[67] Vitality was no defense against a workload that never diminished. Horikoshi's right-hand man in the Zero project, Sone Yoshitoshi, was among the first to be prescribed one-month bed rest for his overwork. Sone's return in August 1940 was immediately followed by Horikoshi's. Obliged to take his long leave of absence from September 1940 to February 1941, he put everything aside and concentrated on convalescence at home. Shortly thereafter, another experienced technician who had designed the Zero's landing gear, chronically suffered from exhaustion. Devoid of rest, he caught a cold that later caused pneumonia and eventually took his life. In 1942, another highly experienced technician in the team died of overwork.[68] The Nakajima and Mitsubishi teams were ill prepared for the war against the Allies.

A modern infrastructure for engineering education was mostly complete by 1924, five years after World War I. By the 1920s, wars abroad probably altered the relationship between engineering education and society in Japan more than the frequently studied link between the military and industry.[69] If each war abroad was a timely "blessing" that transformed Japan's engineering education, the pattern of development was foreboding. The hiatus of expansion between 1919 and 1937 stymied the production of highly educated engineers. This shortage became all the more clear in war-related industries, especially in aircraft companies, and the problem was already observable from the beginning of the war against China.

The national mobilization of engineers beginning in 1938, as well as the massive and rapid buildup of engineering programs, did little to fix the issue before Japan declared war on the United States in December 1941. The limited supply of competent engineers during the 1930s had unintended consequences into the 1940s. The small number of the best engineers from the nation's top universities flowed disproportionately to the navy, which outmaneuvered the army in recruitment. The legacy of this would extend far into the postwar era.

CHAPTER TWO

Navy Engineers and the Air War, 1919–1942

In 1927, a proposal submitted to the Lower House of the 52nd Imperial Diet raised expectations for a stronger aviation capability in both the civilian and military sectors. Toward this end, from February 19 to March 25, a series of discussions were conducted on the need for coordinated efforts by army, navy, and civilian officials. Both political and economic support was necessary, which Prime Minister Wakatsuki Reijirō requested from the ministers of war, navy, education, and communications. He wrote, "the government needs to expand its aircraft operations immediately" on the basis of the following points: (1) the army should reduce its four divisions and use the saving to expand anti–air artillery and air corps; (2) the navy should strengthen its air power without paying too much attention to auxiliary vessels; (3) the Ministry of Communications should expand such infrastructure as airfields and air transportation as well as facilities for night flight; and (4) a Ministry of Aviation should be established to centralize activities by the Ministries of Army, Navy, Education, and Communications.[1]

Wakatsuki's proposal yielded few results, but he rightly included the four ministries in his proposal. The creative output of highly educated engineers, as well as the application of such talents to research and development, were codependent factors aiding the rise of aviation, civilian and military alike. In advanced research and development in aeronautics, the navy excelled and left behind army and civilian researchers by 1942. This outcome was by no means preordained. It was during the 1930s that the navy devised successful plans for building research infrastructure, recruiting graduates from institutions of higher learning, and nurturing able engineers to work on air power development. Meanwhile, both the navy and army engaged in a series of interservice rivalries, weakening each other beneath the surface. The navy won the subterranean war against the army. How and why this was possible is partly explained by the rise of navy air power from the time of World War I.

Building the Infrastructure for Research and Development

The First World War stimulated weapons research at facilities affiliated with universities and military organizations. Most such facilities were either upgraded or newly constructed in an urgent manner and located entirely in Tokyo and its vicinity. The war exposed the vulnerability of Japan's science and technology because its industry had depended heavily on imports of key resource materials from Germany, including chemicals, pharmaceuticals, and precision instruments.[2] Western countries, busy fighting their own war in Europe, stopped selling their aircraft and aero-engines to Japan. The disruptions of such imports and the technological might of the West justified the military's obsession with national security. Army Minister Tanaka Giichi stated in the spring of 1919, "It is urgent [for Japan] to study the unusually fast pace of development in military engineering by powerful nations and to catch up with the progress in the world."[3] Correspondingly, the military empowered its research and development functions to foster homegrown science and technology.

In 1919, the army expanded two preexisting research facilities in hyakunin-chō, Tokyo to advance military science and technology. At the new Army Science Research Institute (Rikugun kagaku kenkyūjo), groups of scientists studied physical and chemical matters to conduct basic research for weapons.[4] At the second key institution, the Headquarters for Army Technology (Rikugun gijutsu honbu), engineers inspected, standardized, and experimented on weapons across various fields, including optics, acoustics, communications, and electronics.[5] The army, awakened by the war of scientific destruction in Europe, began paying attention to the contributions of scientists and engineers at home in waging future war abroad.

The navy also launched a series of initiatives to build research infrastructure for military technology. In April 1918, the Aircraft Testing Laboratory (kōkūki shikenjo) was established for theoretical and empirical research on machinery and materials for navy airplanes, airships, and aero-engines. Japan's first wind tunnel, the one-and-a-half-meter cross section, was built with technical advice from a leading aerodynamicist from the Eiffel Laboratory in France. In 1923, this research establishment and the Navy Testing Station (kangata shikenjo, created in 1908 for testing on models of warships) were merged into the Navy Technical Research Institute (NTRI, or Kaigun gijutsu kenkyūjo) to cover all research for developing warships and aircraft. The establishment soon had four departments for basic research in fields, including electronics, shipbuilding, and aircraft.[6] Actual research activities began slowly, however. The research budget for

the year 1927, according to a research report of that time, was "clearly insufficient . . . [and thus] an urgent meeting [and] an increase in the [research] budget
. . . are expected." The steady increase in the number of research projects created "a shortage in researchers and workmen, and this needs to be alleviated
urgently." Insufficient physical facilities were equally a problem. "The present
buildings are small," the report stated, and the researchers worked in an "utterly
unbearable" research environment and "endured as much as possible."[7] The research establishment became fully fledged by the early 1930s. By 1945, the organization expanded and hosted eight departments for research in a wide array of
fields, including chemistry, psychology, materials, acoustics, and radio waves.[8]

World War I invigorated research and development in military aviation and
expanded the functions of the imperial universities for basic research in science.
The largest beneficiary of this new trend was the field of aeronautics. Compared
to Western Europe, Japan's research in aeronautics was not well established before 1918. Much of Japan's genuine interest in the field was mostly confined to
those affiliated with Tokyo University. Subsequently established at the Faculty of
Engineering, the Study Commission on the Development of Aeronautics (1916)
made little progress in the field. One major difficulty in running this small study
group of six researchers involved its leader, emeritus professor Tanakadate Aikitsu. Although Tanakadate was a superb researcher, he was notoriously forgetful
about many things. He was a weak administrator, and he was ultimately removed
from his post after another senior member and the renowned mathematician
Yokota Seinen sabotaged him. With his patience exhausted, Yokota yelled at him
at a meeting one day, warning the commission that he would resign unless some
appropriate action was taken.[9] This demoralizing work environment changed
somewhat under Yokota's dictatorship. The commission moved forward, but
its function remained nominal with little substantive research. Many of the researchers' activities involved the compilation of manuals that described certain
machinery. Other work involved simply translating materials from foreign journals into Japanese.[10]

Japan's effort to advance academic research in aeronautics culminated in the
establishment of the Aeronautical Research Institute (Kōkū kenkyūjo or ARI).
Political tensions had initially stymied its creation. Originally forming part of
the Faculty of Engineering at Tokyo University, the ARI was the brainchild of
academic scientists such as Tanakadate who argued for the importance of basic,
theoretical studies on aircraft against the Ministry of Education in and out of the
Diet.[11] Developing the field of aeronautics required a national effort; it required
concentrated high-skilled labor and capital for growth, and World War I provided

both. While early flying machines remained crude in form and function, development costs were prohibitively expensive. The more speed, range, altitude, size, and weight new aircraft gained for wider commercial and military usage, the more alarming the costs of experimentation and development became. Costly research and development in the field around the world took shape with funds from national governments. In Berlin, for instance, the German Research Institute for Aviation was established in 1912 as part of the national effort to advance the field. Similarly, in 1915, the federal government of the United States set up the National Advisory Committee for Aeronautics (NACA)—the model of which was derived from British Advisory Committee for Aeronautics (1909)—to conduct aeronautical research. Russia followed suit in 1918 by founding the Central Aerohydrodynamic Institute. Furthermore, in Great Britain during World War I, the National Physical Laboratory became a major center of research activities in aerodynamics.[12]

Japan lagged behind in the field. Alarmed and frustrated, the president of Tokyo University wrote a statement for the creation of the ARI, calling for academic research in aeronautics that could be applied for military and commercial purposes. Yamakawa Kenjirō observed realistically that "not much drastic reform and progress [can be] expected" in aircraft development because the domestic production of licensed foreign airplanes was the best Japan could do at the time. It was "deplorable" to see Japan having only "extremely small" research and teaching capability in aeronautics, while "Western countries had made surprisingly large investments into research and development" and made "remarkable progress." What Japan needed, wrote Yamakawa, was to stop "taking half measures," but to "massively expand" its research capability by "upgrading or newly constructing facilities and buildings" and by "gradually adding full-time researchers and associates," all for "the independence of aircraft manufacturing in the near future" with "technical expertise superior or comparable to that in the United States and Western Europe."[13]

Yamakawa's wish was granted in a roundabout way. Newly set up in 1921 and renewed after the Kanto earthquake of 1923, the ARI expanded its operation through a five-year plan. In 1924, it hosted 144 researchers and their associates in 9 departments, 35 factory workers, 19 administrators, and 4 librarians, totaling 199 employees.[14] The enlarged workforce in the mid-1930s consisted of 339 associates, including 26 full-time researchers, 4 engineers, 49 technicians, 93 craftsmen, and 70 part-time employees.[15] The ARI hosted twelve research laboratories and covered a wide range of areas, including physics, chemistry, metallurgy, materials engineering, aerodynamics, engines, airframes, measuring instruments, and psychology.[16] Various kinds of state-of-the- art equipment aided the research, including a powerful closed-circuit wind tunnel, built in 1930. Three meters in diameter, the wind tunnel was among the largest in the country for pre-1945 research in high-speed aerodynamics.[17]

Probably the greatest asset of this university-affiliated institute was its human resources. Researchers, mostly full-time professors at Tokyo University, could devote their time fully to research without diverting their attention to teaching.[18] Making full use of this system, a Naval Academy professor moved on to the ARI for a one-year research stint beginning in April 1943, in the midst of World War II.[19] From its inception, the ARI continued to link the military and academic communities by hosting a few technical officers sent from both the army and navy every year.[20]

Smoldering Interservice Tensions in Building Air Power

While academic research progressed, the military was responsible for much of the growth of the field from the turn of the century. Japan's first aircraft research organization, the Research Committee on the Military Uses of Balloons (1909), was a joint effort among the army, navy, and academic researchers; however, it made little contribution to basic research due, in part, to the rivalry between the military services. The army led the organization from its inception; its office was located within the Ministry of War, and its budget came mostly from the army, which had 4 million yen to develop air power from 1909 to 1917, whereas the navy's budget from 1911 to 1917 was 750,000 yen, or merely 19 percent of the army's budget.[21] The committee included four civilian researchers and six navy officials. The other twenty members, including the chairman, were army officers. The army and navy officers vied for control over the research agenda, failing to resolve their disagreements in a constructive manner along the way. Unlike the army, the navy did not need research on balloons but rather on carrier-based aircraft. The navy remained unhappy also because it could not use its investment in the committee without the army's approval. The research yielded few results. For instance, the army failed in its mission to use a balloon for observation over Port Arthur in China because the operators suffered from terrible motion sickness in the air. Meanwhile, a civilian researcher from Tokyo University remained aloof from the daily operations of the committee, calling the aircraft a "soldiers' toy." The army, navy, and civilian academics pursued different research goals, and they finally dissolved the commission in 1920.[22]

By 1918, World War I revealed Japan's geographic isolation from Western Europe as the most formidable impediment in advancing the field of aeronautics.

Warplanes in particular were the product of international technology, ideas, and relations. The international exchange of technical information was vital to any nation—and Japan was no exception. During the war, Japan's military engagement in the air was geographically limited to China. The navy's involvement, for instance, was confined to campaigns in 1914 over Qingdao against Germany. Compared to the navy, the army gained more from the war. It obtained 74 military airplanes, 271 aircraft engines, machine guns, wireless communication devices, cameras, and other equipment from Germany—and relinquished what it found useless, such as the Zeppelin airships, to the navy.[23]

Overall, the Japanese military benefited from the postwar sociopolitical landscape in Europe. The end of the war meant an opportunity for Japan and others to recruit foreign engineers and to narrow the gap in the field between the nation and the West. The Paris Peace Treaty of 1919 strictly prohibited research and development in aeronautics as well as the production of aircraft in postwar Germany, and the resulting oversupply of experienced German aeronautical engineers in their homeland became a blessing for other nations. For instance, inflows of German aerodynamicists played an indispensable role in the formation of post-1919 research in the field around the world, especially in Great Britain and in the United States.[24] As a Japanese army officer observed rightly in 1922, the end of the war "shrank military industries of various countries" and produced an oversupply of engineers in the defeated states, including Germany, which could no longer employ them. Depending on their "expertise, experiences, and personalities," the report continued, such jobless engineers could be recruited and "contribute to Japan's military industries." Accordingly, the army ordered its military officers in France, Great Britain, Italy, and the United States to survey qualified individuals and to report back to Tokyo for future recruitment.[25]

Civilian and military institutions soon carried out their own plans to reap the benefits of postwar changes in Europe. The ARI, for instance, sent its researchers abroad and invited foreign experts to promote technology transfer and the subsequent diffusion of scientific knowledge within Japan. World-renowned scholars, mostly from the United States and Germany, presented their research activities at the ARI. During the 1920s, Theodore von Kármán (1881–1963) and Ludwig Prandtl (1875–1953) delivered a series of lectures on theoretical aerodynamics and applied mechanics, which stimulated academic communities in Japan.[26] A more dramatic example involved German expert in aerodynamics Carl Wieselsberger (1887–1941) at the Göttingen Aerodynamic Laboratory. Originally invited by the navy to advance the field of aerodynamics, he became the only foreign researcher officially employed at the ARI albeit part time. From 1923 to 1930, he contributed markedly to designing and constructing two wind tunnels at the site. His engineering knowledge proved equally indispensable outside the academy. At the NTRI, he offered his expertise in designing and building two wind tunnels with the largest cross section at that time in the country. Likewise, the army benefited from his knowledge in the field, when he lectured at the War Department's Aeronautical Technology Institute on multiple occasions. Recognizing Wieselsberger as an important architect of aeronautics in the country, Tokyo University commended him twice during his seven-year stay.[27]

Similarly, aircraft companies in Japan quickly responded to the changing patterns of international migrations of European aeronautical engineers. These civilian companies enticed the engineers with money, employed them, and had them compete with one another to develop the best flying machines. This was part of a managed competition system devised by each military service. On receiving performance specifications from the army or navy, whichever the end user, private firms first submitted their design drawings for a review. The firms on the short list subsequently developed their prototypes, and after each made a test flight, the military would award the production contract to the firm that most closely met its specifications. The decisions, however, were not always impartial. They were often made for political reasons, or on the basis of the personal tastes of the test pilots. Meanwhile, the Japanese aircraft industries remained dependent on foreign engineers, especially those from postwar Britain and Germany. For instance, Nakajima Aircraft Company relied on British and French expertise; Mitsubishi hired German and British engineers; Kawanishi Aircraft Company retained British engineers; and both Kawasaki and Aichi Aircraft Companies employed German engineers.[28] German experts dominated the field in the mid-1920s. At one point, Mitsubishi used Alexander Baumann (1925–1927), Kawasaki Aircraft Company turned to Richard Vogt (1924–1933), and Ishikawajima Shipping Company depended on Gustav Lachmann (1926–1928)—all experienced German engineers—to create prototype flying machines for the army. The most successful Kawasaki aircraft was later modified for mass production, beginning in 1928, as the Type-88 Army reconnaissance plane.[29]

During 1925–1926, both the army and navy adopted this managed competition system from Britain and France, which proved a brilliant strategy to advance the field. The system successfully recreated on Japanese soil design competitions among the air powers in Europe. In addition, it allowed the Japanese military and aircraft manufacturers to observe the competition process firsthand, while remaining physically unaffected in a time of peace. The system essentially turned the war in Europe to constructive use for postwar Japan. It lowered the hurdle of geographical distance and narrowed the technological gap between Japan on the one hand and Western Europe and the United States on the other.

This system was born in a new, solid command structure in the army and navy, each determined to accelerate the pace of technological transformation in military aviation. The most remarkable creation in the field was the Headquarters for Army Aeronautics (Rikugun kōkū honbu) in 1925. While centralizing many administrative roles in aviation, its functions ranged from training personnel, giving technical advice for manufacturing aerial weapons, and administering the overall development of aircraft industries. Useful technical information was actively collected from overseas. The leadership dispatched resident officers to various Western nations with strong air power, namely, Germany, France, Great Britain, and the United States. During their two- to three-year appointments, Japanese military officers studied many subjects in the country of their residence and its neighboring nations. Their focus tended to remain on how to systematically manage weapons development, how to improve methods of weapon production, how to employ natural resources efficiently for war use, and how to manage patents and coordinate wartime mobilization. These officers functioned as engineers as well. They inspected promising foreign-built weapons and materials, negotiated purchasing and licensing agreements, and infused the latest technology from abroad into the Japanese military industry.[30]

At the headquarters, the Technical Development Division (gijutsubu) played a central role in the army's planning, testing, prototyping, inspection, as well as research and development. While promoting the domestic production of aircraft, engineers in this division also coordinated the design competitions among civilian companies and inspected prototypes before awarding production contracts.[31] The establishment carried on the functions of the Tokorozawa Army Flight School (Tokorozawa hikō gakkō, created in 1919), where 68 workers conducted research projects, occasionally designed prototype aircraft, and inspected many products in close collaboration with pilots.[32] The army engineers' effort to domestically produce foreign airplanes and aero-engines did not yield much fruit. For instance, their prototype in August 1928, an experimental bomber, was a costly metal plane with poor visibility from the cockpit and an inadequate defense mechanism.[33]

Somewhat lagging behind the army, the navy took a similar administrative step. Until the mid-1920s, the volume of activities related to military aviation had remained relatively small. As the technology developed and air groups expanded in size, the navy divided the work responsibilities without an effective central command system. In 1927, however, this issue was resolved. Separated from the Headquarters for Naval Warships (Kaigun kansei honbu, created in 1900), the newly created Headquarters for Navy Aeronautics (Kaibun kōkū honbu) centralized many administrative activities in naval aviation for the first time. Thus, naval aviation gained its own independent status. Consisting of an administration bureau, a training bureau, and a technology bureau, the command center supervised the planning and ordinance for weapons development, inspection, and logistics of aircraft equipment; it also handled technical aspects of airframes, engines, and aircraft equipment. This establishment took charge of all air training as well, except for the aerial combat training that remained in various air groups across the country.[34]

Technology transfer in aeronautics, including employing foreign engineers, was successful but costly for the military, and private aircraft companies often bore the cost. For instance, Kawasaki prepared an office for German engineer Richard Vogt during his nine-year stay, and soon thereafter, three Japanese army engineers moved into that space. They asked him numerous questions about every aspect of aircraft engineering, often requesting him to provide detailed answers in writing for later use. Meanwhile, his hefty salary was paid by Kawasaki, which likewise learned a great deal from the foreign engineer.[35] His compensation of roughly 500 to 1,000 yen was very high by the Japanese standards of the time—at least five times more than that paid to a first-year Japanese university graduate in engineering. A newcomer to a more prestigious aircraft company would receive 80 to 85 yen per month, and a graduate from the most prestigious program in the country, the Department of Aeronautics at Tokyo University, would be paid 95 yen.[36] The army was shrewd and gained useful information about state of-the-art technology from abroad. On finding foreign-built airplanes with great promise, the army had private aircraft companies buy and import them.[37]

For aircraft manufacturers, however, hosting foreign engineers was not only costly but sometimes risky, often resulting in poor investments. During the 1920s, Mitsubishi devised mainly three strategies for technology transfer: buying patent licenses, importing sample airplanes, and developing an exchange program for Japanese and foreign engineers.[38] Mitsubishi's expectation was often betrayed in the development of navy aircraft. During 1921- 1924, British engineers—led by Herbert Smith, who had considerable experience in navy aircraft—designed airplanes all by themselves without teaching much to the Japanese engineers working as their assistants.[39] At one point, a British engineer returned a design drawing to his Japanese assistant with the unkind, condescending comment, "HOW TO DRAW A NUT. SEE YOUR SCHOOL TEXTBOOK."[40] Soon after their

departure, Mitsubishi brought in a group of German engineers such as the leading aeronautics professor Alexander Baumann. It was a costly investment for
technology transfer. Mitsubishi personnel had visited him in Germany, where he
had demanded twice as high a salary as the usual pay to a foreign engineer. The
Mitsubishi delegate naturally protested, arguing that not even the minister of the
army could receive that substantial a salary. Mitsubishi reluctantly agreed on the
requested amount, expecting to learn firsthand a great deal from him about aeronautical engineering.[41] His emphasis on the theory of aeronautics, however, did
not yield much concrete return. A series of large, metal-constructed airframes
they designed were costly and unsuitable for production. If anything, his academic inquiries left a lasting legacy at Mitsubishi.[42]

Interservice Rivalry over the Recruitment of Engineers

From 1919 to 1942, the intensity of the competition between the armed services
grew over the recruitment of the best engineers in society. More shrewd and successful, the navy gradually won that war at home; the gain was highly educated
engineers to build navy air power, often at the expense of the army. The technological superiority of the navy later caught the attention of the U.S. occupation
authority. Compiled weeks after the war's end in 1945, an intelligence survey
report states, "It is generally agreed by [Japanese] scientists and manufacturing
who have had contact with both the Army and Navy technical staff that the Navy
personnel is clearly superior, and that the Navy was generally well ahead of the
Army on technical developments."[43] This observation was historically sound, especially in the field of aeronautics, which embodied the interservice rivalry over
the recruitment of engineers.

The intensity of the interservice turf war was by no means demoralizing in
the beginning. Until the 1900s, both services had different engineering needs.
The army tended to focus on artillery and sappers, while the navy invested its
efforts in ammunition manufacturing as well as engine and waterway construction.[44] By the 1920s, the growing need for air power added another dimension to
this technological landscape. Both armed services began to compete rather than
complement each other, each trying to maximize its share in the finite pool of engineers for war in the air. What had developed concurrently was a quiet, subterranean war at home between the army and navy over recruiting engineers while
engaging in external wars from 1937.

Especially in the latter half of the 1930s, the limited availability of engineers
was a problem for both civilian firms and the military services. One solution for
both the army and navy was to turn to Tokyo University's Faculty of Engineering,

a chief production center of military engineers. The armed services employed 299 graduates during 1918–1922, 388 graduates during 1923–1927, 453 graduates during 1928–1932, 394 graduates during 1933–1937, and 700 graduates during 1938–1942. While more engineers outflowed from the university to military research establishments and arsenals, the military dispatched and educated its cadets in the academic setting via patterned flows of human resources. Especially from 1938, both the army and navy sent their cadets to Tokyo University, where they would become ordnance officers upon graduation. There were 30 such cases in 1938 and 121 cases in 1942.[45] Human resources moved in both directions between Tōdai and the military.

By 1937, both the army and navy had devised three similar systems to educate, train, and promote engineers. First, some minor differences aside, each armed service sent graduates from its military academy to imperial universities for a three-year education in science or engineering. The army was particularly systematic. Graduates from the Army Academy, who had finished one-year of field work, took classes in mathematics, chemistry, surveying, and mechanical engineering at the Army Artillery School. Students with high grades subsequently received further education in science and engineering at imperial universities. An overwhelming majority, 177, moved on to Tokyo University from 1900 to 1940.[46]

The second mechanism involved recruiting graduates from civilian institutions of higher learning to produce technical officers (*bukan*) who oversaw civil officers (*bunkan*). The navy's program, originally devised in 1876, selected students through an examination, financed their education with monthly stipends, and appointed the graduates to the rank of lieutenant junior in the areas of arms manufacturing, shipbuilding, and engine construction. After successfully passing a postenlistment examination, the cadets typically received their military training for six months before advancing further in their military careers. The army's counterpart was off to a slow start. Belatedly, in August 1919, the Ordinance of Army Technical Officers (Rikugun gijutsu shōkōrei) opened the window of opportunity for combatant officers and graduates with baccalaureate degrees in science and engineering to become technical officers. This initiative, however, failed to produce an outcome as favorable as that in the navy. During their careers, army technical officers had far fewer chances for promotion than commanding officers in the battlefield. Lacking enticing career paths, officers in the field rarely abandoned their careers for tedious jobs in laboratories that offered severely limited long-term prospects.[47] The army's prevailing culture in favor of "field work" discouraged officers with science and engineering backgrounds from pursuing careers in research and development.

Third, the army and navy recruited graduates from institutions of higher learning as well as military technical schools to produce civil officers, who included *gishi* engineers and *gite* technicians in the hierarchy (Table 2-1). Unlike military officers, at least in principle, civil officers had more stable jobs because they were not accustomed to transfers. They lacked military education and were expected to remain in their given engineering posts for years. Despite their lack of formal higher education, experienced *gite* technicians could advance to the rank of *gishi* engineers after some training at military technical schools. Quantitatively, and somewhat predictably, the navy emphasized the importance of engineering manpower more so than did the army. As of 1937, the army retained 245 *gishi* engineers and 1,079 *gite* technicians; the navy kept 406 *gishi* engineers and 1,152 *gite* technicians, and the total formed 70 percent of the entire *bunkan* workforce as compared to the army's figure, 40 percent. In addition, unlike the navy, the army had only a few minor guidelines about *gishi* engineers and *gite* technicians.[48]

Until the late 1920s, however, the army had the upper hand over the navy in the recruitment of technicians and engineers because the military draft system remained firmly under the army's control. Originally promulgated in 1873, and modified five times before the end of World War II, the Conscription Ordinance strengthened the nation's war-making capability chiefly through the army. In principle, especially before the creation of the full-scale navy, all the mentally and physically healthy male citizens older than twenty years of age were drafted for the construction of the land force rather than the navy. Both the navy and army ministers took charge of the country's military draft laws after 1927, but the navy remained severely disadvantaged in the system because the army conducted the actual administrative work. The navy minister would first discuss the number of needed draftees with his army counterpart, and the army minister, after approving the figure, would send directives to local draft districts to help the navy gain its manpower.[49]

At times, the army's draft system undermined the navy's engineering ability, as in the case of Nakagawa Ryōichi during the early 1940s. This Tōdai graduate and aero-engine specialist exhibited his talent at the Nakajima Aircraft Company from 1937 by creating the air-cooled, eighteen-cylinder, 2,000-horsepower aero-engine, so-called *Homare*, for navy aircraft of 1941–1942. Despite this success, the army drafted him the following year, sending him off to an anti–air artillery unit in the Tokyo suburbs, where he soon caught a cold that resulted in tuberculosis and confined him to bed rest for months.[50] His absence from the technological scene delayed the chronically urgent development of a trouble-free aircraft

TABLE 2-1
Classifications and Ranks in the Army and Navy

Classification	Rank in the Army	Rank in the Navy
Bukan technical/military officer	Colonel	Captain
	Lieutenant colonel	Commander
	Major	Lieutenant commander
	Captain	Lieutenant
	1st lieutenant	Sublieutenant 1st class
	2nd lieutenant	Sublieutenant 2nd class
Bunkan civil officer		Gishi engineer
		Gite technician

engine, especially after 1943. For instance, the Zero Fighter's 1,000-horsepower engine proved severely underpowered in the face of the Allied aircraft equipped with 2,000 horsepower or more, which enabled faster, higher flight in aerial combat. The Japanese navy fighter, however agile, powerful, and useful early in the war, became obsolete and defenseless in the latter half of the air war.

The frustrated navy relied less on the draft system and moved on to increase its number of volunteers, who could operate without the army's interference.[51] The navy's "sales talk"—no sweat and a promising future career—enticed volunteers away from the army. The differences between the military services were clear in the eyes of future cadets. The army typically recruited graduates from junior high school or above, and then appointed them to reserved second lieutenants only after a mandatory, physically demanding military drill—a deterrent for university graduates who aspired to become military officers. In contrast, the navy offered fast promotions. Navy cadets could skip the ranks of seaman and petty officer and rise to the rank of military officer after enlistment. Recognizing the differences, elite Tōdai students tended to gravitate toward the navy rather than the army.[52] A rumor influenced their decision making. Many graduating seniors at Tokyo University favored the technologically savvy navy because it would give them preferential treatment based on their educational background. The army, however, was said to give them "inhumane" treatment despite their education or technical competence.[53]

The navy successfully gained a greater share of highly educated engineers, particularly those from Tokyo University, through its ties with that elite institution. A snapshot of this landscape could be found in the educational backgrounds of the 127 *bukan* technical officers who formed the core of the special recruitment program. Graduates served active duty first as commissioned officers for two years, and thereafter, they could leave or remain in the navy. Among those who stayed active in 1942, 127 in all, the graduates of Tokyo University dispropor-tionately weighted the most—80 graduates, or 63 percent—outnumbering the second largest group (13, or 10.2 percent from Tokyo Institute of Technology), the third largest (10, or 7.9 percent from Kyoto University), and all others.[54]

Behind the navy's successful recruitment of Tōdai graduates were individual efforts. A central figure was Vice Admiral Hiraga Yuzuru (1878–1943), a highly acclaimed architect of Japan's naval engineering and a graduate from the Tokyo University Department of Naval Engineering. After the Russo-Japanese War, this legend in the field chiefly developed the battleships *Nagato* and *Mutsu* as part of the Eighty-Eight Fleet Project and, in 1923, the light cruiser *Yūbari*. During the 1930s, what mattered more than his engineering knowledge was his administra-tive and political skills. After directing the NTRI, in 1935, he returned to his home institution, the Faculty of Engineering at Tokyo University, where he presided after 1938 and helped to build the Second Faculty of Engineering.[55] His active re-cruitment of Tōdai engineering students into the navy earlier than the customary time did not go unnoticed. Commercial enterprise, needing fresh Tōdai gradu-ates, often criticized his action and tried to prevent this early hiring. Nonetheless, Hiraga continued his practice with his popularity among students. Known for his good temperament and interest in catering to students, he believed that young future engineers should be able to imbibe alcoholic beverages during important social activities. He helped provide more beer kegs for the students' consumption at annual graduation parties.[56]

Even before Hiraga's appointment at Tokyo University, the navy had depended heavily on the talents of highly educated engineers. Enthusiastic about technol-ogy, the navy had invested heavily in engineering education and technology transfer from abroad for building warships. Soon after the Meiji Restoration of 1868, the emerging military-industrial complex capitalized on the active interac-tions among the military, heavy industry, and university—and the navy actively supported this tripod.[57] Historically, the alumni of the Tōdai Department of Navy Engineering bonded the navy and the university. From 1883 to 1903, for instance, the navy recruited 34 out of 104 new graduates, or 33 percent, from that depart-ment. It produced only 3 graduates in 1883, but from 1926, roughly 30 graduates per year. Furthermore, during 1933–1935, half of the 88 graduates sought their careers in the navy, which dominated employment (Table 2-2).[58]

The navy selected, recruited, and maintained elite engineers through the in-genuous system method, called the Program of Learned Workers (Yūshiki kōin seido), which conveniently embodied the navy's flexible interpretation of bureau-cratic rules. For financial reasons, in the midst of arms reduction during the

TABLE 2-2
Places of Employment among Graduates from the Department of
Navy Engineering at Tokyo University from 1883 to 1935

	1883–1903	1933–1935
Shipyard	26	23
Other manufacturing	0	(aircraft) 1
Ministry of Communications	20	10
Ship classification society	0	1
Maritime transportation	10	2
Navy	34	44
Others	University: 6	University: 1
	Graduate school: 2	Fisheries Agency: 2
	Others: 5	Enlistment: 1
	Study abroad: 1	Other: 3
Total	104	88

early 1930s, the navy had maintained quotas on the numbers of *bukan* military officers and *bunkan* civil officers, the former earning more than the latter. Without violating the prescribed ratio between the two, the navy created the new category of "learned workers"; once recruited, fresh graduates from institutions of higher learning accumulated experience as factory hands on the floor and were subsequently promoted to *gite* technicians, to *gishi* engineers, and after military training, to *bukan* military officers. This system, unique to the navy, was blessed with more than enough navy, aviation, or technology enthusiasts in the engineering programs at institutions of higher learning. The program started modestly in April 1931, with three university graduates working at Yokosuka Navy Arsenal for aero-engine development. Relishing the taste of this initial success, the navy expanded the program to recruit manpower for warship development, while proving successful especially in the field of aeronautics. By the war's end, for instance, roughly half of the technical officers at the nation's most advanced research center for aviation—the Institute for Navy Aeronautics (INA, kaigun kōkū gijutsushō)—were products of this program.[59]

In the recruitment process, the navy vigorously protected able engineers and technicians from the army's conscription by compiling draft exemption lists. The navy remained fiercely territorial because the army could draft civil officers (*gishi* engineers and *gite* technicians) already working for the navy. For a year, from April 1936 to March 1937, personnel directors surveyed thousands of engineers, technicians, and workers at many navy arsenals as well as at research and educational establishments across the country. The lists showed their names, ranks, and jobs, categorizing them as "indispensable individuals and specially employed workmen [at navy-affiliated establishments] who should not be drafted for war." The exhaustive survey proved crucial for the navy to protect its engineers from the army. At the INA in 1936, for instance, those already enlisted in the navy aside, 538 employees (including civil officers such as *gishi* engineers, *gite* technicians, and machine operators) were officially exempt from the army's draft.[60]

Once the Second Sino-Japanese War broke out in July 1937, the navy acted swiftly and more aggressively to safeguard its engineering manpower from the army by promoting them from *bunkan* civil officers to *bukan* military officers before enlisting them. The navy simplified its official evaluation procedure to expedite the promotion process. According to the new, elusive guidelines, civil officers with "academic qualifications and outstanding achievement" became military officers.[61] Meanwhile, the navy at least nominally maintained a prescribed balance between military and civil officers, 70 and 30 percent, respectively, and 70 percent of the total new engineering workforce recruited from university graduates and the remaining 30 percent from technical schools. When originally planned in 1937, the systematic promotion involved a total of 50 *bunkan* civil officers, including 38 in arms manufacturing, 5 in shipbuilding, and 7 in engine construction.[62] As it turned out, the first promotion in 1939 took shape on a much larger scale, involving 142 individuals.[63] In the field of aeronautics, the navy promoted 15 engineers and 19 technicians to secure them from the army. Concurrently the navy started its school to train *gite* technicians, accepting 40 students every year from 1939.[64]

Meanwhile, the army replicated the navy's recruitment strategy as a countermeasure. In the navy, the period of active duty for commissioned technical officers was set at two years, and after that, they could remain active in the service or leave. Facing this challenge, the army copied this program in July 1939 to obtain more technical officers from universities and special schools.[65] Then in 1940, the army modified its preexisting Technical Officers Program (gijutsu shōkō seido), a means of recruiting graduates from science and engineering programs at institutions of higher learning across the country.[66] As a result, in October 1943, 230 new university graduates joined the Technological Development Division at the Headquarters for Army Aeronautics.[67] The massive recruitment seemed successful, but this new additional workforce was young, inexperienced, and as the next chapter will show, less educated than previous generations.

In the midst of this recruitment battle, the navy often "stole" engineers directly from the army with enticing offers. Technicians and engineers, even when working for the army, could ask in writing for their transfer to the navy, a request army and navy ministers would later approve. In May 1943, for instance, the navy chose, enlisted, and promoted 90 *bunkan* civil officers (61 *gishi* engineers and 29 *gite* technicians) to the positions of *bukan* technical officers at the INA alone. Among them were 17 army civil officers including Kitano Takio and Matsudaira Tadashi, who were both instrumental in the wartime development of aerodynamics. One illustrative case involved Suehiro Takenobu, an army officer and a graduate from the Tokyo University Department of Physics. He received his military training in the army air corps, and in 1935, he was appointed to the rank of second lieutenant. From March 1937, four months before the outbreak of the Second Sino-Japanese War, this engineer worked part time at the INA for aero-engine development. His starting annual salary in the navy, 1,300 yen, was good by the standards of the day. In May 1943, he and 11 other second lieutenants switched voluntarily from the army to the navy.[68] Outwitting the army, the navy tended to obtain the best manpower from draftees and volunteer officers as well.

Neither military service sat still in the quiet war over recruitment while fighting an external war. As a countermeasure, the Ministry of Army issued a decree and made transfer from the army to the navy more difficult than before.[69] The Navy Bureau of Personnel acknowledged this, striking a compromise with the army's counterpart. Requests for transfer from the army to the navy were subsequently forbidden in principle, but many applications were actually handled on a case-by-case basis.[70] To gain more engineering manpower, the undaunted navy flattened its bureaucratic hierarchy by delegating more authority over personnel management to local levels. In February 1944, regional administrators obtained more flexibility in appointing and promoting petty officers in engineering fields.[71] Meanwhile, the army could still exercise its authority over the navy in the draft system. Responding to the army's practice of drafting workers from navy factories, in February 1945, the navy fiercely protected those with more than five years of experience at their places of employment.[72] Overall, the navy outsmarted the army through aggressive, skillful, and clandestine means of recruiting engineering manpower.

The incessant interservice rivalry at home weakened Japan's ability to continue the war against the Allies above water and underwater. Any country with different military services—army, navy, air force, or marines—has experienced some degree of rivalry among these branches. In wartime Japan, however, the intensity reached a point of absurdity, as illustrated in the Type-3 Japanese submarine developed by the army. Its genesis exemplified the lack of interservice cooperation. During the battle of Guadalcanal (August 1942 to February 1943), the navy refused to transport goods to the Japanese army men remaining on islands across the Pacific. Frustrated but undaunted, the army mobilized its engineers to independently develop the ill-fated product at the 7th Army Research Institute, with technical assistance from civilian firms. Forty vessels went into commission by the end of the war.[73]

Such interservice rivalry marred other research project as each service established its own research institutions throughout the war. The army operated no fewer than 21 institutes, including 10 ground force facilities, 8 air force institutes, and 1 fuels laboratory. The navy ran 4 major research establishments. A series of administrative initiatives to coordinate research efforts across institutional boundaries, such as the creation of the Board of Technology, bore little fruit. An Allied occupation team observed rightly in October 1945 that an "almost complete lack of cooperation" between the two services undermined Japan's research capability.[74]

Building Army Air Power: Institute for Army Aeronautics

Compared to the navy, the army lacked an effective plan to train engineers in its fields of research. A key point of reference was the Institute for Army Aeronautics (Rikugun kōkū gijutsushō, or IAA), which played an administrative role without strong leadership or research and development ability. Its primary functions included creating and issuing design requirements, inspecting aircraft prototypes, gathering ideas from aircraft operators, and ordering repairs and improvements—all without generating deposits of engineering knowledge and data that would support aircraft companies.[75] These administrative functions reflected the capabilities of its employees. When the Technical Development Division of the Headquarters for Army Aeronautics was merged into the IAA in 1935, it employed 199 core workers consisting of 83 officers, 25 engineers, 10 warrant officers, and 81 petty officers and junior officials.[76] In October 1942, the organization was divided into eight establishments, hosting 440 engineers and technicians.[77] They devoted their time to basic research, writing proposals, or giving orders, while exerting unwelcome pressure on civilian aircraft firms.[78]

The army lacked technical expertise in research and development for prototyping, generating little information that civilian aircraft designers could freely draw on for advancement in the field. The IAA hosted monthly meetings with private firms to discuss aircraft prototyping, but the most useful, concrete, and latest research data came not from the army, but from civilian manufacturers in the country or from foreign journals.[79] This knowledge gap was hardly surprising at the IAA, which lacked a design department, not to mention a design room. The army's approach marked a stark contrast to that of the navy; as the leader of naval aircraft development, the navy engineers conducted their own research for superior prototypes and efficient workmanship. The IAA engineers devoted their time to supervising administration, production, and above all, inspection, with little or no experience in aircraft designing and prototyping which, by 1936, was no match for those at civilian firms.[80]

The landscape for research and development at IAA was littered with ineptitude in the eyes of Tanaka Jirō, a fresh graduate from Tokyo Institute of Technology. As part of the army's two-year engineering program (tanki gen'eki gijutsu shikan), this new engineer arrived at IAA in April 1939, where only four to five technical officers were studying many aero-engines that awaited inspection. His superiors was unable to give him technical instructions because they lacked knowledge and experience with cooling mechanisms. The newcomer studied the mechanics of the aero-engine by himself, with research reports in English by NACA, the predecessor of the present-day NASA. Soon thereafter, this novice engineer desired to build a quixotically giant wind tunnel, with a length of 120 meters, on the solid foundation of concrete. This proposal earned the army's official approval in 1944. What surprised him even more was the function of the device. Once heavy aerial bombardment by the American B-29s demolished a wooden office building, the long and strong wind tunnel became the shelter and substitute office at IAA.[81]

The army's ability to advance the field waned especially after 1937. In October 1933, the leadership at the Headquarters for Army Aeronautics set a policy to promote the domestic production of aircraft, with less reliance on foreign powers. The army relied more on private aircraft manufacturers at home for research, design, and prototyping, while the practice of outsourcing formed the core of its engineering culture. A milestone event took place in January 1937, when high-level officials formulated a research policy for Air Arms at the Headquarters for Army Aeronautics (Rikugun kōkūhonbu kōkūheiki kenkyū hōshin). The army, as the council concluded, admittedly lacked engineering manpower. Its solution was twofold: purchase good products from civilian firms, and commission all projects of developing heavy bombers and transport planes to the navy.[82] The army retained this policy of outsourcing, leaving little room for joint engineering projects with the navy. During the war against the Allies, the Japanese army used the navy's strength requirements for aircraft design.[83]

The army's policy of outsourcing was good news for civilian researchers, especially those at the ARI. With its strong ties to Tokyo University, the establishment had social prestige but was poorly endowed. Its total budget in 1941 was only 790,000 yen; after the distribution of this meager amount, each laboratory had about 20,000 to 50,000 yen—hardly sufficient for the aero-engine laboratory to buy a functionally operational engine for research. The ARI's purely academic, basic research was a hard sell in the market, and its financial health depended heavily on incomes from outside sources, most notably, the army.[84] At one point in 1938, for instance, ARI researchers worked on five commissioned research projects for the navy and nine for the army; at least for the next two years, the army's projects outnumbered those of the navy with a ratio of 2 to 1.[85]

One such project for the army involved the long-range, nonstop flight that established the world's record. In May 1938, the so-called *Kōkenki* plane took a circular path over Tokyo for 62 hours nonstop and demonstrated to the world what Tokyo University could accomplish. The flight broke the world record at the time set by a French plane, but the path was by no means smooth at the onset. ARI researchers first approached the technologically advanced navy, which, as it turned out, had a similar project of its own and thus declined the request for support. The next choice, the army, welcomed the project, given its future prospect for long-range bombing. ARI researchers played tennis and baseball with army officials, which helped to solidify their support.[86] Subsequently, the army adopted the ARI researchers' proposal, "outsourced" the project back to them, and provided them with needed financial support.

What impeded the army's effort to advance the field was more than its research culture of outsourcing, but its inability to overcome the geographical distance that separated the engineers and test pilots, or its inability to simplify the cumbersome process of inspecting and testing aircraft at different sites. Before 1935 or so, army prototype aircraft—the creation by commercial firms—ordinarily followed three steps before introduction to the battlefront. First, army engineers and then test pilots flew the planes; soon thereafter, experienced instructors at different flight schools conducted more flight tests for a few months, and each school took charge of a particular aircraft type. In this system, aircraft designers, test pilots, and flight instructors were set apart administratively and geographically. Instructors at different flight schools lacked standardized procedures for flight evaluations, but their subjective, individual, and often ill-defined experiences shaped the final, rather arbitrary decision.[87]

Ironically, the army's effort to correct the problem made things worse in an unexpected way. In December 1939, the Headquarters for Army Aeronautics gained a new section, which proved a deceptively effective fix. Following the German model of administration, roughly 500 army officials took charge of reducing the evaluation process from three tiers to two. Accordingly, aircraft prototypes were tested at the two separate establishments, first by engineers at the IAA for basic, initial inspection, and then by test pilots in the airfield at the Headquarters for Army Aeronautics. The fissure between the two groups—engineers and pilots—proved deep; both unfairly asked each other for a "fair" evaluation of a new flying machine.

In the eyes of the airmen, the reports compiled at the IAA reflected the tastes of engineers and, thus, were neither useful nor reliable.[88] Geographically and functionally set apart, neither establishment was the locus for examining the precise cause of the mechanical malfunctioning that showed typically among new flying machines. The improvised "solution" often stemmed solely from an inspection on paper.[89]

In 1942, the division between engineers and pilots became irreparable. In October, as a solution, the army created the Test Division for Army Aviation (Rikugun kōkū shinsabu) to functionally integrate the two groups but to no avail. Pilots soon began to play a more vocal, assertive role in assessing and developing the prototype aircraft.[90] Receding from the scenes, army aircraft engineers devoted more of their time to technical support for the aircraft already in service. The army, once capable of designing and creating prototypes, completely surrendered this function to civilian aircraft firms. Only after this administrative reorganization came the complete demarcation of functional responsibilities: civilian firms for designing prototypes and the Headquarters for Army Aeronautics for evaluating the products through testing flights.[91] By the end of 1942, the army aircraft engineers had been functionally reduced to administrators.

Building Navy Air Power: Institute for Navy Aeronautics

For both the army and navy, concentrating various assets in and near Tokyo was important for advancing research and development in aircraft engineering. The safe, effective, and rapid dissemination of technical information within and between institutions in writing and face-to-face was crucial. The stronghold of the army's research in the field was located in Tachikawa, Tokyo, the navy's in Yokosuka, Kanagawa. Each site had a runway for airborne transportation. Meetings among engineers, held frequently, could thus benefit immensely from the geographical proximity.

Compared to the army, the navy more skillfully pulled resources for research in aeronautics from across the country. This effort was embodied in the creation, content, and use of the INA, the birthplace of a series of highly advanced aircraft from 1935 to 1945.[92] Located in the city of Yokosuka, the INA was the navy's sole research establishment in science and technology for air war. It was the country's finest research organization in the field and played a central role in building the naval air power.[93] To conduct its research, the navy drew its core facilities and personnel to this site from across the country—such as those in the aircraft-related divisions at the Navy Technical Research Institute and Yokosuka Air Arsenal, as well as the engineering workforce from the Navy Arsenal in Hiroshima.[94] At the time of its inception on April 1, 1932, only 113 employees (101 military personnel and 12 clerks) occupied the institution.[95] The research ability of the INA took a few years to mature. Shortly after its creation, the organization earned the unhappy nickname, "empty arsenal," for its meager human and material resources for research.[96]

The INA quickly gained its footing and displayed its ability by creating the Type-96 carrier attacker, B4Y. To replace the earlier B3Y model, the Headquarters for Navy Aeronautics issued a design requirement for a carrier-based attack bomber. But neither Mitsubishi nor Nakajima came up with a satisfactory prototype. When it won the design competition, the INA cleared a path for the civilian aircraft companies to follow. The success stemmed from the navy's strong research ability and facility. Rigorous strength testing of aircraft parts minimized the weight of the biplane, and advanced research prevented various parts of the aircraft from vibrating uncontrollably during its high-speed flight.[97]

The INA almost single-handedly led the navy's pursuit of technological independence in aircraft engineering. Among its multiple functions for research and administration, the main ones included (1) conducting comprehensive engineering research in aeronautics, (2) sharing research results across institutional boundaries, (3) administering projects assigned to commercial firms, and (4) planning future navy aviation.[98] The INA laid the ground for the diffusion of engineering knowledge as the intermediary between the navy and civilian aircraft companies. The navy engineers supported experiments conducted by the civilian firms, a practice which could lead to novel ideas; the navy also acted as an intermediary, relaying useful ideas and test results from one company to others, closing their technical gap in the process. The Aircraft Department, in particular, maintained close contact with the aircraft companies in nearly all stages of prototyping.[99] To support these functions, the INA provided many advanced research facilities. Its tangible assets within the compound included 91 buildings (34 factories, 33 laboratories, 9 warehouses, 15 offices, and others) and a great many machines (985 machine tools, 21 industrial machines, 16 electric machines, 1 woodworking machine, and 20 testing machines). Among the most notable assets were 7 different types of wind tunnels, one of which produced a gust of wind with a maximum speed of 30 meters per second within the 3-meter cross section, the most powerful one in the country.[100] Five months before the Pearl Harbor attack, the INA employed 289 researchers, including senior/junior engineers, professors, and technicians.[101] By the end of the war, the organization had grown into a giant establishment with 14,301 employees in 12 research departments.[102] In November 1945, the Allies' occupation authority examined "a complete range of test laboratories, machine shops, foundries, forges, metal shops, assembly shops, etc." at the INA, concluding that its "functions were equivalent to MAF [Mustin Air Field], Philadelphia, Wright Field [in Ohio] and NACA, Langley Field [in Virginia]," the best research and development centers for air power in the United States.[103]

Esprit de corps was actively promoted at the INA through indoctrination using ritualistic practices. For instance, guidelines for employee education emphasized central ideas, all encouraging patriotism and Confucian ethics: loyalty and sincerity, diligence in perfecting weaponry, engineering service to the nation (extending knowledge, accumulating experience, and honing technical skills to maximize efficiency), strict discipline, and fortitude and vigor. These written codes of work ethics were picture-framed and enshrined at various plants within the INA compound. And employees recited the writing in unison every morning.[104] Music was equally important. Regularly, INA employees sang together and shared their elevated pride in the navy, technology, and nation. The lyrics praised the INA that proudly "bore the destiny of technology on its shoulders." Engineers were "brothers equipped with wisdom and intellect" and devoted to experiment, research, and maintenance of machinery for the nation. "Look up at machines flying in the air with a whirr," the lyrics continued, for "they crystallize our sincere, patriotic service to our engineering nation."[105]

The most remarkable asset of the INA was its research engineers. In their recruitment, its founder and director, Hanashima Kōichi, played a chief role with his strong background in naval engineering. This vice-admiral visited institutions of higher learning across the country in search of the top talent. Dressed in full military uniform, with his chest covered in awe-inspiring decorations, he displayed remarkable sincerity and commitment to recruitment during his visits, and he reviewed letters of recommendation from instructors at imperial universities. His affable, almost self-effacing personality drew future engineers and technicians into the INA.[106]

Manpower in the research establishment was also a product of Tōdai old-boy networks and, more importantly, the geographical proximity between Tokyo and Yokosuka. One means for the establishment to recruit talent was a summer training program in which engineering students gained hands-on experience at civilian firms, arsenals, or research establishments for one month every year during their three school years. In July 1941, 143 engineering students at Tokyo University participated in the program; a distinct minority, only 11, worked at two civilian

A group photo of aeronautical engineers from the Aircraft Department, the Institute for Navy Aeronautics, August 11, 1938.

firms with strong ties to the army, while the rest—132, or 92 percent—worked in one navy arsenal, five navy-related civilian firms, and above all, the INA.[107] There was a strong correlation between the students' places of summer employment and the locations of their educational institutions. Students at Kyoto University, for instance, were trained at aircraft manufacturers near the city, such as in the Hyōgo, Gifu, and Nagoya areas. Furthermore, students at Osaka and Nagoya Universities worked at Mitsubishi in the city of Nagoya. These strong correlations were by no means accidental. Both the army and navy respected the students' preferences, accordingly assigning them as much as possible to the sites of their choice. This system favored the INA, which dominated the recruitment of future engineers from Tokyo University. Among 55 Tōdai engineering students in the summer of 1942, none worked in the army-related facilities; INA obtained 40, or 73 percent, and the rest, 15, worked at a private commercial aircraft company in Tokyo.[108]

The navy's establishment of its research and development center in Yokosuka was by no means smooth. Before the INA was created, the whereabouts of its location was a matter of contention. One group rightly supported Yokosuka because of its ideal location for seaplane development, facing an ocean. The site seemed logistically sound. It would bring airmen's air tactics and engineers' expertise together in this city that already hosted the Yokosuka Air Corps. The opposing group supported Tokyo, emphasizing the crucial importance of the navy's close collaboration with Tokyo University for successful research operations. Also in their view, the aircraft noise of the Yokosuka Air Corps could distract researchers from pursing their quiet inquiries. More serene locations in Tokyo, such as Tsukiji, the site of NTRI, were considered better suited as the research site.[109] Favoring the city of Yokosuka in the end, in May 1931, the navy purchased land strips of roughly 10.8 acres (13,175 tsubo) for a reasonable price, 82,652 yen, from local landowners.[110]

This decision in favor of Yokosuka proved wise in the following years during research and development in, for instance, aeronautical radio technology. Before 1933, the research site for the development of electrical engineering had been the NTRI in Tokyo. Its location, however, turned out to be less than ideal for a series of field experiments that required the use of an airfield. To resolve this issue, the entire research division was relocated in 1933 to the INA in Yokosuka, where engineers later developed a series of new aeronautical radio prototypes and tested them in flight.[111] A venturous spirit pervaded with ample financial resources. Unlike profit-seeking commercial firms, the navy could afford costly, time-consuming research projects that might yield no profit in the end. Such was the case in materials engineering (conducting strength tests for aircraft parts) as well as in aerodynamics (wind tunnel experiments with scale models to observe and solve their vibrations in airflow).[112]

At the INA, aircraft development was possible in a close circulatory loop of feedback among research, prototyping, and field testing. This process of communication required the formation of a small, mutually supportive group of engineers and pilots. The natures of the two professions initially presented a major source of difficulty; engineers were rarely experienced test pilots, and vice versa. In general, the faster the circulation of concrete and accurate information between the developer (engineers at the INA) and the end user (airmen in Yokosuka Air Corps), the faster the introduction of the flying machine was into war use. This circulatory feedback mechanism was crucial in aircraft development. In the process, first, a test pilot felt the flight characteristics of a prototype and observed its acceleration and orientation on flight instruments. The engineers needed to understand his highly subjective, ill-defined experience, with respect to, for example, controllability and stability, while the pilot needed to explain his experience in more concrete, objective, and measurable quantities. Meanwhile, the engineers needed to identify the pilots' needs and translate them into criteria specific to the hardware in development.[113] At the INA, this closed-loop feedback mechanism played out in both the organizational and individual settings. At one level, the engineers at the Aircraft Department developed aircraft prototypes, which the test pilots at the Flight Inspection Department flew. From 1941 to 1943, unlike the army, the navy turned five highly educated aircraft design engineers into qualified test pilots, internalizing the research and test-flight functions at the level of individuals.[114]

Compared to the army, the navy was more successful in leading the field to technological independence from foreign powers. Creating effective technology policy was crucial, and it was a top-down initiative with strong leadership. One chief figure was Admiral Yamamoto Isoroku, the head of the technology bureau at the Headquarters for Navy Aeronautics during 1930–1933. In a war of vast organizations and impersonal strategies, an individual leader's impact was not always decisive, but his role, like his masterminding of the Pearl Harbor attack in 1941, was indispensable for the navy to move irreversibly into the age of aviation. Yamamoto was remarkably free from blind faith in the cult of the battleship that dominated maritime strategic thinking. He remained realistic and cautiously optimistic. He was well informed. Having the information from outside of Japan, he cultivated the potential of air power in modern warfare.

Yamamoto and his replacement, Matsuyama Shigeru, planned strategically for an air power buildup through the "Independent Aircraft Technology Plan" (kōkū gijutsu jiritsu keikaku). This official three-year plan, implemented in 1932, proved highly effective in breaking the navy's dependence on foreign aircraft designs. This uncompromising plan pushed for the domestic production of naval aircraft and engines, all to be based entirely on Japanese designs. Until 1932, the military had emphasized ends over means in technological development; it awarded production contracts to the aircraft manufacturer that created the best products in intercompany competitions, regardless of the nationalities of the design engineers. This old model, however, had revealed major drawbacks by the 1930s. It allowed each manufacturer to hire foreign engineers temporarily as the easiest, fastest means of winning a production contract and earning revenue. In 1932, the navy restructured this system in accordance with nationalistic imperatives. Japan's increasing political isolation in the international arena after the Manchurian Incident of 1931, in which the Japanese army had invaded Manchuria in northeastern China, called for such a new measure. The resulting procurement system began to emphasize means over ends; that is, the designers of all aircraft and engines were to be Japanese engineers.

The new, centralized arrangement called for the pairing of the firms in competition for developing prototypes. The navy awarded the contract to the firm whose prototype successfully met its specifications. The losing firm was to produce its competitor's design, or to produce the engines for the aircraft, as a second-source supplier with the incumbent sense of shame. This system constructively fanned the competitive fire in each firm. It effectively mobilized all civilian companies toward the single goal of developing naval aircraft by using only Japanese aircraft engineers. Soon thereafter, the aircraft industry engaged in competitions, integrated components, and built aircraft designed entirely by Japanese engineers.

Research and development, production, and maintenance were subsequently integrated under the navy's watchful eye. The navy successfully managed the entire process in accordance with the design specifications after a series of discussions about tactical thinking and available technology.[115] The navy's general guidelines of 1933 set civilian firms as the manufacturers of aircraft arsenals as the repairers of the machines. The functional demarcations were clearly drawn. With a strong research capability, the INA pursued aircraft prototyping for experimental purposes, whereas the private firms created aircraft for specific tactical use, such as deck-landing fighters, bombers, and attackers. The civilian engineers and military officials actively exchanged technical information in writing and at meetings. This fruitful strategy enabled the navy to successively introduce a series of highly acclaimed aircraft, for instance, the Mitsubishi A5M fighter Claude, the Mitsubishi G3M attack bomber Nell, the Nakajima B5N carrier attack bomber Kate, the Kawanishi H8K flying boat Emily, and the Mitsubishi A6M Zero Fighter.[116]

The prevailing engineering culture at the INA was an unrelenting pursuit of excellence often at the expense of the top-down order. In building the best flying machines for war, the engineers were entitled to a great deal of flexibility in making technical and administrative decisions, and they freely exercised that power, which the navy allowed to continue within reasonable limits. The top-down hierarchy was not rigidly enforced. As a former navy engineer recalls the first meeting he attended, the participants were treated equally and opined freely about prototype aircraft, despite their different military ranks. Modesty was unnecessary. Their free discussions pleasantly dispelled the stereotype of this new hire who had wrongly believed that the military operated in a strict hierarchy. Once their decision was finalized, however, following it was a matter of course.[117] The engineers certainly formed an integral part of the navy command structure, but they were somewhat set apart from the top-down military culture. Technical knowledge was power—and the engineers knew it well. Their ethos was not to blindly follow military orders but to examine all kinds of questions to find the right ones, and the right answers, in the pursuit of excellence.

The navy's engineering culture left a lasting impression on Tani Ichirō, Japan's leading expert in aerodynamics at Tokyo University. This academic researcher, with considerable exposure to the army's engineering culture, regularly attended study meetings at the INA, often coauthoring research reports with navy engineers.[118] By his own account, what impressed him during a symposium was the depth and scope of the navy's research in his field of study. Much of the technical information—such as analyses of airfield test flights and the collection of empirical studies on scale models of prototype aircraft in wind tunnels and water tanks—was new even to Tokyo University researchers. On the research agenda was how to avoid shock waves once a high-speed flight reached the speed of sound, an issue solved in Japan only after World War II. Study meetings typically left the participants with "homework" for more rigorous analyses. A carefree atmosphere encouraged active participation and frank discussion; seniority had no room in debates; and no clapping of hands, for courtesy or otherwise, followed presentations. These meetings, probably impossible in the army, were an optimum setting for exchanging technical information—and in Tani's view, they were more open, scholarly, and productive than many of the academic conferences he had attended.[119]

A community of private firms and Tokyo University benefited from the somewhat orchestrated diffusion of engineering knowledge in writing and meetings. And the INA set the stage by providing its technical expertise and guidance.[120] Among many types of engineering meetings, one of them focused on practical, technical problems in aerodynamics and hydrodynamics, while another was devoted specifically to issues related to aircraft structure, strength, and fittings.[121] In addition to minutes of the meetings, research results were shared in writing. For instance, about 100 copies of a laboratory report were distributed from the INA to other navy institutions, the marine meteorological observatory, as well as university and army research establishments.[122]

In the world of navy aeronautics, technical information moved not only horizontally and top-down but from bottom-up. In discussions about future aircraft plans, engineering initiatives often came from men of lower ranks to high-ranking officials. The navy leadership rightly observed that young technical officers—who read foreign journals and spent time in laboratories and airfields—best understood the technical circumstances surrounding aviation. Their voices were sought out, listened to, and examined; then, what seemed plausible and useful was implemented. The prevailing atmosphere encouraged even petty officers to often explore, explain, and submit their ideas in writing. Upon reviewing such proposals, appropriate departments of the navy headquarters or research establishments conducted further research, and the promising ideas were shared within the entire navy. This spread of information was possible within the closely knit communication system of the headquarters, Yokosuka Air Corps, and research establishments.[123] The relaxed and encouraging atmosphere was a prerequisite for INA engineers to relentlessly pursue excellence in their research projects.

A series of highly advanced, experimental aircraft born at the site was a product of the navy's successful recruitment of able engineers, their unconventional thinking, and their uncompromisingly detail-oriented approach to engineering—all represented by three Tōdai graduates who had gained considerable experience at the INA before the outbreak of the Sino-Japanese War in 1937: Yamana Masao, Miki Tadanao, and Matsudaira Tadashi. These senior engineers represented the navy's research efforts bearing fruit for war. More specifically, they enabled the navy to own its most advanced flying machines by minimizing their drag resistance, structural weight, and flutter vibrations. A point to note was that streamlining, as well as the reduction of weight and vibration, was among the most common and most important determinants for successful flying machines across nations. The three INA engineers, among others, honed their skills of heuristically solving technical problems and cultivating intuition and judgment along the way. What they saw as the fundamental problem, why they did so, and how to solve it, all formed the basis of research and development at the INA for the D4Y dive-bomber, P1Y twin-engine bomber, and A6M Zero Fighter.

Case 1: Yamana Masao in Streamlining the D4Y Dive-Bomber

In 1938, Yamana Masao led his design team in the 13th Experimental Dive Bomber Project, which resulted in the product known to the Allies as the D4Y Judy. His effort at the INA epitomized the navy's scientific research in the field of aerodynamics. Born in 1905, he was an avid airplane enthusiast in his childhood. He graduated from the Department of Aeronautics at Tokyo University, and joined the navy through its recruitment system, the Program of Learned Workers (yūshiki kōin seido).[124] His considerable experience in minimizing drag resistance bore fruit in a high-speed flight record in the navy. His brainchild, the Judy dive-bomber prototype, was originally designed neither for mass production nor for aerial combat, but solely for research purposes.[125] The aim of the project was to train engineers at the INA, to deepen their theoretical and empirical knowledge about aircraft development through data gathering and analysis. Such a project would have been impossible in the army. It was the navy's way of accumulating technical expertise in designing advanced aircraft and leading civilian aircraft companies in the process.[126]

The navy's concerted effort often went beyond comparable development found in the West. What drove this experimental project was the navy's faith in the tactical importance of dive-bombing against the fleets of its primary hypothetical enemy, the United States. During 1935–1936, a group of INA engineers and technicians visited Western Europe and observed German aircraft manufacturers, including Heinkel, to gain useful technical information.[127] Yamana proved perfect for the mission, given his aptitude for foreign languages; he was fluent in English, and he had written his senior thesis about the stability of aircraft parts at the Tōdai Department of Aeronautics in German.[128] The group's preliminary investigation for six months resulted in the navy's purchase of two Heinkel 118 dive-bombers for further testing in Japan and possible domestic production. Careful analyses of the aircraft at home revealed that its large size, heavy weight, materials, and flight characteristics would not complement the navy's vision. Later, the navy's domestic production plan ended with the crash of one Heinkel airplane during a dive test.[129] The foreign products were inspirational but unsuitable, unsophisticated, and unreliable.

Upon his return, Yamana led the 13th Experimental Dive Bomber Project at the INA. His successful design was a product of the team's relentless streamlining of the prototype on the basis of sophisticated theoretical studies of high-speed aerodynamics. His design policy was single-minded, but it brought navy tacticians' idea to fruition. The performance specifications of the dive-bomber included four components: it should fly faster than the state-of-the-art fighter aircraft in development, the A6M Zero, by 10 knots or 18.5 km/h; it should have superb maneuverability and cruising power; it should be able to carry a maximum bomb load of 500 kilograms;[130] and the structural parts of the aircraft ought to be strong enough to endure high-speed diving, that is, be capable of withstanding a maximum load of nine times the force of gravity, or 9 Gs.[131]

Deep skepticism initially surrounded this experimental project at the INA. These high demands required ambitious research and capable engineers under strong leadership. Yamana was painfully aware of the limited engine power available at the time. The stronger the power, the faster and higher the flight would be. He was determined to create the lightest, smallest aircraft with the least drag resistance—hence, his relentless pursuit of streamlining the project.[132]

Nothing particularly unique stood out in this design philosophy, but *how* it was pursued was unconventional by navy standards at the time. Yamana intuited, and later empirically proved that the aerodynamic design of the German Heinkel 118 was inspiring but scientifically crude. He carefully examined the two-dimensional model of the aircraft's coolant cover in a passing water tank. His experiments revealed that the air flowing into the cooler created a vortex around the external bomb attached at the bottom of the fuselage during climb and high-speed flight.[133] After a series of scientific experiments, Yamana's team created a smaller and more efficient cooler that produced less drag. The result was a far more aerodynamically sound coolant and slim fuselage that internalized the bomb-bay.[134]

After this finding, Yamana questioned what had been traditionally accepted in the designs of foreign and domestic aircraft. His design team paid attention to the navy's then widely used parachute, on which airmen sat in the cockpit. At his request, a navy laboratory created a new parachute that was portable on the pilot's back. With its height reduced by 100 millimeters, the fuselage weighed less, and the smaller cross section gave way to a slimmer, more aerodynamically clean shape.[135] Soon, Yamana began to doubt the navy's predominant use of an air-cooled aero-engine, a product less vulnerable to gunfire and less compli-cated in structure. Since it was easier to maintain than the water-cooled type, it was deemed ideal for naval combat flight. After scientific research in pursuit of the aerodynamically cleanest design, Yamana unconventionally chose the water-cooled type. It minimized the engine's diameter, fuselage's cross section, and drag, with the same power output. Subsequently, Yamana was at greatest pains to figure out the optimum shape and location of the engine. His obsession with streamlining every conceivable part in this scientifically experimental project was at the heart of his dive-bomber that could fly faster than the fighter aircraft.[136]

Yamana's success also owed to a new aircraft design technique that departed from the traditionally used ship-design method. A three-dimensional object such as a ship was heretofore represented in reduced drawings from frontal, side, and plain views. This old method was not immune to misjudgment and inaccuracy during designing at a mold loft, especially when the drawings were enlarged to full size. The requisite adjustments were painstaking, labor intensive, and time consuming.[137] Yamana's team came up with the new, alternative design tech-nique, which reduced the surface of a solid into fairly simple algebraic graphs. Such a mathematical representation of the fuselage, main wings, canopy, and other aircraft parts enabled the team to calculate the skin friction resistance more accurately.

The benefit of this design technique was measurable. Skin friction resistance aside, the scale models of Yamana's design showed virtually no drag in a wind tunnel.[138] The scientific data were astonishing. The surface area under the drag divided by weight—0.144 m²/103 kilograms—was the smallest among *all* the Japanese aircraft of the same generation. This figure was comparable to that of high-speed racing planes in the United States and Western Europe.[139] America's wartime report concluded that this aircraft with its "clean" design "may become the most important carrier-borne dive bomber in the Japanese Naval Air Force."[140] The relentless pursuit of streamlining bore fruit at the INA, and the legacy of this new aircraft design technique outlived the wartime machine. After 1945, Yamana's wartime associates, including Miki Tadanao, resurrected this method in the aerodynamic design of the high-speed rail cars.[141]

Case 2: Miki Tadanao in Minimizing the Weight of the P1Y Bomber

Born in 1909, Miki Tadanao was four years junior to, and one rank below, Yamana, with whom he worked at the INA. Similar to Yamana, anything related to aircraft thrilled Miki during his childhood. His father, a cavalry veteran who had remained unhappy with the army during and after the Russo-Japanese War, steered the young Miki away from joining. After graduating from the Naval Engineering Department at Tokyo University in 1933, he joined the navy's Program of Learned Workers. At the INA, he shared with many engineers Japan's need to create its own independently designed aircraft. As a trainee at Hiro Navy Arsenal, Miki was involved in projects to develop all-metal flying boats and water-cooled engines, all designed by Japanese engineers. By his own account, his education at Tokyo University helped to shape his later career in the fields of hydromechanics and structural mechanics. He kept multiple volumes of class notes he had taken on his favorite subject of study, vessel design and structure. His senior thesis examined ventilation in a ship's hold.[142]

   Under Yamana's supervision, Miki played a central role in designing the twin-engine, long-range bomber, known to the Allies as P1Y Frances. The 15th Experimental Navy Bomber Project of 1940 bespoke the navy's engineering culture that espoused uncompromising effort in prototype research and development, often against seniority. The P1Y bomber realized the navy's ambition that was revealed in the design specifications; it ought to carry on low-altitude assaults, as well as torpedo and dive-bombing attacks, with a high dive speed limit of 648 km/h. Moreover, the aircraft's top cruising speed was set at 555 km/h, faster than ordinary fighter aircraft. The bomber's range, 5,500 kilometers, was too long for a fighter escort, and its speed was fast enough to be able to evade the enemy's assaults in the air. In Miki's mind, the key to success was to minimize the structural weight, frontal area, wing area, and the amount of equipment on board and to integrate all the components in the smallest package possible. The new algebra-based design technique, whose success was proved in the D4Y Judy, formed the foundation of the P1Y bomber's highly refined curvature.[143] America's wartime reporting praised the bomber's "long, slender appearance."[144]

Miki's belief in the importance of the aerodynamically cleanest design had no seniority-based boundaries. When he inspected a prototype of the Mitsubishi Type-1 land-based attacker (G4M or Allied code name, Betty) at the INA, its fat fuselage caught his attention. The inner space in the tail was large enough to carry cannons, a gunner, ammunitions, and fittings as required by the navy's design specifications. The cigar-shaped fuselage embodied the careful design by a senior engineer at Mitsubishi, Honjō Kirō, a graduate from the Department of Aeronautics at Tokyo University. He was eight years senior to Miki with considerably more experience in aircraft design; for instance, in 1926, when Honjō wrote his senior thesis in English about wing's structure by using German sources, Miki was only a ninth-grade student with no exposure to actual aircraft.[145] At one meeting, the quick-tempered Miki sharply criticized the dull-looking, aerodynamically crude design of the tail, which, in his view, resulted in unnecessary drag resistance and structural weight; the inner space of the tail was needlessly large, or rather, big enough to be "a dance floor." Saving no face, Miki insisted on the importance of a clean, streamlined shape that minimized the surface area, friction with airflows, and weight. The meeting ended inconclusively,[146] and Honjō and Miki went their separate ways.

At the INA, Miki's obsessive weight reduction defined much of his design projects from the beginning. In his P1Y bomber, the number of aviators was reduced to three (pilot, observer, and radio/dorsal gunner). The number of cannons was also minimized to save weight and, thus, increase speed. Miki and Yamana had scrutinized the drawing of each aircraft part, checking, for instance, the thickness of the aero-engine cover to minimize its weight. Earning their approval was, as one engineer recalls, "very difficult." At the entrance of the prototype assembly plant was the gatekeeping section, where *all* the aircraft parts were first weighed with scales—and these parts were compared with precomputed weight figures in the original design.[147] After careful structural analysis, numerous holes were drilled in supporting materials to save weight. Plates of all kinds became thinner, and parts of the aircraft used the lightest materials possible.[148] Miki encouraged group leaders individually to pursue a relentless weight reduction in developing the aero-engine, propeller, wheels, and other parts.[149]

The project bore fruit at the expense of the aviator's comfort. "The whole cockpit is exceedingly cramped," concluded an American inspector of the bomber.[150]

The space within the fuselage was so small that the maximum width of this twin-engine bomber, 1,200 mm, was the same as that of the small single-seat, mono-engine fighter Zero. Unable to go back and forth in the fuselage, the three airmen uncomfortably communicated with each other through microphone and earphone.[151] In addition, the bomber used high-pressure rubber tires, the diameter of which was 20 percent smaller than that of the comparable bomber, Mitsubishi G4M heavy-bomber Betty, to minimize the weight of the landing gear.[152]

Case 3: Matsudaira Tadashi in Reducing the Flutter Vibration of the A6M Zero Fighter

The genesis of the legendary A6M Zero illustrates several pivotal factors at play, namely, active collaboration between public and private institutions, the navy's successful recruitment of engineers, and their pursuit of excellence in research at the INA. The aircraft's success, which stemmed from its agility and light weight for dog fighting, was legendary among the Allies during the war, but it was by no means preordained. Its precursor, the 7th Experimental Fighter, ended in technical failure. During test flights, all the prototypes were lost in uncontrollable spin and crashes because they suffered from a major problem with respect to aerodynamics. The prototype aircraft's appearance impressed no one. The chief designer, Horikoshi Jirō, likened the product's crude appearance to "a slow-witted duck"; one navy officer described its incongruous appearance as similar to a Japanese peasant grandmother dressed in Western clothes and high heels.[153] This fighter prototype, however, merits our attention, for it was the first product of the INA and the navy's "Independent Aircraft Technology Plan," the combination of which paved the way for all of the navy's successful carrier-based aircraft from 1932 to 1942.

This 7th experimental project capitalized on a network of engineering expertise in the navy. Upon receiving the navy's performance specifications for the next deck-landing fighter plane, in 1932, Mitsubishi selected Horikoshi as the leader. Anxious and confused, he wondered why such an "immature" 28-year-old engineer had been appointed at all. As it turned out, his lack of practical experience in aircraft prototyping proved advantageous to pursuing an unconventional way of thinking. His pioneering spirit prevailed. As the chief designer, he determined the prototype's basic configuration, which eventually became the single-wing-type aircraft without external supports. Lacking useful data about the configuration, Horikoshi turned to the INA for professional advice through his personal network of Tōdai alumni. Lieutenant Commander Saba Jirō, in charge of the navy's fighter design at the establishment, encouraged Hirokoshi to design the world's first cantilevered carrier-based—not land-based—monoplane. This was a historically revolutionary decision in navy aircraft development.[154]

An explanation is needed for this distinction to illustrate Horikoshi's and the navy's commitment to technological independence. Against the law of gravity, an airplane needs to lift its own weight, including its energy source and pilot(s). Among different sizes of wings, smaller ones are desirable because a substantial portion of the structural weight of a plane resides in the wings. But so-called trade-off factors in airplanes pose a conundrum. Smaller wings will result in a greater load per square foot on the wing (wing loading), necessitating a longer takeoff run and a higher landing speed.[155] Deck-landing aircraft need a particularly slow landing speed and sufficient lift to take off in the short distance available on a carrier. This factor makes lower wing loading indispensable. Thus, aircraft designers worldwide, well into the late 1930s, adhered to the paradigm of biplanes as the predominant form of carrier-based fighters. They had a large wing area (thus, small wing loading), adequate lift for takeoff, and maneuverability for landing in the limited space available on aircraft carriers.

Later manufactured by Mitsubishi and Nakajima, the Zero Fighter was a product of the elite engineering workforce that solved highly technical problems at the INA. A key figure was Matsudaira Tadashi, a former army second lieutenant recruited into the navy. Born in 1910, and having studied hydrodynamics, he graduated from the Department of Naval Engineering at Tokyo University. He received military training in the army but moved to the INA in 1933 to gain hands-on experience in the navy's research environment before the outbreak of the Second Sino-Japanese War.

His wartime accomplishments included solving the flutter phenomenon, which commonly existed in agile, high-speed fighter aircraft. To describe this problem in aeronautical terms, flutter meant a self-induced vibration that could occur in many parts of an aircraft, such as the propellers, engine, wing, tail, ailerons, elevators, or rudder. Flying at higher speeds tended to increase the vibration amplitude of these parts, causing the aircraft to become less stable. In the worst-case scenario, planes could suddenly disintegrate in midair, often leaving no survivors who could explain what had happened. During the 1930s, researchers worldwide placed a greater emphasis on theoretical analysis and wind tunnel tests to predict both the magnitudes and phases of aircraft parts in vibration.[156]

Matsudaira built his career by solving this highly technical problem in the navy from 1933 to 1945. In 1935, a type-96 twin-engine army bomber showed signs of elevators-tail flutter which was, as his investigation proposed, successfully solved after directly connecting the two separate ailerons. The next year, there was a series of flutter casualties involving the type-96 carrier-landing fighter as well as the type-95 army fighter aircraft. As the solution, Matsudaira installed the mass-balance weight, which prevented the aircraft's elevators from fluttering.[157] On earlier airplanes, which could not fly as fast, such weight was not necessary. But it became an indispensable attachment because the fighter aircraft performance improved, and its speed increased, particularly during a steep dive or a dogfight.

The successful development of the A6M Zero Fighter was the high point of Matsudaira's navy career. In 1940, a series of Zero midair disintegration incidents posed a major conundrum. Investigations to pinpoint the exact cause of the Zero accidents reached a dead end because the pilots died almost instantly when the aircraft shattered into pieces in midair. Subsequent studies produced mixed results. Inclusively, previous research had indicated that the aircraft—at least theoretically—would not flutter until it reached a speed of over 750 km/h. This datum, as Matsudaira revealed, was highly optimistic and dangerously erroneous. His painstaking research showed that the prevailing flutter model failed to simulate exactly all aerodynamic characteristics of the full-size aircraft, including the distribution of stiffness, weight, and air-load. He conducted vibration and stiffness tests on the actual aircraft and then incorporated the research data to make a new experimental flutter model for laboratory testing. His new calculations demonstrated that the critical flutter speed for the Zero was only 600 km/h with aileron balance tabs, and 630 km/h without them.[158] This research result was fed into subsequent modifications of the Zero. Until the war's end, Matsudaira's research helped push the speed limits of flying machines with tolerable flutter. He also simplified flutter calculations by developing tables so that engineers could avoid complex calculations. Minimizing the risk of the aircraft's flutter, his research reduced the uncertainties of designing navy aircraft for war.[159]

The fast rise of navy air power resulted, in part, from the bitter interservice rivalry after World War I. The war stimulated the research interests of both civilian and military officials, and it was the navy that excelled by the time of the Pearl Harbor attack in December 1941. It successfully built a research infrastructure, recruited able engineering students from institutions of higher learning, and skillfully applied its skilled manpower to advanced research and development in the field of aeronautics.

The external war that began in July 1937 exposed the limited availability of capable, experienced engineers in society. In the ensuing competition between the army and navy, each tried to maximize its share of the limited pool of human resources, while gradually undermining the other's war-making capability. By

1942, the navy had won this quiet, subterranean domestic war against the army. The prize was highly educated engineers from the nation's top universities and their cognitive abilities for advanced research and development. By 1942, the navy aircraft engineers had prepared the nation's military technology for air war.

The navy's success partially stemmed from the work environment it had created. Its engineering culture balanced top-down and bottom-up initiatives, while its army counterpart chose to embrace outsourcing. Determined to gain technological independence, the top leadership steered the navy into the age of air power led chiefly by the INA. The navy's relentless pursuit of excellence at all costs in minimizing drag resistance, structural weight, and the hazardous vibrations of aircraft parts was possible before 1942, as long as there were more opportunities for than constraints on research and development. Engineers could devote their time and energy to new product development rather than to improvement and repair. At the Department of Materials Engineering at the INA, for instance, student trainees gained useful, practical experience in developing a stronger aluminum alloy with ample time and material resources. Basic, advanced engineering research was in full operation.[160] The atmosphere of victory and optimism remained predominant on the evening of the Pearl Harbor attack.[161] However, this optimism dissipated in the next two years, replaced by pessimism marked by self-imposed constraints in conducting research projects for war against the Allies—resulting in the deaths of kamikaze suicide operatives by engineers.

CHAPTER THREE

Engineers for the Kamikaze Air War, 1943–1945

"Victory is before our eyes [because] divine lightning will annihilate our enemy with a single stroke," wrote Kawabata Yasunari on June 1, 1945. This wartime navy correspondent in the South Pacific later achieved worldwide fame as the first Japanese to earn the 1968 Nobel Prize in Literature. The "divine lightning" on which he reported in wartime *Asahi Shinbun* was the navy's wonder weapon, the rocket-propelled flying machine, MXY7. It was designed and deployed specifically for one-way suicide missions against the enemy fleet on the march. To cheer up his readers, Kawabata wrote evocatively, "Death awaits our enemy fleet at the sight of [this] divine lightning . . . [and] Our enemy is terrified" especially after the "hair-raising" scene of one attacker sinking an American destroyer in one blow. He urged the newspaper readers on the home front to "produce aircraft, produce aircraft [because] our victory by [this] divine lightning is near at hand."[1]

This exhortation echoed one month later among military engineers. Tada Reikichi, a physicist and the head of Army Technical Research Institute (rikugun kagaku kenkyūjo) and later the Technology Bureau, under the supervision of the Prime Minister, defined the roles of scientists and engineers in the beleaguered nation. He wrote, "In accordance with the wills of Heaven and the people, the world of science is responsible for creating weapons for special [suicide] attack that can deliver a death blow [to our enemy]." "Scientists and engineers," he continued, "need to embrace the spirit of special attack . . . and promote science" for this purpose.[2] By the spring of 1945, military engineers were mobilized as the architects of deadly technology for Japan's suicide operation for homeland defense, the strategy widely known as *kamikaze*. The call for the engineers to play a role in the special attack operation was by no means sudden; rather, it mirrored the country's weakening research capability at educational institutions of higher learning as well as at military research institutions, including the newly established Central Aeronautical Research Institute (CARI).

A New Research Initiative for Civil Aviation

A close examination of the CARI (1939–1945) could reveal that Japan's hasty mobilization of human and financial resources failed to produce concrete, fruitful results in the short term. Administratively, this civilian research establishment remained under the Ministry of Communications until April 1943 and then, under the cabinet until the end of the war in August 1945.[3] At least initially, the construction project looked promising, with ample financial capital and the navy's technical support. It epitomized a successful political compromise between military and civilian officials.[4] The proposal, submitted in the spring of 1938, asked for 100 million yen for the first phase of construction; a later proposal requested a total of 300 million yen for the second phase. The diet granted 50 million yen for the five-year construction—which grew to the subsequent eight-year project in 1944 with a budget of 100 million yen.[5] The CARI was expected to host a wide array of basic and applied research projects in the field of aerodynamics, hydrodynamics, aero-engine development, flight experimentation, and materials engineering for civil aviation; meanwhile, the ARI was responsible for academic inquiries and basic research for both civil and military aviation. The CARI leadership planned to obtain a complete set of large-scale, advanced research equipment, including a giant high-speed wind tunnel and many facilities for testing full-scale motors.[6]

The research establishment turned out to be more impressive on paper than in actuality, and it faced two challenges. The first was a shortage of material resources for the construction. Lacking steel, copper, and cement, the leadership begged both the army and navy for their material support, but they were only able to respond in a minimal way. Subsequently, the civilian leadership revised the construction plan, prioritizing the construction of four research facilities above all else. By the end of 1944, only two facilities—a medium-sized high-speed wind tunnel and workshop—had been completed. Other construction projects ended prematurely, including an airfield for test flights, for which the CARI purchased facilities and 1,952 acres of land (2.4 million tsubo) in Ibaragi prefecture. Neither facility was ready for use when the war ended. The original plan to set up two wind tunnels did not materialize; only the foundation of one wind tunnel was complete. By August 1945, the construction project had used up 40 million yen, or only 40 percent of the total approved budget.[7] In the eyes of a CARI researcher, a dire shortage of material resources made any on-site research in 1945 "exceedingly difficult" if possible at all.[8] During the Allied occupation, the Technical Intelligence Group inspected the CARI and concluded that while it corresponded "to the N.A.C.A. [the predecessor of the present NASA] in the United States," it did "not have its size or its prominent position in the aviation affairs of the country."[9] Further, the inspection team concluded that "the entire installation was dirty and poorly maintained" and that the CARI "did not enjoy high priority in obtaining materials."[10]

The second challenge was a shortage of able, senior engineers to supervise on site. By the time the CARI became operational in 1939, a great many experienced engineers had served in the military, especially the navy. The CARI drew in experienced military engineers and stretched them thinly across multiple posts. From 1943, senior engineers from the Institute for Navy Aeronautics (INA)—two experts in aerodynamics and one in aero-engine development—served at the CARI on a part-time basis.[11] At the end of that year, 20 engineers and technicians from the navy, army, and other public institutions were working part time at the CARI.[12] Research projects started slowly in this environment. For instance, the first issues of the CARI research journal appeared in 1942 and included results from laboratory tests and lecture materials.[13] Having noted the wartime shortage of materials and senior engineers for research, the Allies' observers concluded that the CARI "made little or no contribution to the war effort."[14]

At the end of the war, the CARI preserved ample human resources. It started with 196 employees in 1939. By October 1942, this number grew eightfold, occupying research, construction, and administrative departments, with 1,081 workers, including 34 engineers and researchers as well as 129 technicians.[15] This increase was the result of technology policies approved by the then Tōjō Hideki cabinet. One "urgent task" was "to set up research facilities [to develop] aviation and nurture [aeronautical] engineers." In their vision, the CARI was to play a central role in the field with "recent graduates selected from universities and technical schools [who] would undergo leadership training [at CARI]," where "other qualified candidates would receive advanced technical education for two years."[16] By August 1945, manpower had grown to 1,526 workers, with an influx of engineers fresh out of institutions of higher learning who could only contribute modestly to successful project completion.[17] The CARI had fallen short of expectations during the war, but its manpower had great potential, as rightly noted by the Allies' inspection team: "The caliber and extent of equipment for research was very poor," but "the personnel was [*sic*] of fairly high caliber."[18]

Higher Education and Research in Engineering

Like the CARI, the hasty mobilization of human and financial resources produced few positive results at institutions of higher education across the country.

Once the Japanese-American war broke out in December 1941, the Ministry of Education paid more attention to the quantity of engineering students rather than their quality. Tokyo University correspondingly accepted more applicants. From 1927 to 1941, the number of freshmen in the Faculty of Engineering fluctuated between 308 and 482, but the annual quota increased massively to 793 in 1942 alone.[19] With the creation of the Second Faculty of Engineering in 1942, the total number of engineering students at the undergraduate level increased from 875 to 2,049, or by 234 percent.[20] Engineers were mass produced nationwide. In October 1943, the Ministry of Education enacted an education policy as part of the State of War Emergency Initiative (kyōiku ni kansuru senji hijō sochi hōsaku). Accordingly, specific higher commercial schools were converted to special technical schools, and private liberal arts colleges into special schools for science and engineering. The ministry also reduced the student quota in liberal arts programs by 17,000, while raising the quota in science and engineering by 6,000.[21]

The quality of engineering education was compromised at all levels, from the top to the bottom. In November 1942, the Ministry of Education shortened the period of study by a total of six months at colleges and universities, high schools, higher normal schools, special schools, vocational schools, and teachers' training schools across the country. Consequently, engineering students at imperial universities received education for two-and-a-half years rather than three years.[22] At Kyoto University, for instance, by 1943, the war had reduced the total number of students on campus by two-thirds, and the remaining students could hardly concentrate on their studies in the middle of the war. Many of them were later dispatched to factories and hospitals for service, depending on their fields of study.[23] As one physics student later recalled, the wartime urgency reduced the university to a prep school for war. His last six months at the university were spent undergoing physical checkups to join the navy. He thus lacked substantive research experience, and education, in his field of study.[24]

The reduced length of study across junior and senior high schools as well as universities alarmed Hiraga Yuzuru, president of Tokyo University at the time. In response to the cabinet's decision in August 1942, he declared that it was "absolutely deplorable" to see the nation's leadership implementing a policy that could cause "a decline in academic ability." He expressed his view in a 26-page booklet, which he subsequently submitted to the Ministry of Education. The government agency followed one of his suggestions. Accordingly, from April 1943, graduate students at the nation's top 12 universities were to remain exempt from conscription.[25] But the number of these students increased only modestly throughout the war. At Tokyo University, for instance, only 4 engineering students entered the graduate program in 1937, and after a gradual increase, the number of newcomers peaked at 31 in 1941. No graduate students entered the Faculty of Engineering in 1943.[26] The total number of graduate engineering students increased gradually from 17 to 40 during 1937–1940; it reached 61 in 1941 and then peaked at 117 in 1944.[27] This intellectual resource remained misused during the war. "After a few years' experience," concluded a US occupation report, "the men were diverted to administrative work, leaving the actual technical work of the laboratories to inexperienced men who lacked the benefit of having conducted graduate work."[28]

Meanwhile, wartime demands overloaded university faculty, as experienced by Professor Tani Ichirō, the nation's leading authority on aerodynamics. As of May 1940, at the age of 33, this researcher at ARI in Tokyo regularly visited the INA in Yokosuka across 45 kilometers to consult on research in the fields of airfoil design and boundary layers. He also worked part time at the IAA and Army Technical Research Institute, while serving on the research installation committee at the CARI, all located in Tokyo.[29] After Tokyo University established its Second Faculty of Engineering in Chiba prefecture in 1942, his already busy schedule worsened with the additional duties of teaching. He lectured on aerodynamics for 14 hours every week on campus and supervised his two to four students a year as they wrote their senior theses. He also conducted research at the ARI, located 45 kilometers away, for the army and navy to develop advanced, experimental aircraft.[30] His time, energy, and attention remained divided like that of any other university researcher. During the war, concluded a U.S. occupation report, "university staffs undoubtedly constitutes [sic] a pool of untapped scientific ability which far exceeded the talents available within the military and industrial laboratories . . . In comparison with American and Britain [sic], University researchers in Japan were only about one-tenth mobilized for war."[31]

Nor could engineering students concentrate fully on their education as long as their hands were needed elsewhere. Those in the Second Faculty of Engineering at Tokyo University underwent a full course of military drill in the playground—including crawling, shooting practice, and assault exercises—for two hours a week, and after 1944, four hours a week.[32] The students' muscles also seemed useful for agricultural labor. Many Tōdai engineering students were mobilized to farm in Tokyo, Chiba, and Gunma in preparation for a potential food crisis.[33] At one point in 1943, the entire Second Faculty of Engineering was required to grow castor beans for lubricant oil for aero-engines. Those in the Department of Aeronautics successfully produced roughly 45 kilogram and met the demand for that year, while raising pigs and farming a variety of products—such as peanuts, radishes, carrots, eggplants, and tomatoes—in allotted sections on campus.[34] From

May 1944, following the example of the liberal arts students, the science and engineering students lost their draft exempt status and subsequently worked at government agencies and munitions companies.[35]

While the Tōdai engineering students were busy preparing for a food shortage, the threat of air raids posed by the American forces paralyzed much of the engineering research and, above all, education across Japan. At Osaka University, the risk of air attacks dispersed physical infrastructure and students alike into the countryside. Research laboratories were physically relocated to remote areas for safety, while an incendiary bomb attack in June 1945 damaged the remaining research facilities and equipment and the library at the Faculty of Engineering.[36] Because of its location, an even more dangerous campus at the time was that of Nagoya University. The city of Nagoya hosted not only that university but also aircraft-related heavy industries, such as Aichi Aircraft Company and, above all, Mitsubishi Heavy Industries, which had annually produced more than 40 percent of the nation's total output of engines. As the Allies' primary target during the war, the city came under attack 63 times.[37] In May 1945, American air raids torched the Faculty of Engineering building at Nagoya University. The risk compelled the military to issue an order for the students, as well as university faculty and staff members, to halt all of their education and research; for safety, they manually transported their research equipment to 68 remote locations with bicycle-drawn carts and drays.[38] Nagoya University was hardly an ideal setting for studying engineering.

While engineering students lost their opportunities for solid education, they retained something more valuable. Mobilized for research and production for war, they served on the home front rather than on the battlefront, consequently facing far less of a threat to their lives. Probably the best example could be found among students at Tokyo University. From 1926 to 1945, students with the highest death ratios were draftees to the front line, such as students of law (4.1 percent), economics and liberal arts (4.0 percent each), and literature (3.7 percent). Students of science (1.8 percent), engineering (1.9 percent), and science-related subjects (2.3 percent) represented the lowest death ratios among all the adult male groups.[39] Engineering knowledge was a hidden source of safety during the war.

A total of 122 air raids conducted by the Allies in Tokyo dispersed resources for research from the capital to various locations across the country, effectively suspending ongoing engineering investigations. The threat from the air proved especially damaging to the Japanese army's research projects for air power, because they had depended on the geographic concentration of facilities and human resources in and near the city of Tachikawa, Tokyo. The city came under attack 13 times from February 16 through August 15, 1945, because of its military importance as the host of Tachikawa Aircraft Company as well as many army establishments, including an arsenal, a flight school, an air arsenal, and the IAA.[40] Research projects virtually came to a halt at the IAA; *all* of its eight divisions were relocated from or near the city of Tachikawa to the distant countryside across the nation for safety. The first division, originally in the city of Fussa, Tokyo, was divided into three and relocated to Mito city (Ibaragi prefecture), Matsumoto city (Nagano prefecture), and Kyoto. The second division was transferred from Tachikawa to Takayama city in Gifu prefecture, and the eighth division from Tokyo to Niigata prefecture. All the other divisions of the IAA were resettled in the prefectures of Yamanashi, Nagano, Ibaragi, and Saitama. The labor-intensive evacuation to the countryside did not completely free the IAA from the danger from the air. For instance, the Allies' bombs obliterated the seventh division in its relocated city of Kōfu, Yamanashi. Given the dire threats posed by the Allies' bombings, concluded a Japanese army report, research on new weaponry at the IAA suffered "to an extreme."[41]

Other research establishments of the army and navy met the same fate of physical relocation. The Allies' bombing scattered the Seventh Army Technical Research Institute, a locus of research on high-energy physics, including an atomic weapon. Personnel and facilities were relocated from Tokyo to the distant cities of Matsumoto (Nagano), Itō (Shizuoka), and Kanazawa.[42] The Navy Technical Research Institute, located in Meguro, Tokyo, was dispersed even more widely. In May 1945, the Allies' bombing demolished most of its physical facilities, compelling the navy to set up 10 branch stations in rural areas across the nation, such as Shizuoka, Fukushima, Nagano, Tochigi, and Kyoto prefectures. Subsequent research in the new settings remained incomplete when the war ended.[43]

Research and Development for Naval Air Power

From 1943 to 1945, the INA signified the country's declining capability for research and development in military engineering. The eventual breakdown stemmed only partly from the dire threats of the Allies' air raids in 1945. On December 30, 1944, the Office of America's Commander in Chief Pacific listed many military facilities located in the city of Yokosuka. The city was divided into sections, all prioritized in the order of military importance for later strategic bombings. "The most valuable targets were [sic] the YOKOSUKA NAVAL YARD," the report concluded. "Additional targets" included "the principal oil storage area . . . and the major warehouse section." The report listed the INA as "the research and experimental facilities . . . of prime importance to the Japanese Naval Air Forces," but it was placed toward the bottom of the priority list, above hospitals and a former golf course. The INA and other research institutions, the report concluded, "may well be valuable targets during an attack upon this area *following* the destruction of all targets of opportunity [emphasis added]."[44]

Physical damage to the INA was minimal. Gunfire from one air assault destroyed part of a wind tunnel in the compound, and another attack targeting the Yokosuka Air Corps demolished Japan's new jet-rocket fighter, J8M1, modeled after the German Messerschmitt 163.[45] Many air-raid shelters, some underground, were constructed to protect valuable research facilities from such air attacks. By February 1945, the remaining wooden buildings were leveled to avoid fire, and glass windows were replaced with galvanized sheet iron.[46] Meanwhile, the Department of Materials Engineering was geographically dispersed, and physical facilities were relocated to remote lands as a precautionary measure. To transfer one section of the department to Kyoto University, a scouting group arrived on campus in January 1945, followed by 150 employees three months later.[47] Relocating another section of the department to the city of Kiryū in Gunma prefecture was a labor-intensive task that required bicycle-drawn carts to carry vegetables, firewood, and charcoal; in the countryside, navy engineers and technicians deforested three acres of land to farm potatoes in preparation for a food shortage.[48]

Ultimately, what crippled the navy's research and development for air power lay within the organization before the Allies' threat from the air. Especially after 1943, perhaps unknowingly, the navy began to undermine its own effort to introduce advanced aircraft to the battlefront. From 1938 to 1945, driven by an unrealistic expectation, the navy launched a total of 58 official projects that comprised 41 prototyping and 17 aircraft modification projects; only 10 final products, or 17 percent, were deemed effective for war, and 8 products, or 14 percent, were somewhat useful; the rest, 69 percent, were deemed failures.[49] The annual breakdown of these figures could reveal the impractical work schedules followed within the navy's engineering communities, which meant less time on each project. The navy completed and/or adopted 3 to 5 aircraft per year from 1937 to 1939, and 8 in 1940. Fewer flying machines earned the navy's approval over the next two years: 6 in 1941 and 5 the following year. This pace of prototyping increased dramatically to 14 navy aircraft in 1943, or a 280 percent growth in a single year. And the navy created more flying machines in 1944, 16, or 1.33 per month on average! Furthermore, 6 new prototypes were ready for use in 1945.[50] Much of the damage was self-inflicted.

Behind this self-defeating policy was the navy's failure to note the critical issue inherent in aircraft development: time lag. Given the complexity of the product and shortage of experienced engineers, it took five years to complete the entire process, from the conception of the performance specifications to the actual deployment of the final product after training. Planning and crystallizing performance specifications required 8 to 10 months and then an additional 10 to 12 months to complete the remaining processes, such as inspections of mock-ups and wind tunnel models. In other words, it took 18 to 24 months to create the first full-scale prototype.[51] The Allies were among the first to observe the dire consequence of this time lag issue. During their observation in November 1945, the Air Technical Intelligence Group concluded that "when Japanese strategy was forced to take the defensive in 1943–44, it was too late to produce aircraft which were expecially [*sic*] designed for that role."[52]

What worsened this time lag issue was a persistent shortage of senior engineers and technicians, who were already stretched thinly within the navy and aircraft industries. Time became an unaffordable luxury for those engaged in basic, theoretical research—including the navy's research on aircraft flutter at the INA. For instance, Matsudaira Tadashi, an expert in high-speed aerodynamics, had to spend less time on each project after 1942, devoting only six months at most to produce one solid research report, while aiding other projects that strayed beyond their areas of expertise.[53] Many experienced technicians were continuously drafted from the INA and aircraft industry into the army, until the navy leadership devised a countermeasure in February 1945.[54] Meanwhile, draftsmen were conscripted, and it remained very difficult to train them in a short period of time.[55]

What proved more damaging to the navy's aircraft development was its approach to prototyping that underestimated the importance of mass production. In building air power, the navy had successfully carried out its advanced research and development projects at the INA, which remained relatively free from budgetary constraints or technical requirements for mass production. The navy excelled at creating advanced experimental aircraft as both its creator and end user—a strategy useful only while production was needed in a very small quantity. The mass-producibility was at the bottom of the priority list, as typified by the navy bomber, known to the Allies as D4Y Judy. The original plan for this aerodynamically advanced 13th experimental bomber asked for only five prototypes with no intention of mass production.[56]

The research-production conflict, which had remained unsolved in the interest of advancing research, exploded in the subsequent project. Originally, the design team of the 15th experimental bomber, code-named P1Y Frances, had no idea that its creation would be produced later in a large quantity.[57] Converting this experimental machine to a mass-producible, mass-serviceable one was a daunting task that seemed impossible at first, given the complicated mechanics of the aircraft. For instance, obsessed with the weight reduction program, Miki chose hydraulic controlling gear because the alternative, the electric-operated system, required many transmission gears and, thus, added weight.[58] This design decision was risky at the time. And it was reckless in hindsight, because the design team lacked experience in developing hydraulic control mechanisms. The team started from scratch, calling in experts in the field from civilian firms, creating what turned out to be exceedingly complicated mechanisms. More technical complications awaited this lightweight bomber because it could only use custom-made parts. To convert the aircraft for mass production, the design team redrew the assembly diagrams of *all* aircraft parts, making them compatible with the ready-made goods of the time. Meanwhile, the team conducted studies and introduced new standards for complex products in the hydraulic system, such as a pipe joint and V-shaped packing.[59] Buried in this type of work, the INA engineers could hardly concentrate on their research for new prototype designs.[60]

More technical difficulties awaited a civilian manufacturer that had yet to be assigned when the prototype was complete. As it turned out, all the aircraft planning at the INA needed to be manually "translated" into a context suitable to the production capability of the newly assigned, sole manufacturer of the bomber, the Nakajima Aircraft Company. The navy engineers converted the aircraft, which had originally required extensive welding, to the amalgamation of forged, assembled parts; and Nakajima workers modified forged parts by hand or on milling machines on the production floor. The INA engineers extensively used magnesium alloy, which was hardly conventional in civilian aircraft companies.[61] Concocted as an experimental aircraft to advance research, the P1Y Frances remained notoriously complex and difficult to produce in large numbers. Only 1,100 had been produced by the war's end.

Despite these technical difficulties, a fairly relaxed atmosphere prevailed at the INA as late as October 1943. Twice a year, hundreds of workers would compete in ball games (soccer, tennis, ping-pong, and baseball) as well as in martial arts (swordsmanship, archery, judo, and sumo wrestling); the female employees exhibited their skills in flower arrangements.[62] By no means pressed for time, in 1941, the engineers at the Materials Engineering Department enjoyed their basic research by using their state-of-the-art facilities, which were superior to those available at the universities.[63] The engineers in the department were confident and cheerful at the time of the Pearl Harbor attack in December 1941.[64] Their research activities in 1942 allowed creative thinking. They involved a series of experiments to develop substitutes for rubber, nickel, molybdenum, and cobalt. At one point, silk was tested as a substitute for cotton as the material for machine belting and gaskets, and rabbit hair was tested as a substitute for wool.[65]

By the end of 1944, however, optimism at the INA had shifted, first, to pessimism and then to desperation. Its time-consuming basic research became less affordable. In 1943, a new chair of the Materials Engineering Department abandoned all basic research and required immediate applications of research findings. Accordingly, the engineers shifted their research goals in the face of the declining availability of material resources, such as duralumin from Southeast Asia. One of their new projects was to increase the wooden content of all types of aircraft; thus, wood was required to constitute at least 10 percent of a fighter aircraft, 20 percent of an attacker, bomber, and reconnaissance aircraft, and 50 percent of trainer and transporting planes.[66] In March 1945, one partially wooden airplane met this unyielding requirement. Its construction was unbelievably simple; it required minimal steps for assembly, roughly one-tenth of what was required for Japan's most mass-produced aircraft, the A6M Zero Fighter. Meagerly equipped, the wooden aircraft possessed only one weapon: the suspended bomb attached to the belly. Nothing was to be "wasted." The flying machine lacked defensive features, and even the landing gear became detachable and reusable by others for takeoff.[67] The birth of this aircraft, or the wooden coffin with wings, signified the grim fate that awaited young aviators mobilized for home defense. Technology and engineers formed an indispensable part of the wartime *kamikaze* operation.

Technology for the Kamikaze Air War

With its diminished research capabilities, the navy embraced its new defensive policy out of desperation. By design, manned carriers equipped with explosives inflicted two types of damage on their enemy on the sea and ground, the first type was physical damage on the target and the second type, psychological damage on the survivors at the scene. Compared with the army, the navy devised a greater variety of technologies, such as man-driven torpedoes, midget submarines, and self-contained oxygen tanks for frogmen, all for underwater kamikaze operations. This variety stemmed from the navy's mission. Unlike the army, the navy had to defend the archipelago nation from the Allies advancing in the air as well as under and above the ocean. For air defense, both military services quickly converted preexisting aircraft, even slow biplane trainers, for air attack. It was an

easy fix that entailed little additional time, effort, resources, and training of airmen for the operation.

The summer of 1944 marked a watershed in Japan's war history. Until that
point, both the army and navy had witnessed isolated cases of suicidal attacks
from the air and on the ground at the tactical level. Unable to return safely from
the enemy's territory, individual aviators could spontaneously opt for a suicidal
plunge into their enemy target. The suicide tactic, as such, was not deemed a significant tactical element in the Japanese navy before 1941.[68] The final, conscious
decision to survive in captivity—or die in an attack (as tacitly encouraged and
expected)—was left to individual choice. When placed on the defensive in the
war against the Allies, the navy leadership began to make that decision for the
individuals. Suicide attack emerged as a viable option for homeland defense.

A community of military engineers played an indispensable role as the architects of technology for the kamikaze operation, as epitomized in the manned
glider bomb, MXY7—the only aircraft in the world that was designed, developed, and deployed for suicide operations. Its creators at the INA subscribed to
a work ethos not expected of ordinary citizens on the home front. Their aircraft
design was not created for its own sake or in isolation. Designing, as historians
of technology point out, is a social activity directed toward the achievement of a
practical set of goals intended to serve groups of individuals in some direct way.
The performance, size, and arrangement of an airplane, for instance, are direct
consequences of the specific task to be performed. Chief aircraft designers first
translate some ill-defined requirements into a concrete technical problem and
then move on to overall and then component designs.[69] During defensive war, it
might be rational to act more daringly; war could, and usually does, render opera-
tive assumptions of peacetime untenable.[70] So, what initiated and sustained the
engineers' efforts to create this suicide technology for war at the INA?

The MXY7 project signified the disintegration of civilized rules of war within
the military organization, which lacked civilian control. The mechanics of this
single-seat monoplane were disturbingly simple. This small glider—with a 5 meter span, 6 meter length, and 1.2 meter height for Model 11—was attached to the
bottom of the mother plane for takeoff and detached at an altitude of roughly
3,000 meters. Once released in the air, the pilot lost all means of communication. By design, this partly wooden glider lacked technical features for a safe
return, for instance, the fuel capacity for a return flight and a mechanism for
landing on water or ground. Therefore, crashing the aircraft was the only way to
end the flight, and the fail-safe fuse on the warhead was designed to detonate on

touching the enemy target or water. This glider could not be used at night. For roughly 15 minutes in daylight, the pilots would steer the machine with increasing speed toward the enemy targets moving on the sea, especially aircraft carriers. And at least in theory, the power of one explosion on direct hit could sink an aircraft carrier. The first MXY Model 11 carried a 1,200-kilogram explosive in the armor-piercing nose, which constituted roughly 60 percent of the fully loaded weight of the aircraft.[71] This "relatively cheaply and simply constructed" aircraft was "an expendable weapon," concluded an Allies' wartime report. The aircraft earned its disgraceful code name among the Allies: Baka, or "fool."[72]

This technology of war was mostly a bottom-up initiative from the navy battle-front. The originator of the project was Ōta Shōichi, a sublieutenant second class in an airborne transport unit. He had witnessed an acute shortage of airmen at his Atsugi air base as well as the grim air war on the southern front, especially over Rabaul in the Pacific. The only effective way to turn the tide of the defensive battle seemed to be the suicide mission in the air. His rhetoric, as such, sounded tactically rational and practical, yet he had neither the credentials nor the engineering knowledge to realize his idea. Subsequently, he studied guided-missile development in the army, which was not remotely successful. Upon his visit to the ARI, Tokyo University researchers, such as aerodynamics expert Tani Ichirō, drafted drawings in support of Ōta's idea. This plan acquired the technical support and prestige of the university. Shortly thereafter, Ōta took his project to the Headquarters of Naval Aeronautics, and then, in early August, to a meeting at the INA.

His idea for a human guidance system dismayed those at the meeting. Hot-tempered Miki Tadanao angrily yelled at Ōta, calling this idea the height of folly. He and his supervisor initially refused to cooperate, opposing the very idea of using "the precious lives [of their fellow aviators] to compensate for the poor technology [of the time]." Miki believed that the idea was a "desecration of the dignity of engineering."[73] Nonetheless, on August 16, the Headquarters for Navy Aeronautics officially adopted the proposal and ordered the so-called marudai project—the alternate reading of Ōta's name, "dai," was drawn within the circle or "maru"—that symbolically encoded his central role in the project's subsequent development. At the INA, Yamana Masao directed the overall project. Under him were the chief designer Miki Tadanao and other supporting senior engineers.[74] They faced technical and, above all, moral barriers to overcome the creation of this highly unethical weapon, which would end the lives of their own fellow countrymen. Within the INA's institutional culture, disagreements were encouraged historically, actively, and openly because they were considered important for the

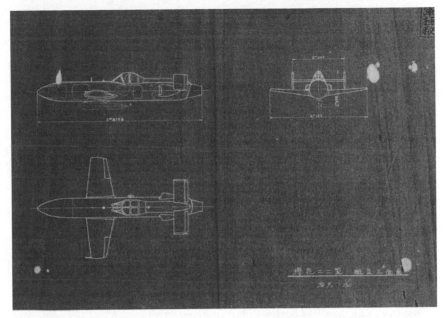

Original design blueprints of the Navy suicide glider MXY 7 Ōka Model 22.

growth of engineering knowledge. But once the decision was finalized after discussions, it was followed by a governing principle and disagreements were effectively contained.[75] By the end of August, the INA engineers were embarking on
the kamikaze technology project.

Their struggles to justify their actions remained unresolved, but the engineers crossed their moral barriers due to at least four factors. First, geographical
distance, real or perceived, was an indispensable factor for the INA engineers in
Yokosuka to support the suicide missions of their fellow aviators. In the summer
of 1944, Ōta's visit to the INA reduced the psychological distance between the battlefront and the home front. His presentation at the meeting, filled with pathos,
conveyed the sense of desperation at the battlefront in Yokosuka, where the INA
engineers felt the everyday plight of the servicemen in the South Pacific for the
first time. The airmen on the front line and the research engineers at home had
rarely, if ever, met face to face. The ARI in Tokyo and the INA in Yokosuka were
safe havens with a curious mix of unreality and alienation. The reality of gory, un-
abated destruction was kept at a distance until America's strategic bombing campaign began in late 1944. In this context, Ōta himself chose to directly appeal to
the navy engineers on the home front, asking for their expert knowledge to assist
his suicide in MXY7 for the beleaguered nation. Only after the meeting with Ōta

did Miki come to realize how tough the military situation was on the front line. As Miki testified, he and others at the INA were "opposed to the idea, but it came from the battle front," and, according to Ōta, "the war could not be won without it [MXY7]."[76] With no other direct communication between the battlefront and the INA in Yokosuka, military engineers were swayed by Ōta's plea and made decisions without full consciousness of the consequences of their unethical actions.

Detachment in Tokyo and Yokosuka from the suicide mission on the battlefront could make it easier to support the engineering, and it could also provide a certain perspective on the operation. The perceived distance could both decrease and increase hesitation toward the military mission. One could draw few firm, verifiable correlations between one's personal attachment to other Japanese on the one hand, and the geographical proximity to them on the other during the war. Arguably, geographical distance neither increased nor decreased the degree of personal attachment to the airmen on the front line so much as it altered its forms of expression. Once the navy had officially moved forward with manufacturing, it probably became easier to support the suicide operation in Tokyo than on the front line, in a less personal, but more effective way. A case in point was the refusal by the head of the INA factory to produce the MXY. Saba Jirō, the chair of the Aircraft Department, who had participated in the meeting with Ōta, criticized his action, saying that the MXY7 was "not what the navy brass on the home front ordered for the aviators on the battle front . . . [but rather it was] what a young lieutenant brought from the frontline for his own undertaking."[77] Ōta's plea for engineers' assistance in his suicide thereby gained momentum.

If the geographical distance between the battlefront and the home front helped the engineers to embark on the project, the same distance prevented them from learning tactically crucial lessons firsthand. The technicians and ground crew in the battlefront airfields could—and did—observe that the MXY7 operation was unlikely to succeed. The suspended bomb added considerable air resistance and weight to its mother plane, which made both aircraft easy, slow-moving targets in the air. When fully loaded with the heavy manned bomb, the mother plane Mitsubishi G4M Betty had to taxi from 500 to 600 meters farther than usual. Its maximum range of 4,700 kilometers was reduced by some 30 percent. In addition, its cruising speed of 170 knots was reduced by about 10 percent due to the additional drag. For the operation, long-range fighter escorts were a prerequisite, but they were deemed an unaffordable luxury by the spring of 1945.[78] The mother plane, with the heavy MXY7 attached, remained defenseless in the air— and this was precisely what the Allies targeted. In June 1945, teams of American engineers studied the captured MXY7 and concluded that "if launched from a distance, . . . [the aircraft] would be vulnerable to attack by [Allies'] fighters before reaching a position to complete effectively its own attack." To deter its threat, "the destruction of the parent aircraft is extremely important."[79]

The second factor that numbed the engineers' moral sensitivity to the suicide destruction had much to do with their professionalism. The immediate problem that the engineers sought to solve was more technical than moral. To win a war, a concrete, quantifiable technical problem was far easier to solve than an abstract and moral problem. The INA engineers lacked the education, and experience, to solve this inherently complex issue. Given the wartime urgency, the plane's wartime use seemed a secondary concern to its creators; in other words, perfecting the technique of war became an end in itself. Miki, and many others in the war, felt compelled to make rational calculations and pragmatic decisions about the relationship between the ends and means. While Miki led the MXY7 design project, he convinced himself again and again that it was his "duty" to produce what was asked of him, without questioning how they would be used; it was a fate he felt compelled to accept as a military engineer.[80] Engineers like Miki had few opportunities and colleagues to alter the course of the momentum in the vast military organization. His moral qualms were submerged while performing highly technical functions.

In addition, the combination of the sense of accomplishment and the urgency of the war was numbing, especially when faced with the most challenging part of the project involving high-speed aerodynamics. The glider's unprecedented velocity could reach 550 knots, or Mach 0.85, roughly the same as the cruising speed of present-day commercial jet aircraft. Neither data on test flights at that speed nor guidelines of any kind were available to the design team. Miki was especially concerned about the possible flutter of aircraft parts in the air. In wind tunnel experiments, the engineers closely observed the aerodynamic characteristics of the glider model as it detached from the mother plane model. The scenes were photographed with a speed camera for close analyses.[81] Many engineers and technicians in the aircraft design division, including Matsudaira Tadashi, who solved the Zero's flutter problem, tested the prototype aircraft in the airfield.[82] When the team of INA engineers successfully finished the first test of MXY's release from the mother plane, one engineer at the scene was "glad," and Miki was "very pleased" with the result.[83] On the completion of a test flight on October 31, 1944, "those involved [in the project] shook hands with each other in joy."[84] As Miki recalled the entire project 30 years after the war, the MXY7 gave "the greatest pleasure [to] the engineers . . . [who] detached themselves from the concept of suicide aircraft" and successfully completed "the all-night works within the short period of a little over two months."[85] The end product epitomized the engineers' ethical qualms submerged by their task-oriented thinking.

The strict military secrecy of the MXY7 project in wartime Japan produced similar numbing effects. The design team was physically confined to two rooms on the third floor of the central laboratory building: one room for the design work and another for resting. Both rooms were strictly off limits to all other personnel.[86] The workmen were not informed of the nature of the weapon they were working on. The wind tunnel was off limits to all unauthorized personnel during the tests, and the entrance to the building was guarded by naval guards specially dispatched from Yokosuka Naval Station. All offices and work areas in the vicinity of the testing tunnel were evacuated.[87] Knowledge was compartmentalized to preserve secrecy. As one INA engineer recalled, "not even co-workers of long standing" could bring up the top-secret MXY7 project in conversation.[88] The military secrecy was self-imposed and internalized within the engineering community. One design engineer maintained, for instance, that the project could not be assigned to private firms. To protect security and the navy's reputation, he went on to say, production had to be completed within the INA in secrecy.[89]

The military secrecy was a product of national and organizational efforts in and out of the INA. For instance, it was illegal to carry cameras or to take photographs in the city of Yokosuka. Tall concrete walls surrounded the coasts of areas near the naval station, and only a few people knew what all the navy facilities looked like.[90] The INA remained secluded within this confined section of the city. Factory workers and administrators routinely pledged to keep the secret, and neither news about, nor work carried out was disclosed to the public in any form. Most of the construction within the compound was also done in secrecy. Disclosures were severely punished in accordance with the Military Secret Protection Law or the Naval Penal Code. Photography, or even sketching, was strictly prohibited at the INA.[91] The production and deployment of the MXY7, the top-secret weapon, remained successfully hidden from the Japanese public. The manned bomb was reported in *Time* magazine, first, on April 30 and then for the next two months; belatedly, after obtaining the navy's approval, Japanese newspapers reported on the first MXY7 mission of March 21 two months later, on May 29.[92]

One secret behind the mission haunted the navy leadership. The originator of the MXY7 project, Ōta, purposefully used his secret, which, once revealed, would traumatize the navy leadership. At the Headquarters for Navy Aeronautics, one chief commanding officer, Itō Hiromitsu, listened to Ōta's plan in the summer of 1944. It sounded "like a project based on considerable studies outside [the navy]." Like many others in the navy, Itō remained convinced that Ōta himself was a qualified pilot determined to use the proposed aircraft to carry out the mission himself, as he declared so resolutely. Itō subsequently "made up his mind to realize the plan of the man [he believed was] representative of the airmen [on the frontline]." Ōta's machinations involved not revealing his own credentials fully—he lacked proper pilot qualifications. Only after the official approval of the project did he receive pilot training, but he failed. Had this fact been known, wrote Itō remorsefully, "he would have rebuked" Ōta for the engineering plan.[93]

Engineers' Autonomous Search: Maximizing the Survivability of Kamikaze Operators

While actively participating in the MXY7 project, the engineers retained their moral sensibilities until the war's end. They tried to maximize the operators' chance of survival in the kamikaze operation. The order of the high command blurred the human-machine boundary: a living airman was himself the guidance system for the machine. Accordingly, for successful operations, a steel armor plate protected both the pilot and glider until the eventual crash. Neither human nor machine was expected to ask ethical questions or make decisions once detached from the mother plane. The aircraft design team redefined these assumptions of the suicide operation and, accordingly, tried to refine the flying mechanism with considerable autonomy. Only the engineers had access to the technical knowledge to do so.

The result was a bailout mechanism for use during a flight. On September 21, the design team compiled a list of important components and specified the weight of each one, including "a parachute of 8 kg." These weight calculations formed the basis of the MXY7 prototype, which included two safety aids as explained in the operational manual of November 1944. The first one was the compartment "behind the pilot's seat for storing a portable backpack parachute." The second safety feature was a jettisonable canopy. This "installation for an emergency exit" was designed for a simple bailout operation in the air: the aviator should simply "push a lever on the right to release the canopy," which would be "blown apart by air flow." In addition, the cockpit panel included a turning coordinator, a means for the pilots to turn around if desired.[94] The combination of these mechanical features made ditching possible in theory and in practice.

These devices symbolized the efforts of the engineers and military tacticians. On October 1, 1944, representatives from the INA and 721 Air Corps, the first unit created on the home front for the new attack operation, discussed several issues, including how to put the prototype to actual use.[95] Paying close attention to the engineers' idea about the bailout, the flying corps asked the design team on November 3 to "conduct further research" on 19 features of the MXY7 trainer version "that functioned poorly." One request was to "move [the lever] for the detachable canopy to a better place"; another was to "make the seat, [with or] without the parachute, more comfortable."[96] Two weeks later, the 721 Air Corps made more requests to 12 navy engineers, including Miki. At the INA meeting, a representative of the flying unit provided a list of "matters that need no discussion during the final inspection," but matters that "require modification from the view of the flying corps." One request was "to make the [manual] operation for canopy release easier" to carry out. Accordingly, the representative continued, the engineers should modify "the next version of the MXY7 trainer" and "confirm the usability" of the safety feature before moving forward.[97]

Incorporating this request, 100 actual flying machines were quickly produced at the INA before November 28. One of them survived the war and attracted the curious eyes of researchers at the Royal Aircraft Establishment in London. Their report noted the safety feature as follows: "The hood slide on rails [are] mounted on the two upper main fuselage longerons, the forward section of each rail being secured by quick release fasteners, so that the hood when in the close position can be jettisoned together with [the] rails."[98]

In December 1944, however, further discussions set back the engineers' efforts. Given the urgent need to mass-produce this Model 11 for homeland defense, the MXY7 modification plan itemized "structurally complicated parts [which] need to be simplified," specifying the mechanism of the detachable canopy. The meeting on December 15 decided that the safety feature was "no longer needed."[99] Subsequently, "the jettisonable arrangement," reported a U.S. intelligence team during the war, became "neither complete nor serviceable."[100] The turning coordinator was removed from the cockpit panel three days later.[101] The subsequent model 22 lacked all the means for the pilots to bail out and survive the operation. By design, its construction was simplified even more than the earlier Model 11 "for mass production."[102] Weight calculations conducted on January 20, 1945, did not factor in the weight of parachute.[103] As the operation manual of May 1945 describes, the canopy was divided into three sections, none jettisonable.[104]

The design team remained undaunted, however. Exercising autonomy in the military project, Miki and others continued their efforts to install a safety mechanism in the subsequent model 43b. Unlike earlier ones that were released in the air from the mother plane before an attack, this particular model ascended from catapults on the ground with the power of an Ne-20 turbojet engine before diving into the enemy target. On March 9, 1945, Miki calculated the overall weight, range, and flight characteristics of the aircraft; for the next 19 days, he conducted more detailed weight calculations, which neither included the parachute's weight, nor suggested the possibility of installing the detachable canopy.[105] For the next four months, the design team considered a mechanism for safe landing to maximize the pilots' chance of survival. The team subsequently worked closely with civilian aeronautical engineers at Aichi Aircraft Company. On April 5, they compiled a weight-calculation report that factored in the parachute weighing of 8.7 kilograms for "the trainer version" of the latest attacker.[106] Subsequent meetings between the navy and civilian engineers resulted in the creation of "Design Specifications for the Model 43b Prototype" on April 26. Its "standard equipment" included the 8.7-kilogram parachute and two other required features: "skid for landing" and "detachable canopy for emergency [exit]."[107] The prototype project explored the possibility of installing these safety devices in the actual flying machine.

On June 23, however, the civilian and navy aircraft engineers met at the Headquarters of Navy Aeronautics to discuss the need to "simplify the construction" of this latest model. The participants paid attention to technical "features that could compromise the aircraft flight performance and practical use," one of which was "skids for landing."[108] Further research and test flights did not materialize in the next month. Subsequently, none of the models 22, 33, or 44 had become functionally operational by the war's end. Overall, the research and development process for kamikaze air war was, neither united, nor sudden, nor monolithic; it signified the engineers' effort and autonomy to increase the survivability of the airmen without compromising the military operation for the beleaguered nation at war.

The MXY7 project set a dangerous precedent for kamikaze engineering projects in the navy and the army until the war's end. From the end of 1944, the navy issued various projects to develop prototype aircraft for the suicide defense operation, baffling the engineers who spent their time modifying preexisting aircraft.[109] The MXY7, for instance, was modified four times in design and/or construction. In the spring of 1945, INA engineers developed another suicide glider specifically to resist the invasion of enemy tanks on the ground. Internally equipped with 100 tons of explosives, the wooden flying machine was to take off from the ground with rocket propulsion.[110] While this machine remained unserviceable, the navy initiated and/or supported six suicide aircraft projects, which further weakened the INA's research capability. Army aircraft engineers pursued at least one suicide technology project for air defense.[111] While the navy launched its last MXY7 attack on June 22, 1945, the army continued its ambitious plan until the last moment of the war. On August 9, 1945, Army Unit 114 was mobilized to fly the long-range, twin-engine reconnaissance bomber Ki-74II one way into the skyscrapers of New York City, which presaged the 9/11 terrorist attack of 2001. A full-scale model of the bomber was completed at Tachikawa Aircraft, and the prototype became serviceable in August 1945.[112] The technical and tactical feasibility of this operation remained questionable, but in any event, the war's end saved the lives of Japanese airmen at the very least and possibly, the lives of New Yorkers as well.

As the wonder weapon for one-way suicide missions, the MXY7 was expected to repel the incoming Allies with the divine wind, or kamikaze. However, the entire MXY7 operation turned out to be a fiasco, as the first attack on March 21, 1945, showed. That day, the entire phalanx consisting of 15 MXY7 gliders, 18 mother planes, and 30 escort fighters came under enemy fire in the air before reaching their targets that were moving toward Okinawa. All 160 Japanese men perished en route, causing no damage to the Allies. From March 21 to the end of the war, the navy dispatched its manned bombs on 10 suicide missions, which sank three destroyers and damaged eight others, while leaving 715 Japanese airmen dead.[113]

Not just these MXY7 combat missions but also the entire suicide operation in the air turned out to be ineffective. The physical damage inflicted on the enemy targets was limited. During the naval battles over the Philippines, Okinawa, Iwo Jima, and its vicinity, the army deployed 1,185 airplanes and the navy, 1,295, for the one-way suicide missions. Among the total, only 244 aviators reportedly accomplished their missions, damaging 358 Allies' ships to various degrees. After all, the hit ratio was only 16.5 percent.[114] During 1944–1945, Japan's self-congratulatory rationale of the operation lacked any verifiable data to support the operation. In the end, the damage to the Allies seems to have remained more psychological than physical. The final plunge was seen as an individual act of lunacy, and the operation itself as an organizational act by an inscrutable enemy.[115]

The MXY7 project caused more psychological damage to Japan after 1945 than to the enemy during the war. The wartime experience continued to haunt those directly involved, and many took their inner struggles to their graves. On August 18, 1945, the originator of the project, Ōta, left a short suicide note, took his airplane, and disappeared into the Pacific. In July 1948, the commander of the MXY7 operation, Okamoto Motoharu, jumped in front of a running train and killed himself. Itō Yoshimitu, who ordered the MXY project at the Headquarters for Navy Aeronautics, survived the war with a painful past; at least for the next 37 years, to use his words, he "tried to erase the painful wartime memory every time it surfaced . . . [and he] avoided war-related publications as much as possible."[116]

The engineers of the manned bomb lived with agonizing questions about the meaning and import of technology, which had failed to bring victory. Yamana Masao, the project supervisor, reportedly called his MXY7 a "disgrace to the engineers" and quietly took his memory to his grave.[117] Miki Tadanao believed that it was "desirable that the attacker be erased from the country's history of aircraft engineering in the honor of navy aeronautical engineers as well as of the entire wartime engineering community."[118]

By the end of the war on August 15, 1945, the eight-year conflict had bequeathed an important legacy that involved the INA. On November 23, 1945, groups of Allied inspectors examined the INA's wartime research on high-speed aerodynamics. "It appears," their report concluded, "that the Japanese were in no way behind on their research in theory of high speed flight." The report went on to state that "they had not actually put their theory into flight tests to prove it bad or good, but this was not due to the holdup in research. The major delay was the development of power plants and . . . air frames for high speed, high altitude operation." Equating the INA to Wright Field in Ohio as "major research establishments," the inspection team recommended that a small 30-centimeter wind tunnel, "particulary [*sic*] of interest" given its "blade design," be transported to the United States.[119]

Lieutenant Colonel Frank Williams held a similar view toward technology transfer from Japan to the United States. He arrived in Japan from the Air Material Command Center, Wright Field, in rural Ohio, to experience "difficulty . . . in finding Japanese reports [because they] were all destroyed per imperial edict about 15 August 1945." Having examined "most phases" of research and development in aeronautics, he concluded that "they [the Japanese] were far behind us." Like other Allies' observation reports, his work also stated that "wind tunnels [in Japan] are far superior to the academic equipment in the U.S. universities . . . [and they] should be shipped to this country [America] rather than being dismantled . . . [they are] of use to government and university laboratories."[120] At least one other high-speed wind tunnel in Yokosuka interested a German engineer who, after the war, moved to the United States and observed the INA for the Allies. Subsequently, in 1949, the "transonic wind tunnel with a Mach number range of 0.7 to 1.34" was dismantled in Yokosuka and transported to the flight-test facility at Arnold Air Force Base, Tennessee. Three years later, the product became "of old design" and was made "available to any educational institution capable of operating it."[121]

The second, and more important, legacy of wartime Japan was the thousands of engineers mobilized against the Allies. After all, the intensity and duration of the war was unexpected. Especially after 1943, the wartime mobilization of engineering education was unplanned and delayed and research and development in military establishments crippled. The hasty, ad hoc mass production of engineering students was no solution to the scarcity of experienced, senior military engineers in and out of the INA. The end of the war on August 15, 1945, had left the battered country with wartime engineering workforce, much of which was effectively contained in the national boundary for postwar reconstruction.

CHAPTER FOUR

Integrating Wartime Experience in Postwar Japan, 1945–1952

At noon on August 15, 1945, Emperor Hirohito of Japan announced the end of World War II in a four-minute, prerecorded radio broadcast, presenting himself as the guardian of those who survived the destruction. "I am terribly concerned about those injured, suffering, or lost in war," he said in a reedy, crackling voice. He went on to ask the nation to "endure the unendurable, and bear the unbearable . . . [for he would] open the path for our future and peace."[1] Millions of listeners found his message somber, even hollow, and eventually accepted it with despair.

This historic broadcast sounded an ominous alarm bell for the nation's aeronautical engineers such as Kimura Hidemasa. On arriving on the Japanese soil, the American occupation authorities completely banned the field of aircraft engineering because of its military applications. Stripped of his career and former prestige, this senior researcher at the Aeronautical Research Institute (ARI) suffered a mental breakdown. After the dissolution of the institution, he survived by selling his camera, golf clubs, and furniture at pawn shops. The sight of his haggard appearance convinced his mother that he would commit suicide. What he called "tear-jerking misery" affected his bladder; his unstoppable urge to urinate every few minutes was later diagnosed by a psychiatrist as symptomatic of a severe stress disorder.[2]

The occupation policy elicited diverse, ad hoc responses from an oversupply of engineers like Kimura, especially those in the field of aeronautics. While their individual experiences varied, their career transitions formed observable patterns of domestic migration in the country. Some abandoned their field of research altogether; some became self-employed; and a great many embarked on second careers as engineers in the civilian sector—moving into the shipbuilding, electronics, agricultural, fishing, and automobile industries. The effect was far-reaching. These career moves benefited important companies such as SONY, Canon, Toyota, Nissan, Honda, and Mitsubishi Motors and, thus, helped the transformation of the defeated nation into an affluent, technology-savvy society.

This domestic migration was part of a broad, historical, and international landscape of the Cold War era. Postwar Japan lacked any major "brain drain" phenomenon between 1945 and 1952, and a series of comparisons with Germany reveal why. Both the examples and counterexamples of this migration pattern highlight the geopolitics in Asia (i.e., among Australia, China, the Soviet Union, and United States), legal and economic impediments, and sociocultural expectations—all of which hindered the exodus of military engineers during the occupation years. Japan's wartime brains were, thus, effectively contained within the country, a situation favorable to its postwar recovery and expansion.

Demilitarizing Military Engineers in Japan

As soon as the war ended in August 1945, thousands of military scientists and engineers became social pariahs, at least temporarily, because, instead of bringing victory to their country, they were deemed partly responsible for its defeat, misery, and shame. Pundits claimed that Japan had lost the war because of its weak technology and industrial power. The defeat very much rendered the former military engineers a dispensable, or perhaps, unneeded part of the newly emerging peace-oriented society. The postwar years constituted a time for social initiatives on behalf of democracy. In this culture of defeat, some viewed science and technology as pawns of militarism and ultra-nationalist ideology. Correspondingly, the institutions of wartime science and technology suffered low prestige.[3]

The end of the war was especially damaging to university researchers who were overwhelmed by a sense of futility. In the summer of 1945, they suddenly lost their wartime research goals and sources of income. For the next few months, their research was barely able to continue given the sharp decline of heavy industries.[4] The two largest investors in wartime research disappeared from the scenes; zaibatsu business conglomerates were disbanded, and the entire military dissolved. No longer able to draw on nationalistic rhetoric for support, scholars in some cases lacked any physical means to carry out research and development. Wartime research facilities, placed under the direct supervision of the Allied occupation as targets for reparations, ceased to operate in August 1945. Only with the authority's approval, which came painfully slowly, could the university researchers initiate or resume any creative research projects, and it was only if they lacked military applications. Moreover, material resources were lacking in society. A severe paper shortage reduced the volume and frequency of academic journal publications, especially after the wartime stock of paper ran out in just one year. The various scientific societies were voluntary associations that relied on membership fees. A lack of paper and money jeopardized their existence.[5]

Already the least paid among educated workers, university scholars were severely deprived financially. The monthly salary for new elementary-school graduates was 200 yen, while that for graduates from women's schools or secondary schools was 300 to 400 yen amid inflation. The pay for full professors at imperial universities was no different, ranging from 200 to 400 yen. This modest amount was a source of envy among assistant professors, who earned 120 to 130 yen, and research assistants, who received only 80 yen. As public servants, such professors could not legally seek other employment for income. Moreover, research assistants had not authored books, hence lacking royalties as a source of extra cash.[6] Unlike engineers who were working in factories, academic researchers in science and engineering lacked the legal means to ask for a salary raise at the time of inflation. A labor strike was not an option. According to the press, their "right to a certain standard of living" remained "under threat."[7]

Some academic researchers successfully adapted to new circumstances by making practical use of their wartime research. A Tokyo University researcher, for instance, used wind tunnel facilities at the recently disbanded ARI to develop a windbreak forest based on wartime studies in aerodynamics.[8] Nishina Yoshio, a renowned physicist who led the army effort to develop a nuclear weapon, applied radiation to plants to examine the process of anabolism. At Kyoto University, a team of biologists, industrial chemists, and theoretical physicists—including Arakatsu Bunsaku, a highly respected physicist—studied the effects of radiation on high-yielding agricultural products. Their research improved plant species, food processing, and sterilization.

Ironically, some weapons for war became a source of inspiration in peacetime. Professors from Kyoto University examined the effects of ultrahigh frequency radio waves—which had formed the basis of a death-ray electric weapon for the war—on the germination, growth, and sterilization of plants.[9] Torigata Hirotoshi, once the head of the chemical laboratory at the Navy Technical Research Institute (NTRI), resumed his wartime research in the field of high-polymer chemistry. At Kyoto University, he used Japanese paper, Japanese isinglass, a sunset hibiscus, and konjak jelly—the materials he knew well because he had studied them in the navy to create the light, yet durable surface for balloon bombs, a weapon that alarmed the US Army on the West Coast during the last few months of the war. From 1951, he carried on his research on artificial silk at the Mitsubishi Rayon Company.[10]

For ex-navy engineers in the job market, the shipbuilding industry was a viable option. As of 1944, according to one estimate, 287,799 factory workers and 45,922 engineers had occupied the industry and produced, for instance, merchant ships of as many as 1,730,000 displacement tons. Their factories had come under assault from the air, but many facilities remained in usable conditions. By 1946, the postwar shipping industry began to show early signs of revival, especially after private companies obtained ownership of three of the five navy arsenals across the country. Thousands of former navy engineers resumed their careers as shipbuilders after a temporary suspension was lifted during the early occupation years. Fukuda Tadashi was such an engineer from the former Kure Navy Arsenal. Using his wartime experiences, he widely promoted the use of electric welding in the postwar shipbuilding industry. Nishijima Ryōji, too, was a naval engineer who once led the construction project of the giant battleship *Yamato* for the war. To a realm formally dominated by gut-feel technicians, he introduced knowledge he had gained in wartime, the block assembly method—a modern means of building ships by assembling and welding various prefabricated sections of different sizes into one framework.[11]

Out of the social transition from war to peace emerged new enterprise, such as commercial giant Sony Corporation, which had a military heritage. Its founder, Morita Akio, was an electrical engineer at the Institute for Navy Aeronautics when he first met his lifelong business partner, Ibuka Masaru, an engineer who developed a direct current amplifier for aircraft at a commercial company. For them and other military engineers, hot-spring inns in remote areas offered a quiet setting amid war to brainstorm about weapon technology such as a heat-seeking missile. To such an occasion the navy often brought wine, which Morita and Ibuka enjoyed during their overnight conversations. Their wartime contact survived the war. About twenty former navy engineers joined them after the war, including Iwama Kazuo, later the fourth president of Sony Corporation. The core navy engineers who aided in its birth included Tokorozawa Keisaburō, an engineer who helped develop the Mitsubishi Zero Fighter for the war. Sony capitalized on various wartime contacts in the field of electrical engineering for postwar technological development.[12]

The optical industry, too, received peace dividends in the postwar society. For instance, Suzukawa Hiroshi—an ex-navy engineer who had developed torpedoes at the Kure Navy Arsenal—developed cameras at Canon. In his view, torpedoes and cameras were somewhat similar because they required both high-precision processing technology and automatic controllers. Joining the Canon design department in 1948, he established basic procedures for research and development as well as for the production of cameras. His 1953 creation, the Canon IV Sb, achieved commercial success in the field.[13]

In a way, the optical industry as a whole was a product of the war. It capitalized on the special procurement after the outbreak of the Second Sino-Japanese War in 1937 and continued to expand into the 1950s and beyond. Hitachi and Fuji Film—today's top commercial companies in the field—had procured optical equipment for the wartime army. The Olympus Corporation could offer another case in point. Its predecessor, Takachiho Optical Company, originally specialized in the production of microscopes and some cameras. But once mobilized for war, the company received technical guidance from the Army Science Research Institute and developed binoculars and other optical equipment.[14]

Agriculture also benefited from wartime developments. A salient example involves rice-planting machinery. This device helped to boost rice production after a former army engineer applied the automatic bullet-feeding mechanism of a machine gun to postwar agricultural machinery. Just as the machine gun picked each bullet from the belt and automatically projected it externally, the rice-planting machine automatically picked up each rice-seeding plant and placed it into a paddy.[15]

The postwar fishing industry also prospered from engineers' wartime experience. For instance, sonar radar technology for submarine warfare formed the basis of a fish finder in peacetime. At the NTRI, and later, at the Maizuru Navy Arsenal, Tsurugaya Takeo had conducted research on an echo sounder, a device to measure the depth of an ocean bed and seek safe routes on the ocean. Research for underwater audio equipment was banned temporarily under the Allied Occupation because of its possible military applications. Once the ban was lifted in December 1949, Tsurugaya's new sonar device proved its value for fishing; it detected the location of a school of fish, and this enhanced the supply of protein-rich food for the postwar population.[16]

Medicine was another beneficiary of the technological conversion of wartime technology to peaceful uses. Former army researcher Hagino Yoshio, who had worked to create television and noctovision, effectively closed the divide between wartime equipment and medical devices for peacetime use. After receiving a medical degree in the postwar years, he created a wide range of medical devices equipped with television screens, including an electrocardiograph, an oscilloscope, and cameras for medical procedures. His other former army colleagues continued their wartime researches and contributed to the development of commercial television.[17]

The civilian application of military technology showed most prominently in the material form of practical everyday technology. For instance, soup ladles that appeared in the immediate postwar years were made from scrapped bullet shells. Pairs of rubber shoes had originally been rubber tires recycled from scrapped wartime military aircraft. Created by an aeronautical engineer, a postwar Mitsubishi bicycle used a light and durable metal alloy, duralumin, which had been employed exclusively in aircraft construction. Some misbegotten ideas that caught the curious eye of the public and the press involved an army tank converted to a mobile home. After the war, many tanks remained unused and, thus, were scrapped and recycled, publicly auctioned, or renovated for everyday use. Pine oil—an energy source collected through wartime mobilization to compensate for Japan's dire lack of aircraft fuel—provided a source of alpha-pinene, a chemical compound used to manufacture a potent insecticide for postwar agriculture.[18] In such ways did the end of the war radically change wartime technologies for everyday use.

Demilitarization and Aeronautical Engineers

After it landed in Japan, which no longer retained strong air power, the Supreme Commander for the Allied Powers (SCAP) issued a series of directives to achieve its first occupation objective: demilitarization. The first order of business, as embodied in Directive No. 1 on September 2, 1945, was to preserve all research and development facilities and information. Allied authority soon banned research fields with direct military applications. At the top of the list were atomic energy, radar development, and aeronautics. On September 18, SCAP strictly prohibited all types of teaching, research and development in aeronautics, and aircraft production, as well as other airborne activities by the Japanese. This unyielding policy, which continued until April 1952, radically altered the technological landscape of postwar society.

SCAP successfully demilitarized Japan partly by focusing on aeronautical engineers, ending their careers in aviation and changing their social lives. The rate of unemployment was thus initially very high among them. According to one estimate, 100,000 engineers and technicians were jobless in the spring of 1946—and a great many of them were aeronautical engineers.[19] One example was engineer Doi Takeo, chief designer of 16 army airplanes, including the Ki-45 type-2 Fighter, Ki-100 type-5 Fighter, and Ki-66 type-3 Fighter. This graduate from the Tokyo University Aeronautics Department spent his time at a job placement office, reading English newspapers while waiting in the lines of job seekers. At first, he earned a living by building wooden carts and wagons. But the deflationary economic initiative of 1948, the "Dodge Line," eliminated even this temporary job.[20]

Socially stigmatized for having contributed to the war effort, former aeronautical engineers of necessity explored "new" fields. One such navy aeronautical engineer Watanabe Saburō became self-employed; using his wartime skills, he built farming equipment to extract vegetable oil for cooking.[21] To make ends meet after the war, the chief designer of the Mitsubishi Zero Fighter, Horikoshi Jirō, worked on machinery and tools such as the lawnmower, thresher, and refrigerator.[22] Nakagawa Ryōichi, Nakajima Aircraft Company's chief designer of the renowned 18-cylinder aircraft engine Homare, supported himself in a similar manner. He built an electronically operated bread-making machine, a tire pump for bicycles, an engine for fishing boats, a film projector for theater use, a diesel engine for agricultural machinery, and a sewing machine.[23] The dissolution of Nakajima produced many job seekers such as Nakagawa. Some of the company's former aircraft engineers took purchase orders from the Allied occupation forces and made coffins out of duralumin, a material used exclusively for aircraft construction. This light, durable, and precious metal proved useful for the containers; the US military placed deceased servicemen in the duralumin coffins and flew them back to their homeland.[24]

The accumulated knowledge of aeronautics for war seemed painfully useless in peace-loving society, as in the case of Itokawa Hideo. This professor at Tokyo University had developed the Ki-43 army type-1 Fighter and the Ki-44 army type-2 Fighter at Nakajima Aircraft Company. At the war's end, he suffered from severe neurosis, which initially led him to a doctor's office, and then, under the physician's sponsorship, to employment in Tokyo University's Faculty of Medicine. His new job was to suture the tissues of postoperative patients, although he did not possess a valid medical license. He soon created medical equipment that mechanically assessed the effect of anesthesia given to a patient on the surgical bed. This medical breakthrough ended a then prevalent practice among anesthesiologists; before his invention, a surgical assistant had repeatedly called out a patient's name by the bedside, until the subject under anesthesia could no longer give a vocal response.[25]

Among business operations, the automotive industry could reveal some important patterns of technology conversion across a few countries before and after the end of the war. In the face of war, some European nations had capitalized on their strong automobile industries. In Italy, Fiat had produced engines for cars and aircraft during World Wars I and II. In Great Britain, Rolls Royce Limited had built powerful engines for automobiles and military aircraft during World War II, including the legendary Supermarine Spitfire. Mercedes-Benz had made similar war contributions in Germany as the manufacturer of automobiles and airplane engines. These European countries shared a history of converting engine technologies for automobiles and aircraft before 1945, wherein flows of engineering knowledge were bidirectional between industries.

Because Japan did not have a similar pre-1945 history, the Allies' complete ban on the aircraft business had a fuller, more sudden impact on the Japanese automobile industry than anything of this sort in Europe. The draconian policy in Japan unleashed a massive, powerful, unidirectional flow of engineers from the defunct aircraft industry to the still nascent automobile business.[26] In the process, the former aircraft engineers brought a needed dose of practicality to the automobile design process. Toyota Motor Company, for example, experienced this via Hasegawa Tatsuo, who had been a senior aeronautical engineer at Tachikawa Aircraft Company. This designer of the army Ki-94 interceptor was among approximately 200 engineers who moved from Tachikawa to Toyota by the end of 1946. To his surprise, the preexisting community of automobile engineers had long adhered to relatively primitive engineering design techniques. Automobile engineers had seen no need to calculate the strength of the physical components in use; they lacked the means, or standards, to compute loads on many parts of cars in the design stage. Hasegawa introduced a new way of designing buses and trucks, using the monocoque body structure in his design of the Toyota BW bus.[27] Originally used in the German aircraft of the 1930s, this structural style could disperse the entire load on the sides, roof, and floor. His design team later launched a series of projects to build "smaller, faster, and better" automobiles. One of these creations was the Sports 800—a lightweight, two-seater sports car with an aerodynamically clean and oval shape because of its use of the monocoque structure. Soon, Hasegawa's design experiences bore fruit in the first generation models of commercially successful automobiles such as the Publica, Celica, Carina, and Corolla.[28]

The dissolution of Nakajima Aircraft Company also acted as a catalyst for empowering the automobile industry. Among many ex-Nakajima engineers who moved to Fuji Heavy Industries, the parent company of today's Subaru, was Momose Shinroku, the designer of the carrier-based navy reconnaissance aircraft C6N. In his new job, he infused his wartime experience of controlling tare weight into his car design projects. The Subaru 360, an outcome of this process, was one of the most popular two-door passenger cars in Japan in the 1960s. Despite its compact size, light weight, and small 360cc engine, the car gave Japanese drivers enough inner space with the monocoque body structure. Its charming design earned it the nickname "lady bug." Considered the "people's car," this affordable, mass-produced vehicle was a commercial success. Sales continued for twelve years from 1958 to 1970.[29]

Similarly, the end of Nakajima Aircraft was an opportunity for Honda to build its infrastructure for research and development. From the summer of 1945, scores of aero-engine developers flocked to the nascent company from the army and Nakajima Aircraft Company. Among them was Kudo Yoshito, a graduate from the Department of Mechanical Engineering at Tokyo University who helped to create the turbo-jet engine Ne130. One major difficulty that the ex-Nakajima engineers faced was determining, as one put it, "how low [they should] decrease their engineering expertise in level" when designing automobile engines. Kudo later became the first director of the Honda Research and Development Company—the birthplace of many highly advanced engines for motorcycles, passenger cars, racing cars, and airplanes.[30]

Likewise, Nissan Motor Company embraced former Nakajima engineers, allowing them to pass down their engineering values from generation to generation. As they found themselves out of work, many Nakajima aircraft engineers migrated to Fuji Precision Company, which merged into Nissan in 1966. One such engineer was Nakagawa Ryōichi, the chief designer of the wartime 18-cylinder engine, Homare. At Nissan, one junior engineer received his tutorage in developing various automobile components for the commercially successful passenger-car series, Skyline. "Most of what I [had] experienced after the war," Nakagawa reminisced, "was based on my ideas and experiences from [wartime] aircraft business."[31] Compared to the engineers in other industrial sectors, former aeronautical engineers such as Nakagawa showed an "incomparably more serious" attitude toward technological innovation. For these engineers, developing highly advanced, agile aircraft that could destroy enemy targets in the air, on the ground, or in the ocean was crucial. Before the eyes of the junior engineer under his tutelage, Nakagawa displayed "the extremely meticulous and scrupulous care taken to ensure the most accurate construction" of all components in a mechanism. Furthermore, this engineering style was "the most important lesson prewar aircraft engineers gave to the postwar Japanese auto industry."[32]

Among all the Japanese automobile companies, Mitsubishi Motors Corporation gained the most from the leadership skills of former aeronautical engineers. This was possible because Mitsubishi, however crippled it was during the occupation, preserved its core engineers in safe havens. One stark example involved the design team of the A6M Zero Fighter project. Horikoshi Jirō, the chief designer, was relocated to one of Mitsubishi's subsidiaries, Yoshimi Manufacturing, which operated independently after the dissolution of the business conglomerate.[33] Sone Yoshitoshi, Horikoshi's right-hand man in the Zero Project, settled at a Mitsubishi subsidiary after 1945; at Mihara Train Manufacturer, he played a vital role in developing the running gears for passenger trains. Tōjō Teruo, the second son of wartime Prime Minister Tōjō Hideki, and a key member of the Zero Project, moved to a plant in Kawasaki. Unemployment was rare among Mitsubishi engineers. Kubo Tomio, the chief designer of the army reconnaissance plan Ki-46 and army fighter Ki-83, built motor-driven tricycles at the Mitsubishi Mizushima plant after the war. Some Mitsubishi wind-tunnel technicians moved to a shipbuilding laboratory in Nagasaki. This practice of transferring engineers in the Mitsubishi networks continued until the end of the Allies' ban on research and development in aeronautics. By 1961, upon invitation, all the aforementioned engineers returned to their wartime employment at Mitsubishi Nagoya Plant. Soon afterward some engineers built their careers beyond the field of aeronautics. In the 1970s and 1980s, Kubo Tomio (1973–1979), Sone Yoshitoshi (1979–1981), and Tōjō Teruo (1981–1983) consecutively served as the presidents of the Mitsubishi Motors Corporations.[34]

Retention of Wartime Brains

When placed in the international context of the Cold War, these individual cases together illustrate the absence of a significant, systematic "brain drain" among former Japanese military engineers during the occupation years. A few rare cases notwithstanding, aeronautics offers a stark example of this pattern. Supporting evidence could be found, for instance, in the student directories of the Department of Aeronautics at Tokyo University. Established in 1918, the department stood at the pinnacle of elite professional education in the field. It almost single-handedly trained future aeronautical engineers in Japan; the university was the only institution to offer such a program until the mid-1930s. A series of student directories from the department list the names, career paths, and contact information of all the post-1923 graduates. According to the 1973 directory, the department produced 435 graduates by the end of the war, and among them, only 2—1 in 1940 and another in 1942—established their careers outside of Japan (both in the United States). In other words, 99.5 percent of the alumni remained in the country.[35]

This remarkable absence of a brain-drain pattern among the Japanese engineers invites explanations of factors that operated at individual, organizational, national, and international levels. Germany, for instance, offers a case for comparison. The engineering communities of both Germany and Japan shared similar fates. Before 1945, both Germany and Japan aggressively pursued territorial expansion, exploiting the human and natural resources of foreign lands that fell under their empires. During the ensuing occupations, both countries lost industrial and research capacity as well as their independence in technology sectors with possible military applications. The Allies dismantled the wartime aircraft industries of Germany and Japan, ending education, research, and development in aircraft engineering. In the case of Japan, the relative lack of a major brain-drain pattern had examples and counterexamples—both pointing to the importance of geopolitics, economic and legal barriers, and sociocultural norms that discouraged the emigration of Japanese engineers. Overall, Japanese engineers had less of a chance to leave their homeland than did the Germans to leave theirs. The emigration of engineers from Japan was far less common than that from postwar Germany. Examining how and why things did *not* take place is more difficult to explain than how and why things *did* take place, but one could gain a new perspective about the former question by exploring the latter.

In retrospect, geographical location could help explain the emigration of Japanese engineers but only to a degree. Sociologists have empirically noted a strong inverse relationship between the volume of migration and the distance traveled; simply put, the greater the distance, the lesser volume of migration.[36] After the colonization of Hokkaidō, countries near Japan were ideal places for emigration. Imperial Japan's former territory in northeastern China, Manchuria (occupied by Russia after 1945), had attracted young industrialists and farmers from Japan, especially in the 1930s.[37] In China after the summer of 1945, the Nationalist and Communist regimes vied for more control in the ensuing civil war until the formation of the People's Republic in October 1949. If not Russia, China probably retained the largest number of Japanese engineers and technicians, many of who contributed to the post buildup of both countries.[38]

The Chinese civil war makes it difficult to ascertain the total number of Japanese engineers and technicians in China proper, but some records offer insight. At the end of 1946, according to a US estimate, more than 90,000 Japanese remained in the country (including Taiwan and Manchuria). Not all of them were skilled workers or technicians, and in fact, many were family dependents. The Nationalist government conducted a similar demographic survey around the same time, concluding that the number of Japanese technicians (excluding those under Communist control) was slightly above 14,000. The Nationalist regime confiscated Japanese-owned enterprises and human resources with the collapse of the Japanese empire. This strategic move paid off. Japanese expertise was necessary in postwar China because the departure of the Japanese would interrupt regular work or technology transfer. About a quarter of all Japanese technicians in post-1945 China, many of whom were administrators or economists, worked in local factories. Many of the skilled workers held positions in hospitals, schools, and government agencies in China, and many continued to engage in a wide range of specialties, including the textile, railway, medical, and mining industries.[39]

The physical structure of Japan and its island status did not help the emigration of military engineers after the summer of 1945. Historically, Japanese had migrated to distant lands across the ocean such as North America (Mexico, Canada, and the United States, especially Hawaii and the West Coast), the Caribbean, South America, Australia, and New Zealand, but leaving the archipelago posed far greater physical and mental challenges than simply walking overland out of, for instance, Germany. Japan could not provide the relative ease of the emigration portrayed in the musical and film *The Sound of Music*, in which a large Austrian family and its female tutor walked together to the contiguous, politically neutral country of Switzerland. It was virtually impossible for Japanese military engineers to experience the life transition of, for instance, Claudius Dornier. Born in Germany in 1884, this aeronautical engineer moved to Switzerland in 1947 for a job, returned to Germany in 1954 when the Allies lifted a ban on the German aircraft industry, and died in Switzerland in 1969. Geographical proximity was a key factor in his movements. To express the emigration from 1850 to 1950 as a percentage of population increase, the emigration rates of Western European countries—England (75 percent), Italy (47 percent), Germany (24 percent), Denmark (22 percent), and France (6 percent)—surpassed Japan's rate of 1 percent.[40]

International geopolitics further complicated this picture. Japan was not situated like Germany, a country surrounded by the Allies. The nearby occupiers, Britain, France, and the Soviet Union, actively evacuated German specialists for postwar development, often from under the noses of the Americans. No such countries took control over Japan because the United States single-handedly occupied the Japanese islands. For exodus seekers, relocation to nearby countries was rarely the best option in part because of Japan's colonial past. A former Japanese army engineer experienced this firsthand. His wartime job was to develop deadly biochemical weapons in Manchuria, a task also pursued in the infamous Army Unit 731. At the war's end, he missed his opportunity to be repatriated to his homeland, and he wandered into China after the summer of 1945. He finally settled in Hong Kong, where he hid his wartime past and lived incognito throughout the subsequent decades.[41] Apart from China and perhaps Russia, it

was only the United States that could absorb military engineers in any significant numbers. But the American occupation authority set up "directives [which]
contained instructions relative to the transporting of Japanese nationals in the
United States military aircraft," decreeing that "no Japanese nationals will be
transported outside of Japan proper" unless "required for the accomplishment
of urgent occupational matters."[42]

Post-1945 emigration into the United States was highly contingent on historical precedents, which worked against the Japanese but worked for the Germans.
The influx of German engineers into the United States was not new. After the
end of World War I in 1919, several prominent German scholars in fields related
to aeronautics resettled in the United States. The Treaty of Versailles dissolved
German air power and prohibited the production and importation of new aircraft. Many aeronautical scientists looked for job opportunities abroad. It was an
opportunity for the United States to enhance its military capabilities with technical help from German experts, thereby demilitarizing Germany at the same time.

Historically, the United States was determined to catch up with European
advancements in aerodynamics and thus welcomed the wartime "brains" with
proven records of excellence in certain fields. For instance, Michael Max Munk,
a highly acclaimed German aerodynamicist, was responsible for the first systematic measurements of airflow around wing sections, known later as Göttingen
profiles. In 1920, he accepted an invitation to the United States. His employer
was the National Advisory Committee for Aeronautics (NACA), the predecessor
of the NASA. At the headquarters, his method proved useful for theoretically
predicting airfoil lift and moments. Shortly thereafter, he worked for different
commercial companies such as Westinghouse, Brown Boveri, and the Alexander Airplane Company. He also taught at the Catholic University of America in
Washington, DC.[43] Notable precedents like this reduced the cultural and political
resistance to the post-1945 immigration of German engineers.

The presence of earlier German immigrants in the host country could help
the flow of information about various opportunities abroad, encouraging a "brain
drain" from post–World War II Germany, whereas the absence of such individuals could impede emigration from post–War II Japan. Commonly, the availability of relevant information—such as the difference in wages between the home
country and target country, economic and political conditions of the target country, and immigration restrictions in the target country—could help shape the size
and attraction of communities overseas. Individual contacts in the target nation
could play a vital role in the outflow of immigrants. A network of individuals who
shared a common place of origin could attract migrants with the same traits,

resulting in chain migration. Past emigration could encourage present emigration. The mass media and other channels of information flow could also considerably influence the volume and direction of a migration.[44] In this framework, the German engineers and scientists of World War I could supply information to the engineering communities of their origin via formal and informal networks, reducing any uncertainties associated with the post-1945 migration. This option was not available to Japanese technicians and scientists. Before 1945, with very few exceptions, no such Japanese community existed in the United States or elsewhere.[45]

In trying to advance the field of aircraft engineering, the United States had good reasons to look not to Japan but to Germany. American wartime intelligence had noted the highly developed nature of German science and technology, aiming to exploit these resources against the Japanese following the end of the war in Europe. Noticeably, American research lagged behind German research in rockets and guided missiles, synthetic fuel, jet engines, and high-velocity aerodynamics. Capturing scientific and engineering know-how from wartime or nearly postwar Germany was an urgent task, which lasted into the 1950s. Conducted in military secrecy under the guise of national security, the ensuing operation— Operation Overcast/Paperclip, later replaced by Project 63—was very successful. Individual examples included Werner von Braun, the chief aerospace engineer for the German V-2 Rocket and, after the war, for ballistic missile development in the United States, which led to the Apollo 13 space program. Another example includes Adolf Busemann, a leading figure in the high-speed aerodynamics of wartime Germany. He moved to the Langley laboratory at the NACA and helped to create the swept-back wing for high-speed flight early in the Cold War.[46]

These two military engineers were among 2,756 German specialists who resettled in the United States shortly after the war. From 1948 to 1952, the United States Air Force brought in 1,044 German scientists and engineers, or 40 percent of the total of 2,627 people; the army integrated 866 Germans (33 percent), followed by the navy (425 Germans, or 16 percent), and the Department of Commerce (260 Germans, or 10 percent). The active recruitment also bore fruit in the long run. About 90 percent of German specialists hired from 1945 to 1952 stayed in the United States permanently.[47] In the case of Germany, the visible hand of the US military was pivotal in the "brain drain" and technology transfer.

However, the Allied occupation of Japan failed to develop the idea of relocating scientists and engineers from Japan to overseas countries. The architect of the audacious plan was Brigadier John O'Brien from Australia, the first chief of the Scientific and Technical Division in the Allied occupation forces. Upon his arrival in November 1945, he built a system to supervise and administer science and technology in occupied Japan. His keen interest in the question of reparations formed the basis of an ambitious proposal wherein he aimed to permanently re-locate wartime assets—advanced research projects, as well as families of notable scientists and engineers—from Japan to Allied countries. This reparation plan was designed to eliminate the war-making potential of Japan.[48]

O'Brien's proposal was short-lived, however. His superior, Major General William F. Marquat, who had served with General Douglas MacArthur in the Pacific campaign, rejected it for at least two reasons. First, O'Brien was not an American; he was an Australian, who was heading a unit in the Allied occupation gover-nance and, hence, lacked political clout in the hierarchical bureaucracy domi-nated by Americans. He received neither a salary, nor access to various services provided for American staff members in Tokyo. Second, Australia and the United States valued the importance of reparations differently. O'Brien—the chief of the Australian survey team serving in, and paid by, the Australian army—was determined to pursue various reparation questions. This was understandable in retrospect. Many Australians had feared the Japanese military, as it could ad-vance into their homeland as they had done in the Philippines. Their concern for the Japanese underlay the Australian Scientific Mission, whose purpose was to visit Japan and make claims for the most desirable reparations items. However, generally speaking, the military services of the United States had far less interest in reparations partly because overall they did not seem impressed with Japan's wartime research.[49]

While the story of the United States in this regard remains unclear, only a very small number of Japanese engineers seem to have been able to provide the Americans with knowledge useful in the Cold War.[50] One revealing case in-volved at least a dozen former army engineers. Their wartime jobs had included forging foreign paper currency and various documents, including identification papers issued by the Chinese Communist Party as well as by the North Korean and Soviet military forces. By the spring of 1950, they found employment in the Yokosuka US military base, where their wartime expertise supported America's Cold War efforts in East Asia. The end of the Allied occupation in April 1952 opened the door for their migration abroad. To continue their clandestine opera-tion, their place of employment moved from Yokosuka to San Francisco. From June 1952, they lived in suburban, civilian apartment complexes with their rents paid by an unknown US government agency. The disbanding of the project team in 1960 returned nine members to Japan; two, however, remained on the West Coast. With his Japanese wife and child, one member became a naturalized US citizen, relocated to Washington, DC, and worked for a government agency for the next 20 years.[51]

Apart from this seemingly exceptional case, the cultural backlash against Japanese nationals in the United States did not create a healthy, ideal, or supportive environment for flows of information or individuals across the Pacific. Racial discrimination remained tenacious. As historian John Dower shows, wartime racism against the Japanese was more intense in degree and malicious in nature than that against the Germans.[52] This barrier had legal and historical grounding. Introduced in 1924, the Johnson–Reed Act limited annual immigration from particular countries. The quota system fixed the number of German immigrants at 25,957 per year and effectively prohibited the inflow of Japanese immigrants. According to the legislation, nominally, the annual quota for Japanese immigrants entering the United States was 100. But the restrictive quota system used the euphemism "aliens ineligible for citizenship," constructed "Asiatic" as a racial category, and effectively excluded Japanese immigration until June 1952.[53] The Immigration and Nationality Act of 1952 finally lifted the ban against Japanese immigrants. The new, more salutary law assigned an annual quota of 185 to Japan, whereas the annual immigration from Germany after 1953 had averaged at 24,500.[54]

Similarly, Japan's wartime sentiment diminished to a degree between 1945 and 1952. An underlying factor was repatriation of Japanese from Japan's former colonies although this aggravated the already acute food shortage. To solve both the population and food problems, Nakasone Yasuhiro, a newly elected member of the House of Representatives, insisted in 1949 that Japan should revive its emigration policy for the exodus of roughly 1.6 million people a year. Popular only in the abstract, this idea soon mirrored a social desire for emigration to the United States. According to an *Asahi Shinbun* survey of public opinion conducted in November 1951, 28 percent of the respondents favored emigration, which was greater than the percentage of advocates for birth control (24 percent) and those for both emigration and birth control (14 percent). Among this 28 percent, more males supported emigration than females, and more members of the urban population than their rural counterparts expressed the same point of view. Furthermore, the most desired destination was Brazil (21 percent) followed by the United States (18 percent).[55] This social sentiment aside, anti-American sentiments lingered especially among the senior military engineers (see the Appendix), who had built their careers in the war against the United States. Living in the former enemy country during or even after the occupation was, according to two anonymous interviewees, not only "unthinkable" but "detestable."

During the occupation, any Japanese seeking foreign travel by ship would experience at least two barriers. The first one was economic. Ordinary Japanese citizens had neither lawful access to hard currency nor any other adequate financial support for their departure. Typically, the higher the cost of movement in money, time, or energy, the less likely it was that a migration would occur. Migration cost could be higher than any conceivable wage gain for those without recruiters, friends, or relatives at the destination. The longer the distance to the target country, the lesser the resources available, such as money and other types of necessary assistance.[56] In Japan after the war, travel costs were prohibitively expensive with the exchange rate fixed in April 1949 at 360 yen to 1 US dollar. Even after the end of the occupation in April 1952, converting the Japanese yen to the US dollar was exceedingly difficult.[57]

The second barrier that stood before those Japanese wishing to go abroad was the legal restrictions at home. In the first half of the occupation, Japanese could not obtain a passport and visa unless they had a relative abroad who could provide a written affidavit of support. The visa was a central tool of migration control in the United States. Japanese applicants wishing to visit the United States had to sign a statement swearing that they were not Communists; in addition, they had to have fingerprints taken and pass a mandatory medical examination. Applicants with symptoms of the widely spread disease, tuberculosis, were denied. Only after 1949 did government-sponsored, study abroad programs become more available—such as Government Aid and Relief in Occupied Areas and the Fulbright Program—but their selection procedures remained strict, thus precluding many scientists or engineers from pursuing their chances overseas.[58]

What played an indirect and important role in removing these economic and legal barriers was gender. The ordinary Japanese citizens who had the best chance of emigrating to the United States were not the male ex-military engineers but the young brides and grooms of the US military service members. By World War II, national border control replaced other factors, including distance, cost, and local institutions, as the main obstacle to international mobility. Families and personal ties were among the most important dynamics in the decision to move. While World War II sent a large number of US personnel to military bases in Europe and Asia, family connections to those service members provided their brides and grooms with a strong incentive to emigrate. Once engaged or married to US citizens, foreign nationals could enter the host country often with their dependents, irrespective of the fixed immigration quotas assigned to different countries. During the occupation years, for instance, 8,381 Japanese females married US service members. Each year from 1946 to 1965, partly because of the war bride/groom factor, the number of female immigrants entering the United States far exceeded the number of male immigrants.[59]

Among all the Japanese males, renowned scientists who could share their theoretical studies in writing were the most likely to obtain the three prerequisites to go overseas: money, a passport, and a visa. University researchers had a good chance of passing this set of hurdles, especially if they had individual contacts abroad. Nobel Prize laureate in nuclear physics, Yukawa Hideki, exemplified this tendency. He was among the few scientists who received invitations from universities and research institutes in the United States via personal contacts. With support from theoretical physicist J. Robert Oppenheimer, Yukawa stayed at the Institute for Advanced Study in Princeton before joining the Physics Department of Columbia University. Ordinarily, a Japanese specialist wishing to go abroad would try to find a position in foreign countries by directly writing letters to relevant researchers in the fields in question. Then their hosts would guarantee to pay for travel and living expenses of the Japanese researchers. Having cleared these barriers, 265 Japanese students temporarily stayed in the United States during 1949–1950.[60]

Aside from personal contacts, sociocultural expectations could aid or hinder an exodus of Japanese engineers. For instance, Japan's first female physicist, Yuasa Toshiko, immigrated to France via her individual contacts, in part, because she sought a laboratory research culture that could accommodate her research needs.[61] Probably more so than their German counterparts in their pursuit of a career overseas, Japanese engineers faced more cultural restraints at home, partly due to family expectations associated with birth order.[62] At the risk of oversimplification, later-born males in urban areas seemed to have a better chance of emigrating than did first-born males in rural areas; in cultures that espouse filial piety, first-born sons (especially only sons) were apparently more likely to stay with their aging parents and take care of them, while later-born sons were relatively free from the familial expectation. Compared to younger siblings, first-born males would bear the weight of sociocultural, and often legal, obligations as the head of the household, which could include his own family, his parents, and often, his spouse's parents. Cases of adoption aside, in Japan that codified primogeniture until May 1947, firstborn sons tended to inherit tangible assets and carry on their family names, businesses, and other duties. For instance, among 18 interviewed former military engineers whose family backgrounds were accounted for, 9 were the firstborns, and 13 were the eldest sons in the families (Appendix).

Religion could also play a key role. One major duty expected of the eldest sons was to financially support, and periodically and physically pay tribute to, their

Mahayana Buddhist temples in which their cremated ancestors are entombed. For key religious anniversaries at their temples, eldest sons played a key role as the registered guardians of their ancestors. Except for one case involving a Protestant Christian family, all the interviewed former military engineers were from urban areas (see Appendix) and confronted this issue in one way or another during the 1950s. Their spatial mobility remained culturally bound.

Engineers' geographical mobility was often a function of age. Almost always, the ages of senior, experienced military engineers as of the year 1945 ranged from the early thirties to early forties. As expected, those in this age group were married and tended to support their families with young children, and often, aging parents. This age group was at least several years older than the median ages of all the male immigrants to the United States during 1946–1952, which ranged from 25.0 to 29.9.[63] These family dynamics could become complicated in a foreign environment. Wartime engineers could read reports and journals written in foreign languages such as English, French, or German, but conducting daily conversations in English was a different matter altogether. Former military engineers typically lacked the experience of living abroad for an extended period of time; in fact, none of the informants (see Appendix) even went overseas before the end of the war. Relocating their family members abroad would have demanded daily communication outside the home in a foreign language—and this was hardly the best option for many household heads.

One counterexample could explain this cultural landscape. The case involved Koyama Akira, a former army engineer and a later-born male, free from any expectation to succeed as the family head. His expertise was derived from producing counterfeit foreign currency, especially that of China, during World War II. He was 27 years old in 1952, when he was invited to immigrate to the United States. Like many others like him, he cut all kin ties when he left his home communities and arrived in the host community as a minority. Frequent return by ship to his ancestral home, *furusato,* was physically and financially impossible.[64] A wide range of problems awaited his wife and child, who joined him later and became naturalized citizens of the United States. The legal barrier was far easier to overcome than cultural, everyday issues, including language, social customs, food, and child education.[65]

From 1945 to 1952, the emigration of military engineers from Japan was far less visible than emigration from postwar Germany. Noted counterexamples aside, the Cold War geopolitics in East Asia pushed Japanese engineers to remain in their own country. Meanwhile, economic and legal barriers stood formidably before those seeking to go abroad. Sociocultural norms pulled them inside the national boundaries. Lacking the means and paths to pursue an exodus, former military engineers were contained in the country and migrated domestically into the civilian sector. Their absence from the post-1945 emigration in any significant numbers ensured that their wartime expertise was passed on exclusively in Japan. The benefits of their transition were apparent in the shipbuilding, electronics, agricultural, fishing, and above all, automobile industries. By the end of the Allied occupation in April 1952, former military engineers lacked incentive to emigrate because they had firmly reestablished their second careers in Japan. One government agency that creatively cultivated wartime human resources was the Japanese National Railways. The process, however, was very contentious from the beginning, but in a roundabout way, it eventually bore fruit in the development of high-speed rail service in the following decades.

CHAPTER FIVE

Former Military Engineers in the Postwar Japanese National Railways, 1945–1955

"In winning the war, the victor is only too likely to assume that he is superior to the vanquished in every respect, not only materially, but also intellectually, ethnically, and culturally," warned Dr. Fritz Zwicky, a Swiss émigré and an expert in jet propulsion at the California Institute of Technology. His appraisal from November 1945 followed a three-week tour of Japan. His report about Japan's wartime technology was remarkably free from wartime stereotypes that remained in the United States. Referring to "intellectual potential and intrinsic technical skill," he wrote that "the Germans were very impressive, [and] the Japanese are also more impressive than commonly is believed." While noting "remarkable copy work" in Japanese products modeled after foreign turbojets and superchargers, he observed, "Rocket-propelled missiles were visualized and experimented with at an earlier date than in the United States . . . and supersonic wind tunnels were installed several years in advance of the construction of similar equipment in the United States."[1]

The end of World War II released thousands of able military engineers, including those who had impressed the occupation authority with their technical expertise, into the Japanese National Railways (JNR). This development during the occupation (1945–1952) was a blessing in disguise for Japan. Only in these years did various engineering groups mingle with one another, setting the stage for the successful marriage between former military engineers and preexisting rail engineers for the postwar development of high-speed rail service. To illustrate these points, we begin by examining the cultures of defeat in society and in the railway industry after the total war. In the years of postwar reconstruction, former military engineers helped the JNR to provide a safer, more comfortable riding experience for millions of passengers who had grown weary of war and death. These engineers began to reshape their fields of study, their engineering communities, the JNR, and, to an extent, society.

The Japanese National Railways after World War II

When the war ended on August 15, 1945, most public railway facilities in the country were urgently in need of repair. The damage, both severe and extensive, included 682 breaks in main rail tracks and 361 breaks in side-rail tracks. As many as 85 tunnels and bridges as well as 465 stations and other buildings were in poor condition. Aerial bombardment and gunfire had been particularly intense in urban areas during the war, devastating 100 bridges in 50 locations across the country. Roughly 15 percent and 13 percent of steam and electric locomotives, respectively, had been heavily damaged. Also in a terrible condition were passenger carriages and freight wagons; overuse and low wartime manufacturing standards posed serious safety concerns to the riders. In 1947, only 47 percent of 18,000 coaches were considered functional. Meanwhile, demobilized soldiers and evacuated schoolchildren in the countryside returned home on trains.[2]

The urgent need to repair the physical damage remained unfulfilled for a time because Japan as a defeated country lacked material resources. Wood, which was needed for housing, was in short supply. To fix damaged crossties, for instance, in 1946, the JNR needed roughly 34.7 million cubic feet of wood (5,450,000 *koku*), but it obtained only half of that amount.[3] Steel was not available either. According to one estimate, at least 150,000 tons of shaped steel products were required per year to repair physical infrastructure and locomotives, but in 1946, the JNR obtained 46,000 tons, or merely 30 percent, of the needed quantity. It was only after 1950 that the JNR obtained the bare minimum amount of steel parts for railway construction.[4] Moreover, a severe shortage of coal—the chief energy source of motive power in the railway industry—hampered logistics. Heavily damaged by air raids and torpedoes, maritime transportation essentially halted at the end of the war. A major campaign to save coal began in September 1946, but in December 1946, passenger operations were reduced nationwide by 16 percent.[5]

In the defeated society, the poor physical condition of the rail service caught the attention of the press. New censorship by the Allied occupation forces after the summer of 1945 defined the media, replacing the heavy censorship of the wartime media as organized by the militaristic regime. After that year, the media were no longer an instrument of war; nationalistic, evocative themes—such as "Sacrifice for the Emperor"—that had fanned the wartime fires were gone. Now, the press expressed the views of many ordinary consumers and civilians, articulating such uplifting themes as "peace" and "coexistence." A society that had survived a destructive war at a considerable cost had acquired a keen sensitivity to life and death issues. This development, a culture of defeat, reflected the reorientation of the individual and collective values in Japan that had appeared after August 1945.

The press began to carry reports about the everyday rail service that endangered the lives of riders. Such reports would not have appeared widely, if at all, in the wartime media. One such incident occurred in December 1945. An infant died of suffocation on its mother's back in a highly congested commuter train in Tokyo. When the police charged the mother with involuntary manslaughter, readers discussed safety in rail service as a social issue. Ordinary citizens—a female journalist, a male student, a mother, and others—wrote on the matter in a leading newspaper, *Asahi Shinbun*, and generated a major public dialogue. The police soon withdrew the indictment under intense public pressure.[6]

Especially in urban areas, many trains defied social expectations of safe and comfortable postwar travel. The average occupancy rate per car was 140 percent in 1946, but would often increase to 400 percent on major urban lines during rush hours in coaches that regularly carried 300 riders, despite the legal limit of 70.[7] The level of public frustration with rail service was high. At the end of 1948, the majority of coaches had reached or exceeded their average life span of 30 years; they were structurally weak assemblies of wood and steel with bare seats, loose nails, excessive noise, the malodor of "a pig farm," leaking roofs, and closed windows that obstructed light, air, and the use of emergency exits.[8]

Millions of ordinary citizens, who were weary of war, hailed its end because it liberated them from the death that had previously surrounded them; however, they quickly learned that a major threat to their lives still lurked in the daily rail service. One serious danger was the combustibility of the wooden railcars. In November 1945, for instance, a cigarette fire burned rapidly out of control and eventually reduced six partially wooden cars to ashes. Sixty-five passengers were injured, and eight died in the ensuing inferno.[9] The fragility of the wooden construction was equally dangerous. In June 1946, the centrifugal force of a congested commuter train running on a curve in Tokyo pushed passengers through a broken wooden door and off the bridge, into a river below, killing five passengers.[10] Train accidents—such as collisions and derailments—were common across the country. In February 1947, four wooden cars at the rear of a train were dislodged from the rest on a sharply declining curve and fell off a cliff. This staggering accident injured 495 passengers and killed 184.[11] Evocative headlines in major national newspapers—such as "Unprecedented Catastrophe!" and "A Thousand Casualties"—alerted readers to the latent danger in daily rail service.[12] The press published articles on this tragedy that symbolized the highly risky nature of the rail service. "Doors fell off, congested trains derailed and often fell off cliffs," one newspaper derided.[13] An editorial denounced the railway's corporate leadership, demanding that the railroad companies as a whole address this tragedy.[14] In 1949, the Civil Transportation Section of the Allied occupation—the supervisor of all railway operations in Japan—recognized the public's frustration over the danger of riding in the fragile and combustible wooden railcars, insisting that they be removed from service.[15]

A series of deadly accidents in the summer of 1949 signaled a new level of alertness. The president of the JNR disappeared mysteriously one day, and his body was later found in pieces on a stretch of railroad tracks. His active involvement in a heated labor dispute at the time included massive layoffs of 95,000 railroad employees over a period of time. His death, either a suicide or, more likely, a homicide, preceded a naked terrorism directed at rail service. In July, an unmanned train crashed through a station and into a residential area in Tokyo, killing 6 citizens and injuring about 20 more. The press widely reported this unprecedented accident. *Asahi Shinbun* focused on an excited man on the scene who ranted against the JNR's recent policy supporting massive layoffs and its use of the wooden, "problematic" car model 63 in its rail service.[16] Shortly after the incident, the police arrested about a dozen Communist Party members for sabotage. The following month, several disgruntled JNR employees in the labor union reportedly sabotaged a track and derailed a fleet of 630 passengers, injuring and killing several of them.[17] Investigators and the media witnessed the disfigured, partially wooden cars on the scene. The fatalities carried political implications, thus alarming the Civil Transportation Section. On August 19, the authority instructed the JNR to report all of the "accidents" aiming to damage the Allied occupation.[18]

From 1945 to 1952, during the occupation era, the JNR faced the daunting and urgent task of providing safer rail service to millions of passengers amid mounting sociopolitical pressure from the left. On arrival, the Supreme Commander for the Allied Powers (SCAP) endorsed a series of liberal labor union policies for better wages, benefits, and working conditions. Consequently, leftists grew in number and in bargaining power, especially in state-owned services such as the JNR. The National Railways Workers' Union, formed in March 1946, altered the size and demography of the JNR employees. According to a February 1947 agreement between the union and the JNR leadership, employees obtained 8-hour workdays, 1 weekly day off, 20 days of paid vacation per year, 5 holidays per year, and menstrual leave for female workers. These new standards reduced the total number of working hours among the existing workforce and, to compensate for the loss of time, increased the total number of employees by nearly 20 percent.[19]

The JNR workforce continued to expand in the defeated nation. Thousands of demobilized soldiers and repatriates returned from Japan's former colonial empire of Taiwan, Korea, and Manchuria. By the end of 1946, 164,000 former employees of the JNR had returned to their jobs. The JNR absorbed 10,000 other workers who left the dissolved South Manchurian Railway and Korean Railways at the end of the war. As a result, the total number of JNR employees doubled from 300,000 in 1939 to 600,000 in 1948.[20]

Where Did the Former Military Engineers Go? The Japanese National Railways and the Railway Technical Research Institute, 1945–1952

The end of the war was, in fact, a blessing for the JNR. Peacetime leadership strengthened the core engineering workforce, welcoming highly educated engineers from the army and, above all, the navy. According to the employee directory of August 1950, at least 267 new employees with baccalaureate degrees joined the JNR after the war, and the majority of them—at least 181 engineers, or 67 percent—shared a background in military engineering (Table 5-1).[21]

This postwar infusion of former military engineers had its genesis during the war. Acting partly from a sense of obligation to absorb them, the JNR upper management had planned ahead for the defeat that seemed imminent by early 1945. They held meetings before the war's end, discussing impending issues for the JNR and the Ministry of Transport, such as how to transport millions of returning repatriates. The leader of the government railway operations at the time was a bureaucrat, Horiki Kenzō (1898–1974)—the director of the Railway Board within the Ministry of Transport during the war, and subsequently, a member of the Upper House of the Diet and a councilor in the Ministry of Health, Labor and Welfare in the 1950s. At a morning meeting on August 15, 1945, Horiki emphasized the duty of the JNR to absorb not only railroad engineers returning from Asia, including Manchuria and Taiwan, but also wartime military engineers in

TABLE 5-1
Wartime Background of 267 Employees at the Japanese National Railways

Wartime Background	No. of Employees
Army	48
Navy	95
Central Aeronautical Research Institute	24
Others (aircraft related)	14
Conscripts	24
South Manchurian Railway	35
Korean Railways	14
Others (railroad related)	13
Total	267

the homeland.[22] After Emperor Hirohito announced the end of the war, Horiki's goodwill effort formed the backbone of the JNR recruitment strategy. This outcome would not have been possible in postwar Germany, where the Allies divided and directly administered the defeated nation. In Japan, the Allies kept its pre-1945 government system and indirectly managed the rail service. The Ministry of Transport operated the railways under the supervision of the Civil Transportation Section of the SCAP General Headquarters.[23]

Most of the former military engineers were "outsiders" to the preexisting management structure of the JNR and ordinarily lacked chances for promotion in Tokyo. Among the 267 new recruits with baccalaureate degrees as of 1950, only 17 engineers—or 8 percent—resettled in the JNR headquarters. The rest—92 percent—remained outside of the central management in Tokyo. They worked at JNR's subsidiary institutes, or railway factories, or bureaus located in geographically remote areas.[24] This geographical diffusion of engineers across the country benefited the employed and employers alike. Local factories gained the expertise of military engineers for repairs and maintenance. Food shortages were less acute, and housing conditions were better in remote areas than in urban centers. Bureaus in the peripheral regions—such as in Niigata, Sendai, and Moji—were able to employ an ample number of former military engineers and support their families.[25]

The integration of "wartime brains" into the JNR was a major gain for the Railway Technical Research Institute (RTRI), a locus of the conversion of wartime technology to civilian uses. Historically, this had remained the only research establishment of the JNR. It began in 1907 as an experimental station with 38 members, who chiefly tested such materials as cement, bricks, and volcanic ash. Located in Hamamatsuchō in Tokyo after 1927, the RTRI acquired its official name in March 1942 and hosted roughly 400 workers throughout World War II. By this time, they were engaged in a wide array of research and development projects; at one point, the research system comprised about 40 laboratories, each specializing in key aspects of technological artifacts. One laboratory, for instance, focused its research entirely on the body structures of different vehicles—steam locomotives, electric trains, passenger cars, freight cars, special utilities cars, and automobiles. Bureaucratically and financially, in June 1949, the RTRI became a subsidiary organization attached to the newly formed JNR under the Ministry of Transport.[26]

During the Allied occupation, the RTRI actually accommodated too many wartime engineers. From 1944 to 1947, hundreds of them flocked to the institution, whose total employees quadrupled from 380 to 1557. This massive expansion was short-lived, however. The Allied occupation authority purged thousands of "militarists" and "ultranationalists" from the public service sector (including the RTRI), allegedly because they had contributed to the war effort. Under this intense sociopolitical pressure, the RTRI lost 821 workers, or 55 percent of its total employees, in 1949 alone. A chief designer of the legendary giant battleship Yamato, for instance, perforce left the RTRI and sought employment elsewhere. Kagayama Yukio, the JNR president from 1949 to 1951, tried to preserve the RTRI's research capability but to no avail. The monetary pressure on the JNR, then operating at a deficit, remained intense. The Science Council of Japan, for instance, discussed the possibility of privatizing the RTRI to alleviate the financial burden. In June 1949, Prime Minister Yoshida Hitoshi proposed reducing its research activities and the size of its workforce by as much as 30 percent. In April 1950, 111 employees and 7 laboratories were transferred from the RTRI to the newly established Transportation Technology Research Center. By 1951, the total number of RTRI employees had bottomed at 512; afterward, it increased slowly.[27]

Despite the fluctuations in numbers, many highly educated former military engineers remained at the RTRI as its core workforce. In 1950, there were 549 workers, including 125 new recruits with baccalaureate degrees in engineering or natural science. Of this additional workforce, 37 former navy engineers constituted the largest group, followed by the second-largest group, 29 new university graduates. The 28 who had built their careers in the railway industry were a distant minority group that included one engineer from the South Manchurian Railway and one from the Korean Railways. Sixty-four individuals, a substantial portion of the new workers, had been wartime military engineers and most of them shared an aircraft-related background (Table 5-2).[28]

Table 5-2 illustrates the importance of institutional and personal linkages for knowledge transfer in the immediate postwar years. The 11 Central Aeronautical

TABLE 5-2
Wartime Background of 125 New Recruits with Baccalaureate Degrees

Wartime Background	No. of Workers
Navy	37
Army	13
Central Aeronautical Research Institute (CARI)	11
Others (aircraft related)	3
South Manchurian Railway	1
Korean Railways	1
Others (railroad related)	26
University faculty	4
New university graduates	29
Total	125

Research Institute (CARI) engineers in the table moved from that agency to the RTRI. These two research institutes were familiar to each other during the war because they shared some administrative personnel. When the CARI ceased to operate on August 15, 1945, the Ministry of Transport became its governing body, managing the defunct establishment as well as the RTRI. Appointed to direct both institutions on September 1, 1945, Nakahara Juichirō further solidified the connection between the two through which 1,220 workers—or 80 percent of the 1,526 workers in total at the dissolved CARI—migrated to the RTRI by April 1946.[29] This personnel transfer was smooth on paper. Officially, the aeronautical engineers were affiliated with the CARI until December 30, 1945, and then with the RTRI from the very next day.[30] The RTRI also inherited the wartime knowledge base. It included 2,200 books from the dissolved navy, as well as 1,000 books and 2,000 back-numbered journals from the CARI.[31] The accumulated experience and knowledge of the wartime years found their niche in the postwar environment.

Less transferrable and successful than human resources were tangible assets, a chief source of postwar reparations as extracted by the Allied occupation forces. Seven months after the war, the opportunistic RTRI management drew up an expansion plan that would put to use buildings as well as 229 acres of land (281,000 *tsubo*), both occupied by the former CARI. The reality, however, differed from what they envisioned. Vigilant over such sources of indemnities, the occupation authority allowed the use of only half of the buildings and 100 acres of land, thus derailing the plan. Dismayed but undaunted, in September 1948, management came up with a more comprehensive five-year plan that would integrate various research facilities. Stymied by the Civil Transportation Section of the Allied occupation, this plan had failed by 1950.[32]

In other times, the end of the war provided an advantage to RTRI, which was determined to collect machinery and facilities from across the country. Management acquired only 24 pieces of machinery from 1937 to 1945, but from 1945 to 1953, they obtained 211, averaging more than 26 pieces a year. Many tangible assets that escaped the fate of wartime reparations were relocated at the RTRI. From the disbanded army, for instance, the RTRI obtained storage facilities from the Second Army Rail Regiment and buildings from the Army Material Arsenal. After some renovation, the wartime facilities became part of a railway testing laboratory for the engineers to test railway tracks, soil, and concrete. One-story buildings and 1,429 acres (43,220 *tsubo*) of land were available use by the RTRI.[33]

More windfalls came from the defunct navy. For instance, the RTRI management gained the right to use a giant tester from the Allies' General Headquarters and then the Ministry of Transport. This machine was the only one in the country that could check the fatigue strength of ship hulls, locomotives, and aircraft parts during the war. In 1952, this 25-ton equipment was transported in pieces from its original site at the Navy Technical Research Institute (NTRI), where highly advanced radar technology had been developed during the war. The RTRI's ambitious effort to obtain navy facilities extended across the country. A case in point was the Testing Station of the Kure Navy Arsenal, located about 1,200 kilometers from Tokyo and only 24 kilometers from the obliterated city of Hiroshima. Granted its privilege, the RTRI temporarily administered assets at this distant site, such as plants, property, and land, on the behalf of the Allied occupation authority. From April 1946, RTRI engineers used the facilities for experimenting on railcars, bridges, rail tracks, and other large structures. Once unshackled from the wartime reparations in 1952 and from the Ministry of Finance in 1955, all the testing facilities were available to the RTRI, including a one-story building on 9.5 acres (271 *tsubo*) of land and additional space of 14 acres (426 *tsubo*).[34]

Smoldering Tensions within and outside the Railway Technical Research Institute

The integration of wartime research facilities and engineers into the RTRI sharply divided the engineering community and rarely did the outcome of the integration seem salutary to all concerned. Within the JNR and the RTRI, former military engineers were both "newcomers" and "outsiders." Aeronautical engineers, however "elite" they might have been in the military during the war, were stripped of their elite status in the peacetime railroad community. At times, wartime railroad engineers openly exhibited hostility to former military engineers. At an RTRI job interview, for instance, one engineer from the Institute for Navy Aeronautics (INA) was branded as *"kokuzoku"*—a traitor to the country—for having belonged to the military during the war. Later, he declined the offer with resentment.[35] Many railway engineers scornfully called former military engineers at the RTRI *"shinchūgun,"* an occupation force.[36]

Why such "outsiders" joined the RTRI was not visibly clear at first to the pre-existing community of railway engineers. At least for some time, the former wartime engineers had neither the necessary research equipment, nor a set of tasks to complete in the immediate aftermath of the war; they spent time by cutting grass and planting potatoes in the backyard to weather the food shortage of the postwar years. Despite this work style, their prerogative included free rides on the JNR trains. Those who spent two to three hours traveling each way to the RTRI essentially became commuting farmers at the expense of the establishment.[37] A disgruntled railroad engineer took his grievance outside of the establishment, criticizing the "new research and development system":

> With no experience in the field of railway engineering, some new boys came [to the RTRI] by invitation simply because they had the title of engineer. Some sections went too far. These engineers formed the center of research and tailored it to army or navy styles, a development detrimental to the traditional spirit and history of railroad engineering. The inconsistent policy of [RTRI] expansion, and a lack of order in each enlarged section disgracefully created similar laboratories. . . . The expansion seemingly resulted in the formation of a strong system for research and development, but in reality, the system became shallow and weak. . . . This is because the opportunistic leadership held certain preconceptions [about military engineers], pursuing expansion hastily and capriciously. The policy resulted in nothing but the opening of new laboratories, enormous expenses for research facilities, and bad feelings against the new researchers.[38]

This highly critical article appeared in a technical journal and was read widely within the railroad community. This type of criticism was pervasive even within the Ministry of Transport. "The 1,500 research associates [mainly consisting of former military engineers] and the massive research system at the RTRI," one lamented in a Ministry journal, "contribute very little to JNR's business operation."[39]

The smoldering tension within and outside of the RTRI resulted in part from management's alleged favoritism for wartime military engineers vis-à-vis wartime railway engineers. The management in fact appointed 14 former military engineers to important laboratory leadership positions. In their appointments, which were never justified publicly, benefits outweighed risks at the JNR that suffered from widespread labor unrest at the time. Even General Douglas MacArthur was on alert when faced with the leftist threat to a fragile social stability. In July 1948, he instructed Prime Minister Ashida Hitoshi to prohibit strikes by public servants, government railway workers among them. While JNR employees retained the right to collective bargaining,[40] the RTRI remained remarkably free from the sociopolitical turmoil. The voices of a few identified leftist engineers rarely attracted much support. Having shared senior rankings in the military, the 14 laboratory leaders tended to lean more toward the right than toward the left, offering a counterweight to the leftist movement within the JNR.

As of 1950, there were 30 laboratory leaders, who could be divided into the camps of former military engineers and wartime railway engineers. Compared to the former group, the wartime railway engineers as a whole were less educated but vastly more experienced in the railway industry. Excluding one rail engineer from the wartime South Manchurian Railway, this group consisted of 15 laboratory leaders who joined the JNR before 1937. All of them had gained at least 13 years' experience in their technical fields by 1950. Half of those in the leadership positions were craftsmen; lacking baccalaureate degrees, they had mastered their skills through steady on-the-job learning. Based on the seniority-based pay system, wartime railway engineers with college degrees were the highest paid among all the 30 laboratory leaders. Two such graduates from Tokyo University received 174,000 yen per year, which was 21 percent more than the second-highest salary paid to former navy aeronautical engineers.[41]

The opposing camp, consisting of 14 former military engineers, was more "elite." They migrated from the former navy, army, or the CARI, joining the RTRI in the first few years after the war's end. This group included 10 engineers from the navy, 2 from the army, and 2 from the CARI. Former navy aeronautical engineers were the highest paid among them. While one laboratory leader from the defunct CARI earned the least at the RTRI, three senior engineers from the former INA earned 14 percent more a year. As of 1950, none of these 14 laboratory leaders possessed more than a few years of experience in the railroad community. What compensated for their lack of experience in the rail service was their elite education; that is, *all* of them had earned baccalaureate degrees in engineering or natural science from former imperial universities, and half had graduated from Tokyo University (Table 5-3).[42]

The ages of the 14 former military engineers reveal more commonality about their wartime past. The oldest one was born in March 1908 and the youngest in January 1916, and their ages as of 1950 ranged from 34 to 42. *All* of them joined the military research establishments *before* 1938 at the latest, gaining considerable experience in their fields of research throughout the wars in China after July 1937 and later in the Pacific after December 1941. These senior researchers had led group projects before the end of the war. Highly educated and experienced, these former military engineers launched a series of opposition movements,

TABLE 5-3
Wartime Background of RTRI Laboratory Leaders as of 1950

Wartime Background	No. of Lab Leaders
Japanese National Railways	15
Navy	10
Army	2
Central Aeronautical Research Institute	2
South Manchurian Railway	1

some quiet ones, against tradition-bound and seniority-based cohorts of railroad engineers at the RTRI.

Former Military Engineers in the Japanese National Railways

While the increasing prominence of the former military engineers in the RTRI was often contentious, by 1952 they proved indispensable in the JNR. Train accidents created a niche in the JNR for these engineers. By 1955, these "newcomers" to the preexisting community of railway engineers had solved many problems, gained professional recognition and established their autonomy in the JNR; and their strength was the engineering knowledge they had accumulated during the war. One major technical problem in the JNR involved vibrations—a phenomenon that endangered the lives of everyone who rode trains. The Rolling Stock Vibrations Laboratory, in particular, could illustrate how the former military engineers challenged the seniority-based engineering community for railway service.

At the center stage was former navy engineer named Matsudaira Tadashi, who eventually developed the air suspension system of the running gear on Shinkansen railcars in the early 1960s. Visibly, but quietly, he challenged the status hierarchy in the RTRI from the time he joined it in the fall of 1945. Born in 1910, he was a 35-year-old engineer with a baccalaureate degree in marine engineering, a field one would expect to be of use in postwar society. He was multilingual as a result of his years at Tokyo University, able to draw information from marine engineering, aeronautics, and railway engineering that had been published in Japanese, English, German, or French. At the RTRI, he actively recruited wartime colleagues from the INA. Matsudaira's pre-1951 research group included at least six former navy engineers and two army engineers. The growing influx, however, divided the Rolling Stock Vibration Laboratory into two camps, the former wartime military engineers and that of the wartime railway engineers.[43]

Matsudaira's cohort differed from the other half of the laboratory workforce, that is, the group led by Musashi Kuraji. Joining the JNR in 1923, Musashi was 11 years older than Matsudaira.[44] Musashi was an expert in the field of vibrations of rolling stock and had earned a doctoral degree in 1940 from Tōhoku University. His expertise helped the NTRI during the war. At the RTRI, he had directed the work of at least six railway engineers from the wartime era; collectively, they dominated earlier research and development in the field.[45] His 22-year career as of 1945 marked a sharp contrast to Matsudaira's complete lack of a proven record of activity in the field.

Seniority was the prevailing norm at the RTRI. When Matsudaira convened a round of first self-introductions in the laboratory, Musashi condescendingly said that the newcomer read his publications on the vibration phenomenon. This self-confidence was unfounded in Matsudaira's view. By his own account, he found "surprising" what he saw as a dire lack of useful studies about the kinetics of rolling stock when moving at a high speed.[46] Known for his quiet and gentle manner, Matsudaira never opposed Musashi in the open, at least not initially. He quietly rejected the senior researcher and pursued his own research. As early as April 1946, Matsudaira's research group developed a mathematical model for investigating the vibrations of bogie cars.[47] Meanwhile, Musashi's group continued to conduct similar research projects in response to official requests from the JNR headquarters. The two groups shared the common task of solving vibrations of rolling stock, a chief source of discomfort and safety in long-distance travel. But they exchanged little, if any, technical information.[48]

Matsudaira's initiative, separate from that of Musashi's group, was possible because the JNR leadership supported the former navy engineers' effort to solve the vibration issue in rail service. From December 1946 to April 1949, for instance, many engineers from across the railroad industry discussed the vibrations of running gear at high speed.[49] The first of the six study-meetings, held in a relaxed atmosphere at an *onsen* hot-spring inn, included 26 engineers; among them, *all* of the 11 speakers, including Matsudaira, had been wartime aeronautical engineers.[50] This initiative was strategic. Its organizer, Shima Hideo, specifically chose these former aeronautical engineers as the presenters at the first meeting. The subsequent meetings hosted more wartime railway engineers, including Musashi, to discuss the vibration problem.[51]

In the midst of rampant vibration-related train accidents, Matsudaira rose to prominence in the engineering community as a technical problem solver. A turning point was a devastating derailment that occurred in July 1947. Adhering to seniority, Musashi chiefly investigated the accident upon official request, while Matsudaira played a supporting role at the site.[52] Musashi's approach to the train vibration phenomenon remained largely empirical, focusing on the physical infrastructure of the transportation service. The vast majority of the investigators at the site concluded, wrongly in retrospect, that worn-out rail lines that had survived the war caused the train vibrations and accidents. In this prevailing view, the summer heat had expanded the steel rail, caused it to buckle, and resulted in the derailment under investigation.

Matsudaira disagreed with tradition-bound railway engineers, especially with investigators of the accident site dispatched from the JNR Bureau of Transport.[53] Relatively new to the field and thus untainted by preconceptions, he questioned the assumptions shared in the JNR community. The result was groundbreaking.

The former navy engineer ascribed the cause of the accident not to the railway infrastructure but to self-induced vibration in the railcars. As he concluded, the vibration manifested in aircraft and trains was essentially the same. At higher speeds, parts of the train vibrated more—as in aircraft flutter—and became less stable. The only difference was the source of energy: air causing the aircraft vibration, as opposed to direct contact with the rails inducing train car vibration. This thinking was the product of his experiences at the INA, where Matsudaira successfully solved the flutter problem that caused the Zero Fighter to disintegrate in the air in 1940 (Chapter 2).

Matsudaira's hunch proved correct. He embraced reproducibility as the core of his following research in a controlled laboratory environment. At the RTRI, he demonstrated the self-induced vibrations of railcars cheaply and repeatedly before the curious eyes of the former wartime railway engineers. To prove his theory about the cause of the 1947 derailment, Matsudaira developed experimental models to test the vibrations of rolling stocks. The train cars, built to one-tenth scale, demonstrated their fluttering at a speed of roughly 50 km/h.[54] This method had its genesis in his navy years, when he examined the physical movements of scale model aircraft in wind tunnels. His research data attested to the self-induced nature of a railway car's lateral vibrations and, thus, the danger inherent in this phenomenon. This research soon elevated him to the status of a leading authority in the field. When the RTRI hosted a two-day workshop in April 1948 that addressed 20 major technical issues in the rail service, it was Matsudaira— not Musashi—who addressed the topic of train car vibrations for 40 minutes.[55]

Matsudaira embarked on a mission to redefine the field of research more openly once Musashi retired in March 1949. With the dissolution of Musashi's leadership, the camp of wartime military engineers and that of wartime railway engineers merged into one under Matsudaira. He became more vocal soon after, openly pointing out inaccuracies in Musashi's formulas, clarifying basic issues about the vibrations of bogie cars.[56] During his career, Musashi had developed a series of mathematical equations for understanding the mechanism of a railcar's vibrations, but they were too complicated and, thus, practically useless. In 1952, leading the entire laboratory with 14 other researchers, Matsudaira revised and simplified Musashi's mathematical equations for practical, wider usage.[57] By 1953, through a series of tests, he diminished the lateral vibration of improved cars to half that of conventional cars at 65 km/h. The freight car that he had improved operated safely at a speed of 85 km/h.[58] By the end of the occupation era, former navy aeronautical engineer Matsudaira had reshaped the field of vibration studies in the rail engineering community.

Vibrations of artifacts, mobile or otherwise, were a pervasive conundrum that needed the expertise of additional former military engineers in the postwar JNR. Shinohara Jinkichi, a wartime expert in shipbuilding at the NTRI, examined the characteristics of vibrations in piers. His theoretical studies of vibrations in concrete bridge bearing were useful within the JNR. His wartime colleague, Hashimoto Kōichi, physically examined the structures and vibrations of bridges across the country.[59] A group of former military engineers, including one wartime expert in the topic of vibrations and the metal fatigue of aircraft parts at the CARI, examined the relationship between vibrations and fatigue of steel rails at the RTRI.[60] Developing the running gear and infrastructure for high-speed rail service was among the earliest research fields in which the civilian application of military technology was partially complete.

Former Military Engineers in Materials Engineering, Acoustics, and Wireless Communication, 1945–1955

Within the JNR, the civilian applications of wartime technologies were successful in several other fields of study. An illustrative field of research was materials engineering—a field that formed an important part of the successful Shinkansen project in the 1950s and 1960s. Before 1945, research on steel for the rail service stagnated because the JNR lacked both material and human resources; meanwhile, the military was excelling in this field.[61] What changed this landscape was the postwar infusion of former aeronautical engineers into the JNR community like Satō Tadao, leader of the Steel Casting Laboratory at the RTRI. Born in 1908, this graduate from the Department of Metallurgy at Tokyo University had built his wartime career at the INA.[62] A senior researcher in the navy, he specialized in metal fatigue, and his research projects involved highly advanced experimental aircraft. One such project aimed to explain why a few of the earliest Zero Fighters disintegrated suddenly in midair.[63] To check metal fatigue, he also conducted load tests on the ailerons of the experimental bomber, which became operational and known as P1F Frances.[64] His wartime research helped to create heat-resistant steel for Japan's first gas-turbine engine, the Ne-20.[65] Showing the devastating effects of repeated loads on metals, his projects formed an essential ingredient in designing military aircraft.[66] He joined the RTRI shortly after the war and helped to solve many accident-related issues in the JNR. For instance, he examined the metallographs of the damaged bearings, and in 1950, he observed many cases of unusually overheated metals under high pressure. His subsequent research on metal alloys was vital in the development of high-speed rail service in the years that followed.[67]

Wartime research for developing decay-resistant wood bore fruit immediately in the postwar period. After the summer of 1945, as the JNR needed more durable wood for crossties as part of its repair service, former navy engineer Yamana Naruo's wartime research offered one solution. A graduate from the Department of Forestry at Tokyo University, Yamana had built his career by developing wooden military aircraft at the INA. His effort in this area was substantial. A diminished importation of metal resources—especially of light but strong metals, such as aluminum—encouraged the Japanese navy to investigate the potential of wooden flying machines. In this situation, Yamana developed durable laminated wood.[68] His wartime creations included wooden propellers fabricated by mechanically pressing layers of wood into one,[69] as well as the bomber and trainer D3Y1—a wooden version of the navy Type 99 bomber.[70] He also established standards for controlling the quality of manufactured wooden propellers.[71] His wartime efforts ended in both success and failure—all crucial for developing wooden crossties in the postwar JNR, where the damaged crossties that had survived the war caused derailment. In 1946, Yamana organized research projects to improve the durability of such wooden ties.[72] Subsequently, he tested various types of wood, warning that Kyūshū pine trees, used widely theretofore, proved less durable than previously recognized.[73] The following year, he developed a crosstie with pieces of compound wood that were glued together.[74] In 1953, based on his wartime experience, he used high- and low-frequency electric waves on crossties for preservation. The gluing process improved after the crossties were dried with electrical heat, and then soaked in cold creosote oil.[75]

The field of acoustics at the RTRI also benefited from the influx of former military engineers from the navy. In the Measurement Laboratory, the goodwill efforts of its leader, Hirokawa Genji, proved vital. He joined the RTRI in 1930 with a degree in physics from Tokyo University. His tasks included measuring the noise levels of running trains in urban areas, a source of social discomfort in the early 1930s. Amid the ensuing war, this expert in acoustics worked not only for the RTRI but also for the NTRI, bridging two institutions. His expertise showed great promise in the Acoustics Department of the navy establishment, which devoted part of its research to minimizing the noise emitted by Japanese warships to prevent the Allies from detecting them. Hirokawa, at one point, measured the level of noise generated from the engine room of a warship. In the immediate aftermath of the war, his laboratory at the RTRI absorbed engineers from the dissolved NTRI. Various noise meters, previously used in the navy during the war, also found use in his laboratory. Under Hirokawa's leadership, rail engineers and former navy engineers measured the level of noise in buses, electric trains, coaches, dining cars, sleeping cars, and locomotives.[76]

Similar to the case of acoustics, wireless communication advanced little if at all in the JNR before 1945. Railway engineers explored the field, but their works remained essentially experimental. During the 1930s, railway engineers tested wireless telephony for communication between running trains, albeit with little success, as it functioned only within a distance of 1 kilometer. In 1932, the JNR developed a wireless telephone with ultra-short 30 mega-cycle bandwidth, but the practical use of the device was severely limited.[77] Observing the utility of the idea, the military mounted various types of wireless telephones in army tanks and airplanes for communication. The navy was ahead of the army overall in developing radio communication for aerial operations over vast territories. While the military excelled in wireless communication, its use in the civilian sector remained marginal at best, in part, because of wartime fears that potential foes could listen in on the uncoded communication system about matters of national security, such as military logistics within the country.

The end of the war released former navy engineers, unleashing changes in the field. Foremost was the Wireless Communication Laboratory at the RTRI led by electrical engineer Shinohara Yasushi. Born in 1916 and a graduate from the Department of Physics at Kyoto University, Shinohara had accumulated his experiences at the NTRI, where radar systems were developed.[78] The end of the war was fortunate to him and the field because it ended many regulations that had governed wavelengths for wireless communication. At the RTRI, Shinohara devoted part of his research to medium-frequency wavelengths for better communication between running trains.[79] During 1946–47, there was a plan to develop medium-frequency wireless communication on the Tōkaidō line, but as it turned out, extraneous noise was too loud and obnoxious.[80] Undaunted, Shinohara led a laboratory team that included, among others, the next leader, Maruhama Tetsurō, who had migrated from the dissolved INA.[81] Inheriting the research and laboratory leadership, Maruhama later focused his work on microwave devices for wireless communication in the JNR. In 1950, he developed a wireless communication system that used 4,000 mega-cycle electric waves over the 80 kilometer distance of the Tsugaru Strait between Honshu and Hokkaidō.[82] In the project, the magnetron tube—an electric vacuum tube previously developed at the NTRI for generating high-voltage microwave energy and wartime radar—proved useful. With it, the field of wireless communication advanced for the first form of communication in Japan using a super-high-frequency circuit.[83]

Wartime research was by no means a panacea for all technical problems in the JNR after 1945, and in some cases, it did not pay commercial dividends. The research and development of the gas-turbine engine at the RTRI could illustrate this point. This method of aircraft propulsion, however promising during the war, lost its niche under the official ban on the aircraft industry by the Allied occupation. Some wartime engineers who had developed the Ne-20 axial-flow turbojet for the Nakajima Kikka—the Japanese version of the German Messerschmitt 262—resettled at the RTRI. The research institute also inherited research facilities, such as the air compressor and combustion tester, for developing the aircraft engine from the CARI.[84] The jet engine for locomotives was a tantalizing idea in the eyes of the wartime experts from the dissolved INA. It could operate on low-quality fuels and without water. And the structure of the engine was barely an obstacle. Nonetheless, the subsequent research and development at the RTRI came to a complete halt in April 1950, when the research facility was physically transported to the newly established Transportation Technology Research Center.[85] Research in the field of jet engines had little practical value for the JNR as it electrified its train operations.

Individual cases of partially successful technology transfer were abundant in postwar Japan. A representative case involved the field of materials engineering, in which developing more sophisticated and durable metals remained a crucial research topic. During the war, Kokubu Kinji had conducted chemical experiments, such as cementing silicon onto light alloys for aircraft engines.[86] Joining the RTRI in 1945, he embarked on a series of research projects to create a better contact strip, the part of the pantograph that touches trolley wires to collect electricity. This was an urgent task in the JNR. Carbon metal plating in the contact strip often broke apart and cut the overhead wires, causing a serious safety problem. For the contact strip, Kokubu created bronze metal alloys that were more stable but, as it turned out, less durable than his ideal.[87]

A closer investigation of this modest success could highlight why some former military engineers, such as Kokubu, failed to bring their years of research to fruition by 1955. Born in 1918, this graduate from Kyoto University joined the CARI in 1940, which was still a fledgling research institute that had become operational only a year earlier. Kokubu was like hundreds of other research associates during the war. Despite publications in technical journals, he remained a junior researcher who lacked substantive experience in the army or navy during the most fruitful research years, which were *prior* to the Pearl Harbor attack in December 1941. The CARI became less promising in the course of the war. Especially from late 1944, many researchers at the establishment lacked material resources sufficient to conduct physical experiments. As a junior engineer from the dissolved CARI, Kokubu was paid the least of more than 200 former military engineers at the RTRI.[88] What mattered was when—before or after 1937—and which research establishments they joined before the end of the war.

Former military engineers cultivated a new niche in peacetime. Once remobilized by the JNR, they played a vital role in the reconstruction of the war-torn rail infrastructure. Central to this development was the RTRI. The engineers successfully solved technical problems, stepping into the void created by train accidents that threatened the lives of millions of ordinary citizens. The technical expertise of the former military engineers was a tool to cultivate what were, for them, new fields after World War II, such as the vibrations of mobile and immobile artifacts, materials engineering, acoustics, and wireless communication. During the Allied occupation, the engineers began to reshape their fields of study at the RTRI and the JNR.

In the defeated nation, the basic sociotechnical structure for the civilian application of wartime technology was nearly complete within 10 years of the war's end. Those conditions were ideal in retrospect. Neither ideological nor religious differences split the RTRI irreparably in the Allied occupation years; for instance, union activism was remarkably absent at the establishment. Matsudaira's laboratory revealed the malleability of the postwar engineering community. While his laboratory was initially divided into the camps of wartime military engineers and wartime railway engineers, the schism had narrowed by 1952. The tension, however reduced, did not disappear in the RTRI. From the mid-1950s, research and development for high-speed rail service advanced almost single-handedly at the RTRI, owing much to former aeronautical engineers who radically altered the organizational culture of the JNR through a series of quiet opposition movements.

Opposition Movements of Former Military Engineers in the Postwar Railway Industry, 1945–1957

From September 21 to 28, 1957, the Japanese rail industry carried out a series of high-speed test runs between the cities of Hiratsuka and Fujisawa, both located about 50 to 60 kilometers away from Tokyo. It was a joint project between the Japanese National Railways (JNR) and the private railway company Odakyū. The experiment involved state-of-the art rail technology as embodied in the so-called Romance Car SE 3000, which was essentially a wingless airplane in design and construction. The success of its test runs dazzled more than 100 observers. On September 25, the train ran smoothly at a top speed of 130 km/h. According to a national daily newspaper, this first test run showed "a very good chance of success." The following day, the train successfully established the world's high-speed record on the narrow gauge: 143 km/h. As *Asahi Shinbun* wrote, "our dream of a super-express train came to a reality" with this "light of dawn." On the last day of the experiment, the "new and powerful" high-speed train renewed its high-speed world record, reaching a top speed of 145 km/h.[1]

Behind this technological success were power relationships in the national, regional, local, and laboratory settings that changed drastically after Japan's unconditional surrender to the Allied forces. During and after the occupation, many local citizens in the city of Yokosuka and the resort town of Hakone welcomed the social conversion from war to peace. Some even initiated the movement at the grassroots level. In a roundabout way, this reorientation of sociocultural values in the Kantō region gave birth to innovative, nonweapons technology as embodied in the high-speed rail project of the Odakyū Romance Car. Furthermore, defeat radically reconfigured engineering communities in the rail industry. The resulting cultural tensions between preexisting rail engineers and former aircraft engineers ran deep, as typified at the Railway Technical Research Institute (RTRI). As each group pursued its ambition, two high-speed rail projects proceeded in parallel and in competition, following no preordained vision of postwar Japan.

A full view of the Romance Car SE3000. Engineers carefully inspected the vehicle after it successfully established a world record of 145 km/h for its high-speed test run in September 1957. Photo courtesy of Odakyū Electric Railway Co., Tokyo, Japan.

Demilitarizing the City of Yokosuka, August 1945 to April 1952

During the Allied occupation, social expectation for peace was encapsulated in the city of Yokosuka that had wartime navy facilities and engineers. It was where probably the most dramatic demilitarization occurred along with three other cities: Kure in Hiroshima prefecture, Sasebo in Nagasaki prefecture, and Maizuru in Kyoto prefecture. Before the summer of 1945, these four cities had been heavily militarized, as the hosts of massive tangible assets for military use. For instance, land for military purposes covered 25,160 acres, or 3.1 percent of the country's military-owned land (804,954 acres). The defeat, however, drastically altered the landscape of the four cities. By 1976, for postwar industrial growth, factories and warehouses were occupying the land once owned by the military; schools and public parks occupied the land to accommodate a postwar population increase.[2] Among the four cities, Yokosuka experienced the most radical transformation of its landscape. From 1945 to 1976, military land was converted for peaceful use in 199 projects, which covered 1,533 acres; this was more than 1,129 acres in the city of Kure (170 projects), 780 acres in Sasebo (144 projects), and 950 acres in Maizuru (109 cases).[3]

Yokosuka was disarmed immediately after the arrival of the US occupation forces. The first wave of American troops came to Atsugi air base on August 28, followed by a flotilla of destroyers off the coast of Yokosuka. On August 30, 13,000 marines came ashore at the naval port, and a flying corps of 4,000 arrived in Oppama and Taura districts—strongholds of Japan's naval air power.[4] The sense of occupation was pervasive in the city. There was an eerie silence, and the advance of the occupation forces into the urban center was peaceful because the day before, on August 29, the city government had prohibited local residents from attacking the Americans on the march and ordered them to stay indoors.[5] The occupation authority quickly closed and seized military facilities, checked property lists as prepared by the Japanese military, and disarmed the city. Machinery and equipment were removed from vacant buildings for requisition. Guns, cannons, and bullets were reportedly abandoned in the ocean. Torpedoes and depth charges were detonated in open areas. Aircraft in the Oppama airfield were disarmed, stripped of their propellers, and destroyed by fire. The battleship Nagato, still fully operational, was demolished as the target of a nuclear detonation test in Bikini Atoll in July 1946.[6]

It was remarkable that this most heavily militarized city in Japan adapted to the subsequent peacetime era quickly and almost voluntarily after the end of the war. Local citizens developed a very adaptive strategy that benefited both themselves and their occupier. Their audacious plan—to lure private enterprises into the land previously occupied by the military—made sense to both parties. It was politically logical to the occupation authority that was determined to demilitarize and democratize Japan. The plan made economic sense to the Yokosuka residents who wanted to use the former military property for the postwar revival of the city.

Conversion advocates launched a bottom-up initiative only one month after the end of the war. On September 15, the Yokosuka City Recovery Committee—with 30 members, including the mayor—was newly formed to discuss various ways to revitalize the city. Three months later, a series of meetings created the Yokosuka City Recovery Plan, which signified the city's struggle to translate the military facilities into a source of hard currency. Five of the seven sections in the plan called for the conversion of former military property to civilian use. Outlining the city's revival, the proposal emphasized the need to develop peacetime industry, commercial business, harbors, tourist facilities, educational institutions, residential areas, and transportation facilities. The Yokosuka Navy Arsenal, for instance, was to be used for commercial shipbuilding as well as for food processing and canned food manufacturing. The INA was to be used to make building materials for the construction industry. Institutions once used for military education, such as the Army Artillery School and Navy Mine Research Center, were to be converted to colleges and other institutions for education and research. Yokosuka Navy Hospital was to become a hotel for international tourism. Former military land, such as the Army and Navy Drill Grounds, was to be transformed into parks and fields for recreational sports such as baseball, tennis, and golf.[7]

This conversion plan was audacious because it indirectly asked for the transfer of all the former military facilities to the city of Yokosuka. This transfer process first required the release of requisitions from the occupation authority and, then, approval from the central government of Japan.[8] Within a few months of the military defeat, the city of Yokosuka announced its commitment to peace. This amounted to the city's quiet, passive resistance to the occupation authority as well as an effective, tacit request for the early withdrawal of the Allies from Japan. The benefit of Yokosuka's challenge seemed to outweigh the risk.

The recovery plan also represented an urgent fix for the city's social problem. On October 25, 1945, the vice minister of public welfare in Japan sent a circular notice to Yokosuka that stressed the importance of converting air-raid shelters to accommodate war victims and repatriates. The central government temporarily allowed the free use of facilities and land for local public organizations, housing administrations, rental housing associations, and war damage support groups.[9] This rapid conversion for the public good set an important precedent for the years to come.

Apart from the social need, economic necessity weighed perhaps more than any elusive enthusiasm for democracy and peace. Local residents aimed to preserve such assets as they had for recovering the city's revenue, that is, land, property, plants, and equipment, which had survived the war relatively unscathed. Yokosuka had come under direct attack only twice during the entire war. Compared with the cities of Kure (1,252 acres affected) and Sasebo (438 acres affected), Yokosuka remained relatively free from major air assaults.[10] America's first air attack on the city was brief and light; led by Lieutenant Colonel James Doolittle on April 18, 1942, the air raid damaged only the power generator room of a submarine tender in repair at the dock. After three years of tranquility came America's next assault on July 18, 1945, which heavily damaged only the battleship *Nagato* resting in the navy dock, facilities in the navy arsenal, and some urban areas.[11] Many local residents remained puzzled as to why Yokosuka, the most heavily militarized city, had not come under heavy air attack. During the war and even to date, there have been rumors that the US military had spared Yokosuka on purpose, so as to use the navy facilities to maintain America's military presence in postwar Asia.[12]

A substantial number of tangible assets of the former Japanese military survived the war with little or no damage. As of August 15, the property of the defunct navy consisted of the Naval Station Headquarters, the INA, nine educational institutions, two hospitals, one prison, and the Yokosuka Navy Arsenal,
including five research labs.[13] The army and navy together occupied about 4,636
acres (5,700,000 tsubo)—all immediately taken over by the occupation authority.[14] Compared with Tokyo, Hiroshima, and Nagasaki, much of which were leveled to the ground by nuclear or incendiary bombs, Yokosuka endured far less
physical damage and civilian casualties—which appears to have encouraged the
local residents to accept the occupying military forces.

Yokosuka's conversion from war to peace unfolded before the eyes of the
thousands of former military engineers who remained in the city. Those who
feared the occupation forces had simply moved out to live with their relatives
in distant prefectures.[15] The dissolution of the Japanese military had driven the
working-age population out of Japan's most heavily militarized city and its vicinity. The diaspora included female volunteer soldiers and factory workers originally from remote areas.[16] On August 26, all but 150 employees at the Yokosuka
Navy Arsenal were dismissed.[17] In the city under siege, convenient handymen,
rather than former military engineers, were in high demand to build accom-
modations.[18] Among the unneeded talents were aeronautical engineers such as
Yamana Masao living in the nearby city of Kamakura. During the war, he had
shared an academic appointment and lectured on the aircraft fuselage structure
in the Tokyo University Aeronautics Department. He had concurrently designed
the dive-bomber D4Y Judy; he was also in charge of developing the twin-engine
dive-bomber P1Y Frances as well as the rocket-propelled kamikaze glider MXY7
Ōka at the INA. Soon after the end of the war, he left for his hometown of Atsugi
to take up farming to feed his family and to alleviate the acute food shortage of
the time.[19] Yokosuka had attracted a substantial working-age population of people
like Yamana before the war, but its population declined sharply from 298,132
people in February 1944 to 202,038 people in November 1945, or by 32 percent
in 18 months.[20] The city had lost the population, military, and industry, which had
been viable sources of revenue.[21]

In the economically distressed city, many remaining military engineers lived
in fear. Miki Tadanao experienced this firsthand. With Yamana, he formed the
core of the navy's effort in research and development. This graduate from the Tokyo University Department of Naval Engineering had built his wartime career as
a senior aeronautical engineer at the INA from 1932 to 1945. His group had been
devoted to designing highly advanced, experimental aircraft on the basis of both

theoretical and empirically proven knowledge in high-speed aerodynamics. His wartime accomplishments included upgrading an earlier model of the legendary A6M Zero Fighter, assisting in the design project of the D4Y Judy dive-bomber, and designing the twin-engine dive-bomber P1Y Frances and the rocket-propelled kamikaze glider MXY7—all of which were deployed in one-way suicide missions as part of homeland defense during 1944–45. News of Japan's defeat on August 15 meant the end of his navy career—and it marked the beginning of his fear of the occupation forces.

The defeat drastically altered Miki's daily life. No longer busy with the urgent tasks of designing navy aircraft for war, he unfortunately gained the luxury of time. He would spend his days taking his daughter to a nearby river; while at home, he was visibly apprehensive at the sight of an unknown visitor, thinking that he might be captured by the occupation forces. His fear was not groundless. Within three months after the war's end, the Allies' Air Technical Intelligence Group had thoroughly interviewed his wartime colleagues, including Tani Ichirō and Matsudaira Tadashi, about their research projects.[22] The Allied occupation authority was keenly interested in Miki. The inspection team wrote, "Japanese design and development was good but not outstanding . . . [and] nothing of a revolutionary or startling nature was in the progress with the possible exception of OKA 43 [MXY7 model 43]," the latest version of the MXY 7 kamikaze glider. "Much can be learned," the report concluded, "from Japanese practices and experiences particularly if time and patience are given to interrogate responsible officers and technicians."[23] More overwhelming than Miki's fear of the occupation force was his acute sense of wartime guilt, remorse over the fact that he had been chiefly responsible for the creation of the manned bomb. Dejected, he followed the social custom of shaving his head to shed the past, but this action gave him no relief. Two months after the war ended, he expressed the burden of this guilt in the following personal memo:

> As I am writing [this entry], I cannot help thinking about the feelings of many young, beautiful patriots who perished [in the kamikaze missions] like the petals of cherry blossoms. Their surviving families must be cherishing the memory of the deceased. The war was no good for us. The tasks left for us as the fortunate or unfortunate survivors of the fighting are, to respond to the noble sacrifice [of the kamikaze airmen], to walk the road beset with many hardships, and to put all of our energies into the reconstruction of our homeland. Our [ultimate] job then is to actualize a beautiful Yamato Japan, a symbol of harmonious Eastern culture in the world by removing all the weapons forever.[24]

His dispiritedness was lifted only after he followed the advice of his Christian wife. Subsequently, he was baptized and officially joined the RTRI, both events coincidentally occurring on his thirty-sixth birthday on December 15, 1945. That day marked his spiritual conversion and, to a degree, epitomized the country's technological conversion from wartime to peacetime.

His city of Yokosuka mirrored this conversion. The local residents began to lure private enterprises such as private railcar manufacturers. Although citizens' lives were disrupted temporarily in the summer of 1945, they resumed as the city hosted more jobs. In February 1946, the Committee for Converting Yokosuka Naval Port was formed to investigate the scope and prospect of converting military facilities to peaceful use. Responding to this local initiative, in July that year, the Ministry of Finance decided on the Conversion Plan for the Main Military Facilities in Yokosuka, which assigned specific tasks to districts in accordance with their strengths. For instance, the Yokosuka navy-port district was to provide commercial harbors, industry, and educational institutions. Kurihama district, with a fishing port, was to accommodate facilities for fisheries and education. Ōgusu and Takeyama districts were to host educational facilities, including housing, for agricultural development. The city successfully attracted a total of 144 commercial enterprises.[25]

By 1947, the dynamics of Yokosuka's conversion were set fully in motion. The occupation authority released from requisition some areas used by the INA during the war. By September 1951, this wartime establishment had been replaced by 22 commercial enterprises that used 169 buildings and employed 1,799 workers on property covering roughly 72 acres (88,020 tsubo).[26] Similarly, Fuji Automobile Company moved into the area previously occupied by Yokosuka Navy Flying Corps during the war. From August 1948, Fuji assembled parts and repaired trucks, jeeps, and trailers for the US military; moreover, during the Korean War (1950–1953), the company employed up to 7,000 workers to meet its high demands.[27]

A series of initiatives in Yokosuka received the attention of the newly formed diet. In the first election for the Upper House on April 17, 92 of 250 seats were dominated by Ryokufūkai affiliates. This political faction strongly supported the transfer of former military property from the central government to local residents and commercial enterprises. In the first Upper House diet session on December 7, 1947, Itō Yasuhira, together the unanimous approval of the Upper House Finance Committee, backed a local plea to lower the sale price of the transfer. The Ministry of Health and Welfare, according to the petition, aimed to "increase national revenue" by selling the previous military assets, but the price tags were "way too much, beyond the bounds of possibility." Unless reduced, he claimed, the high price would leave "a profound[ly] negative influence on support, education, and work for rehabilitation" in the local communities of Yokosuka.[28]

In 1950, the diet gently pushed the cabinet into considering a bill for further demilitarization. Heated discussions focused on the Former Navy Cities Conversion Law. Its "purpose," as stated in Article 1, was to "convert the [once heavily militarized] cities of Yokosuka, Kure, Sasebo, and Maizuru into port cities for peacetime industries, and to make a contribution to the ideal of peace-loving Japan."[29] According to one advocate in the Upper House, it was "extremely important for the state government to give as much support as possible" to the citizens of the four cities "in accordance with the new [pacifist] Constitution of Japan that renounced war forever." He added that the local residents had "strongly desired to make a solemn peace declaration, [were] determined to eradicate the military past from the [new] cities and [sought to] restart as port cities with peacetime industries."[30] Soon, the Lower House of the diet took up debate over the bill. One supporter asserted that the local residents of the four cities "are enduring miserable unemployment." He went on to say that they were "against war more intensely" than the rest of the population; their "thinking to maintain peace no matter what" lay at the heart of the citizens' resolutions and the genesis of the bill.[31] This voice among others gathered support from the Liberal, Socialist, and Democratic Parties, and despite the Communist Party's opposition, cleared the way for legislation at the national level.[32]

Subsequently, Yokosuka, Sasebo, Maizuru, and Kure were legally transformed into peace-oriented cities. On June 4, 1950, with a very high turnout of 69 percent, the law was approved in Yokosuka by an overwhelming majority, 91 percent of voters. The other three cities shared similar results.[33] By the resulting legislation, city governance took responsibility for converting former military property to peaceful use. This new law, enacted on June 28, aided the transformation of Yokosuka and the three other cities. There was a national sense that the country was boldly marching forward after the occupation era.

Engineering Workforce in the Rolling Stock Structure Laboratory at the Railway Technical Research Institute

The experience of the almost bottom-up conversion of Yokosuka was typified by former navy engineer, Miki Tadanao. By 1957, he played a key role in altering the preexisting culture for research and development in the railway industry. In December 1945, he joined the RTRI—the JNR's only research and development establishment—and led the Rolling Stock Structure Laboratory. Within the RTRI, as of February 1956, this was the largest laboratory with 23 research associates.[34] From around 1950, the RTRI had contributed to the design and construction of streamlined, lightweight rail vehicles of Odakyū Romance Car and the Shinkansen, both of which successfully established high-speed world records in 1957and 1963, respectively. Its creations, however, caused tension in the relations between the RTRI and the JNR headquarters, which soon spilled over into the railway industry. Miki's laboratory could offer a useful case study about how two engineering communities collided quietly, albeit notably, in the development of high-speed rail service.

Underneath the simmering tensions was an unusually high concentration of former aeronautical engineers in the Rolling Stock Structure Laboratory. Their leader, Miki, was an aeronautical design engineer by training who voluntarily and independently designed railcars on the basis of his accumulated knowledge and experience in the navy. He was among many ex-aircraft engineers who started new careers in this laboratory. In 1950, Miki had 14 research associates, and at least 7 of them—or 47 percent of the entire laboratory workforce—had been introduced to the field of aeronautics by the end of the war. Most of them had actually designed army, navy, or civilian aircraft at one point during the war. Two exceptions notwithstanding, many did not know each other before joining the RTRI, and apparently, none encouraged each other to come to Miki's laboratory. In the next three years, almost coincidentally, more former aircraft engineers joined his task force. By 1950, an overwhelming majority of 73 percent shared a background in aeronautics.[35] Their presence dominated Miki's laboratory of 15 researchers, as shown in Table 6-1.

Although a newcomer to the field, Miki was able to assert his authority in the laboratory. The major sources of his leadership included his educational background, age, and above all, former military rank. Unlike many others, he was a graduate of Tokyo University; he was 41 years old as of 1950—the oldest in the laboratory; and he was a former technical lieutenant-commander—the most senior rank among all the former military engineers at the RTRI. Miki's wartime work as a navy aircraft engineer, such as P1Y Frances, was a source of awe among the ex-aircraft engineers, who formed the majority in the laboratory. His leadership was also sustained by the military culture to which at least nine former military researchers had been exposed. This was a culture that cherished strict, austere, and hierarchical relations. The rest of the laboratory workforce lacked a military background, but fit in well in the laboratory culture. As the table shows, at least four of them had just graduated from their respective educational institu-

TABLE 6-1
Rolling Stock Structure Laboratory at the RTRI in 1950

Last Name	Educational Institution (Field)	Year Hired at RTRI	Wartime Affiliation	Wartime Job Description
Miki	Tokyo University (navy eng.)	1945	Institute for Navy Aeronautics	Aircraft design
Akatsuka	Industrial Academy (mechanical eng.)	1946	Institute for Navy Aeronautics	Aircraft design
Itonaga		1946	Central Aeronautical Research Institute	Aircraft design
Takabayashi	Industrial Academy (aeronautics)	1946	Central Aeronautical Research Institute	Aircraft design
Ono	Kyoto University (physics)	1945	Institute for Army Aeronautics	Aircraft design
Hasegawa	Tōhōku University (aeronautics)	1946	Institute for Army Aeronautics	Aircraft design
Nishimura			Army Technical Research Institute	
Matsumoto			Army Technical Research Institute	
Nakamura (k)	Tokyo University (aeronautics)	1946	Nakajima Aircraft Company	Aircraft design
Nakamura (h)	Tokyo University (mechanical eng.)	1945	Student at Osaka Army Arsenal	Metal fatigue analysis
Mizushima				
Tanaka		1947	Student	—
Nakae	Industrial Academy (mechanical eng.)	1949	Student	—
Hayashi	Tokyo University (mechanical eng.)	1945	Student	—
Yoshimine	Tokyo University (mechanical eng.)	1940	RTRI	Structural strength test

tions; they were the most teachable and malleable, especially given Miki's strong personality. Neither his stature nor frail physique was imposing, but he was irascible, uncompromising, and domineering. He tended to yell at the slightest frustration, which required his associates to adjust to his disposition.

The predominance of the former aeronautical engineers created a distinct minority in the laboratory. Yoshimine Kanae, a graduate from the Tokyo University Department of Mechanical Engineering in 1940, was the only researcher who joined the RTRI *before* the end of the war. He was among the best educated, but the age-based seniority system prevented him from pursuing the laboratory leader position. He was nine years younger than Miki. Yoshimine's expertise lay in conducting physical tests and assessing the structural strength of railcars.[36] In sharp contrast to all the former aeronautical engineers in the laboratory, he completely lacked knowledge or experience in the field of high-speed aerodynamics.

Quietly and resolutely, Yoshimine and Miki pursued their own ways. Yoshimine minded his own business, conducting a series of tests on railcars; Miki, an expert in aerodynamics, maintained a noninterference attitude toward him. This uneasiness remained until Miki retired from the RTRI in 1963.

Similar tensions simmered in Miki's laboratory on the periphery and at the JNR headquarters. At the JNR, the Mechanical Engineering Department had traditionally taken charge of designing, building, and repairing railcars and locomotives before the war years. The department was subdivided into sections, each taking full responsibility for clearly defined tasks. After a series of organizational restructurings, the department consisted of four sections in August 1952: (1) the Factory Section, (2) Locomotive Section, (3) Repair Section, and (4) Rolling Stock Section.[37] The engineers in the Rolling Stock Section engaged in the privileged task of designing railcars and/or approving the final product after inspection at the peripheral RTRI.[38] This hierarchical relationship remained in place partly because the RTRI's financial health depended on the JNR headquarters. In 1956, for instance, the RTRI processed 1,625 projects, 77 percent of which were commissioned by the JNR headquarters.[39]

The engineering workforce at the Rolling Stock Section qualitatively differed from Miki's group of former aircraft engineers in at least two ways. First, as of 1950, the engineers at the headquarters were overall less educated than Miki's laboratory associates. As the employee directories of the time show, only 1 of the 12 senior engineers in the design unit had a university education. The craftsman's mentality was the prevailing norm. Second, the Rolling Stock Section completely lacked expertise in aerodynamics and was, thus, very ill prepared for the coming age of high-speed rail service. A counterexample involves an engineer who graduated from Tokyo University's Aeronautics Department in 1946. Soon afterward, however, he was officially transferred to Miki's laboratory, which, for a time, meant that there were no university-educated railcar designers in this section at headquarters.[40]

Engineering Cultures in the JNR Headquarters

Although this talent description changed somewhat after 1950, it mirrored the engineering culture of a railway industry that promoted study abroad for improving railcar designs. The JNR and commercial manufacturers often sent their engineers to the United States and Western Europe for extended observation. This managed, costly technology transfer was useful for catching up with the West, but it proved less suitable as the technology gap narrowed. Historically, however, this technology transfer strategy had proved successful for developing lightweight, all-steel railcars.

Before the postwar technology transfer, lightweight, all-steel railcars were rare in Japan's public railways where wooden construction dominated.[41] One could suggest at least several technical, cultural, and social reasons for the persistence of the industry's use of wood. First, wood was reasonably durable and light. It was a suitable material for passenger cars because the wooden, lightweight construction imposed minimal loads on the railroad's physical infrastructure. Rails and bridges, for instance, could last longer with a fleet of light wooden cars. Second, the wooden construction was dominant in much of the railroad community because it was largely unchallenged over time. Rail service began shortly after the Meiji Restoration of 1868, but after that, there was little to no variation for decades. The first passenger car in the Japanese rail service was a part-wooden product imported from Britain. The length and weight of the passenger railcar had more than doubled by the turn of the century, but the car design remained unchanged for years.[42] Third, wood was readily available in Japan—a mountainous country, abundantly endowed with the natural resource. Thus, the wooden car was relatively cheap to make and easy to shape. Wood's malleability enabled craftsmen to create the ⌂-shaped roof for air intake and ventilation, a style that was gradually replaced by the arched roof during the 1930s.[43] Carpentry craftsmanship for building wooden houses underlay such a construction style. Moreover, many passengers favored wood rather than steel. Steel cars tended to absorb and emit more heat, thus not particularly suitable in the hot summer climate without air conditioning or in the cold winter climate, even with a stove inside.[44]

After the war, most wooden passenger cars were in poor physical shape and needed urgent repairs. The lack of structural integrity and strength of the coaches sounded an ominous knell to riders in many ways. In 1945, millions of passengers rode cars that had survived the war and commonly lacked glass windows, futon cushions on seats, and adequate lighting inside.[45] Riders heard the wooden floor and roof squeak because loosened nails could no longer hold all of the parts tightly together. Water often dripped from the roof when it rained.[46] Passengers openly grumbled about the uncomfortable ride, equating the coaches to pig farms due to their odor and lack of cleanliness.[47] Meanwhile, the perennial shortage of necessary manpower or money for maintenance remained unsolved. In fact, the majority of passenger cars (5,949 cars) in service during the immediate postwar years had been manufactured during the 1910s and 1920s.[48]

The war had disrupted the flow of engineering knowledge from abroad, and only after 1945 could Japan's railway industry freely capitalize on such information. The industry depended heavily on the peacetime international framework in which useful technical information could flow safely, especially from Western Europe. The Eurasian continent was fertile ground for the active, formal, and informal exchange of technology across national boundaries; it was also grounds for intercontinental commercial rail service. In 1954, for instance, several European countries—France, Belgium, the Netherlands, Luxembourg, West Germany, Switzerland, and Italy—cooperated in developing a new cross-border rail service known as the Trans Europ Express. The project aspired to offer a high-speed, luxurious ride across the continent without the need to change locomotives or otherwise stop at national borders.[49] This vast international project, and incidental diffusion of knowledge as a result, was not an option for the geographically isolated nation of Japan. Within this milieu, Japan's railway industry needed a strategy. The JNR and manufacturers, thus, dispatched personnel to gather technical information from abroad, developing technical tie-ups in the rail industry between Japan on the one hand and Western Europe and North America on the other.

But not all the players in the Japanese railway industry could afford the costly technology transfer. Only the JNR headquarters and a few profitable private companies could pay all the necessary costs. Lacking funds for technology transfer from abroad, the RTRI mobilized its engineers to gather any information possible about foreign rail systems. Their efforts rarely bore fruit during Japan's occupation because much of the information was scarce though nontechnical in nature. One engineer, fully determined to learn more about rail service in the world, described his effort in his diary in 1947:

[To obtain any useful information] I had to go to the [distant] Ginza district [in Tokyo] once or twice [a month] to know the first sale date [of the journal he needed]. The day before the sales, I prepared two lunchboxes [one for breakfast and one for lunch the next day]; and on the day of the sale, I left home at 4 o'clock in the morning to catch the first train bound for Ginza. My heart beat fast on the way. I dashed from Yūrakuchō station to Kyōbunkan [where the magazine was sold], and then relaxed at the end of the waiting line. Then I heard the clock chime at 7:30 am, 8:00 am [and every half hour] until 9:30 am. [Well, again] I got up at 3:30 am and waited 6 hours to get a copy of LIFE magazine.[50]

The JNR's effort to develop its first all-steel, lightweight coaches capitalized on an international trend in the field. Building lightweight railcars was among several important topics discussed among the participating countries, including Japan, at the International Railway Congress in 1947. The host nation, Switzerland, was a strong advocate for lightweight rolling stock because of its economic benefits. Understandably, some countries with steep slopes in mountainous areas valued lightweight cars because less structural weight would mean less energy and cost required for the service. The overall financial savings from placing lightweight electric rail in service caught the attention of the JNR, which was operating in a deficit at the time. Its engineers then actively observed the state of the advanced technology in the West.

One such individual was Hoshi Akira, an engineer at the Rolling Stock Division of the JNR's Department of Engineering. This graduate from the Tokyo University Department of Mechanical Engineering entered the JNR in 1942 and built his career as a railcar developer after a brief stint in the army during the war. Dispatched by the JNR, he stayed in Western Europe for a year to study the technology for constructing lightweight railcars. From Switzerland, he excitedly wrote in a postcard that after four months, he "was finally able to lay his hand on data and reports about load tests on steel rail cars." The structure of the Swiss railcars bore a "striking resemblance to that of aircraft [fuselage]." This finding encouraged Hoshi. It also evoked "respect" for the RTRI that conducted a series of physical load tests on railcars—tests that were "by no means inferior to how it was done" in Switzerland. After his return to Japan, Hoshi presented detailed information on the railway industries and their lightweight vehicles in Switzerland, Germany, France, Italy, and the United States.[51]

Active technology transfer from abroad like this burgeoned into two design characteristics of light, all steel railcars in Japan. The first was the wide application of the monocoque construction in Japanese train cars. The introduction of this design marked a significant departure from the traditional style for building housing that had heretofore prevailed in Japanese railcars. Until this new application was implemented, the roof had not been considered a vital part of the entire structural integrity in Japan; rather, it had been meant solely to protect passengers from sunshine or the elements. The introduction of the monocoque structure from abroad added another functional utility to the roof. As a result, the central struts inside railcars—for instance, the wartime car model 63—became unnecessary by 1955.

Another was a new strain gauge developed on the basis of technology from abroad that invigorated Japan's rail industry after the war. Used to evaluate the aircraft propellers or analyze the screws of a submarine during the war, a strain gauge found its niche in railroad engineering after 1945. An engineer at the RTRI, with a wartime background in army aircraft development, studied different types of strain gauges imported from the United States and developed a highly versatile strain gauge in 1949.[52] The postwar railroad community benefited immensely from this innovation. After its introduction, the device became a standard feature in conducting various load tests—bending, twisting, and compression—related to the development of passenger cars. With such analysis, computing structural strength was no longer such a formidable task.[53] The engineers, including Hoshi, were able to introduce the monocoque construction widely and successfully.

These technologies formed the basis for all the newly developed steel railcars after the immediate postwar years. The coach model naha-10, introduced in 1955, exemplified this success. Based on a light, steel railcar developed in Sweden, the product benefited from all of the technology transfers. The coach consisted of steel from floor to roof, but was 31 percent lighter than the comparable, then mass-produced model suha-43.[54] During the conversion of JNR railcars from wood to steel, the postwar technology transfers rendered wood obsolete. In 1955, Japan became the first country in the world to complete such a technological transition on a national scale.[55]

Opposition Movements Developed by Former Aeronautical Engineers in the Railway Industry

By the mid-1950s, the weight factor occupied much of JNR's thinking. Many of the technical difficulties in developing lightweight vehicles involved the weight analysis; the key issue was to save as much structural weight as possible without sacrificing strength and safety. The advanced stress analysis and testing of the strength of the materials were indispensable in the postwar railway industry. Before the introduction of composite alloys into the field, the conversion of wood designs to steel naturally increased the strength of the body structure. However, it also added weight, which then damaged the strength and durability of the physical infrastructure in the rail service, such as tracks and bridges. Given this unyielding technological parameter, minimizing the structural weight of the train car became an important part of the railcar design project. Of all of the structural weight, the upper body of the car occupied the largest proportion and much of the attention of design engineers. On average, the upper body and interior equipment constitute roughly 40 percent of the total car weight; the running gear, about 30 percent; and the rest, roughly 30 percent.[56]

All the data collected at the RTRI formed the base of information for the JNR headquarters in developing new steel railcar designs. Meeting the JNR's expectation, for instance, Miki's laboratory at the RTRI engaged in labor-intensive research projects. These tasks included (1) developing a vibration gauge for railway vehicles, (2) basic research on the vibrations of springs and the trucks of the running gear, (3) examining the physical impact of running railcars on rails, (4) developing a lightweight coach, and (5) improving ventilation in coaches and electric railcars.[57] In September 1946, Miki studied the structure and strength of the railcar body, the distribution of stress, as well as the fatigue aspect of welded metal parts. He conducted time-consuming tests on many vehicles, including an automobile, large trailer bus, moha-80 railcar, moha-63 railcar, and some coaches that had been damaged at accident sites.[58] Miki and the other ex-aeronautical engineers were known for their mathematical skills. The RTRI obtained Japan's first giant electronic calculator, Bendix G20, in 1962,[59] but until then, the portable steel-made Tiger arithmometer had been the tool of choice for laborious, time-consuming calculations.

Aside from calculating figures, meticulously controlling the weight of a railcar was a business all too familiar to the former aeronautical design engineers. The lighter the structural weight of the aircraft, the faster, the farther, and the higher it could fly. All the safe aircraft designs required sound theoretical studies and empirical data. Ex-aircraft engineers such as Miki had proved their ability to handle highly complex physical forces in three dimensions within an uncompromising weight envelope. The accumulated wealth of knowledge in aircraft design engineering was indispensable for developing railcars within the same, unyielding weight restrictions. Therefore, Miki and the other former aeronautical engineers gained a reputation for their ability to conduct the advanced stress analysis and testing of the strength of materials. The self-supporting monocoque body structure of steel railcars of the 1950s was the culmination of their expertise in structural analysis and aircraft designs before 1945. The technical knowledge, already internalized, had become part of the engineers by the end of the war. In developing the monocoque, lightweight, all-steel coach cars, perhaps because of pride, the JNR neglected to fully utilize the domestic deposit of wartime aeronautical engineering.

Furthermore, the craftsmanship in wartime aircraft development proved valuable in the postwar railroad industry. Only after the war was spot welding, previously used in aircraft, introduced in railcars; until then, house-building techniques were prevalent—and rivets or often nails supported the structural integrity of passenger cars. The new technique from overseas removed the redundant crossing of steel plates and unnecessary heads, consequently reducing weight by 10 to 15 percent without sacrificing the structural integrity of the railcars.[60] In addition, for further weight loss, a corrugated metal sheet employed in wartime aircraft was used on the floor of the postwar railcar. This type of sheet had covered the surface of the main wings on aircraft until some detailed studies pointed out the surface roughness as a major cause of drag. As the speed of flight increased and necessitated care as meticulous as flush rivets, the corrugated metal sheet supported the inner structure of the wings.[61] This wartime technique to increase structural strength resurfaced in a newly developed niche after the war.

While the JNR headquarters looked abroad for useful information, former aeronautical engineers in Miki's laboratory at the RTRI looked mostly within themselves. After all, the ex-military engineers were able to make technical improvements in many design features based on their own wartime knowledge and experience at home. In fact, except for one associate who visited France, none of Miki's workforce went overseas. The leading figure in this landscape was Miki. Identifying himself as a "designer" instead of as a "researcher" before and even after 1945, he single-handedly designed a series of streamline railcars against various forces of conformity in the JNR. Actively, independently, and resolutely, Miki sought opportunities to capitalize on his wartime experience in designing aerodynamically sophisticated, lightweight aircraft. This thinking was borne out of his wartime past as much as his postwar frustration. He had spent years at the peripheral RTRI as a human calculator for the central headquarters to design what he saw as outdated, needlessly weighty railcars. By his own later account, he was "minding his own business [of designing]." This self-identity, however innocent, proved subversive in the preexisting power structure of the JNR in which, historically, the RTRI on the periphery had nothing to do with the designing of vehicles. Miki's strong-willed personality formed the core of a series of opposition movements that were a welcome salvo among the other former aeronautical engineers in his laboratory who shared the frustration. For them, technology was power—it was a means to alter the preexisting center-peripheral hierarchical structure of the JNR. The corollary was a degree of organizational friction in high-speed rail projects.

Miki had passion and energy, but he needed resources for his high-speed rail service projects. What supported his initiatives was a rule issued in March 1941 that effectively allowed the RTRI to accept requests from the public or civilian sector and to co-opt their research projects as long as duties assigned by headquarters were not impeded.[62] Within this framework, designing railcars, for him, irrespective of the central design section, required commercial niches—and his

designs could only yield fruit outside the JNR power structure. Miki's two oppositions from 1951 to 1957 signified divisions within the JNR and among commercial
railcar manufacturers in the society that had successfully accomplished postwar
reconstruction after defeat.

First Opposition by the Former Aeronautical Engineers

The first major opposition took shape in early 1950 when Miki's laboratory began
to develop a monorail car independently of central direction. This engineering
success drew on a German development of the 1930s. The so-called Rail Zeppelin, *Schienenzepplin*, mesmerized and deeply impressed Miki. This experimental
Nazi product had been developed parallel to new breakthroughs in the aircraft
and automobile industries during that decade. The streamlined railcar, reminiscent of the Zeppelin airships, established a high-speed world record in June 1931.
Equipped with a BMW engine (600-horse power) and a four-blade wooden propeller in the rear, the aerodynamically refined car ran part of the stretch of land
between Hamburg and Berlin within 98 minutes at the speed of 230 km/h.[63] Little, if any, technical information about this vehicle was available in the Japanese
railway or aircraft industry at the time. Its success remained a mysterious black
box. Miki and other ex-aircraft engineers at the RTRI could observe it in writing
and pictures, but its inner mechanisms remained unknown to them.

With his curiosity aroused, Miki capitalized on this experimental German
project. In his thinking, an aircraft propeller and engine in the nose could pull a
railcar on a monorail track from Tokyo to Osaka across 550 kilometers within 2.75
hours at an average speed of 200 km/h. He calculated the structural weight, drag
resistance, necessary horsepower, and projected performance of the wingless airplane—an entire process with which he had had been familiar from developing
military aircraft during the war. Miki used his wartime resources for his project.
When he prepared a blueprint of the product and a 1/30 scale model, the drafts-
man was his associate from the prewar years—a former aeronautical engineer
who had helped Miki at the INA throughout the war. Miki used a notepad previously issued by the navy during the war, something he could feel comfortable
with for calculating the numbers for his design.

Using four German technical sources, his plan for the "rail plane" took shape
in the suspended-type mono-railroad system. He compared American, Japanese,
and German aircraft engines and then selected a 560-horsepower aero-engine
from Daimler-Benz. The motive power, with a 2.6-meter-diameter propeller,
could theoretically pull a light, streamlined railcar carrying 63 passengers (later

reduced to 46) and 4 crew members at a top speed of 250 to 270 km/h. The vehicle was 20 meters long and 2.5 meters wide, and the structure weighed 12 tons.[64] The project showed great promise at least in theory. Major national newspapers, such as *Mainichi Shinbun*, soon reported about this idea. *Nihon Keizai Shinbun* printed an article with the sensational heading, "New Plan for a Bullet Train in the Air: Tokyo-Osaka in 2 1/2 Hours." [65]

This theory gained modest success in an unexpected setting. While the JNR ignored this idea because it seemed unrealistic, the RTRI benefited from this misperception. The president of the private corporation Seibu Railway Company read this article and was more impressed by its optimistic outlook than its technological utility. He showed interest in operating the product within the Toshimaen amusement park as a children's vehicle. This project stimulated Miki in two ways. First, although it increased his workload exponentially, it was his own design project beyond the reach of the JNR central headquarters. Second, the project affirmed his unfaltering commitment to using technology for peace, never for war. Subsequently, his laboratory took charge of developing the vehicle and infrastructure with commercial companies, including Hitachi. The "Flying Rail Car" was introduced as a ride in Tokyo's Toshimaen amusement park. By April 1951, the size of the railcar was reduced from 20 meters to 9 meters in length, and the top speed of operation from 270 km/h to 10 km/h. Suspended from above, the vehicle did not transport 63 passengers from Tokyo to Osaka, but it did carry 20 children on a circular track with a radius of 30 meters. Its power source was not a German aircraft engine, but a 220-voltage generator that motorized wheels. Its dummy propeller in the nose lacked any mechanical function, but it was stylistically appealing. The vehicle was lightweight, owing much to a series of rigorous structural studies based on knowledge drawn from wartime aircraft engineering. To minimize the danger of derailment, the railcar was equipped with three fail-secure devices.[66] This vehicle based on wartime technology was immensely popular, and it was a pleasant metaphor for peace.

This success marked the beginning of similar projects, all solidifying the ties between the RTRI and the private railway company Odakyū outside the JNR power structure. Buzz about the "cute, little flying car" for children reached Odakyū by mid-1953. Its director, Yamamoto Risaburō, subsequently asked the RTRI for assistance in replicating Miki's idea in a different amusement park. The resulting product, designed by Miki and manufactured by Hitachi, was essentially the same in dimensions, motive power, suspension system, and carrying capacity as his earlier vehicle. After a series of surveys in the field, the "streamlined, monocoque-structured" car carried 20 children on a circular, monorail track with

The Flying Rail Car in the Toshimaen amusement park in Tokyo, 1951. The body structure of this children's vehicle capitalized on wartime technology.

a radius of 140 meters that surrounded a baseball park.[67] In 1955, Miki helped to design a roller coaster train for the Kōrakuen amusement park in Tokyo.[68]

In Miki's eyes, these vehicles for children embodied the demise of militarism and the importance of peace. Moreover, these products of wartime military technology and defeat alleviated the mourning of their creators and played a cathartic role for Miki and his associates. The innocent laugher in the amusement parks affirmed Miki's commitment to the construction of peaceful culture and society. He deeply deplored Japan's movement toward remilitarization at the time. The navy was steadily revived in his city of Yokosuka. The outbreak of the Korean War in June 1950 expanded the Maritime Safety Agency of Japan as a staunch ally of the United States amid the Cold War that escalated progressively in Asia. In April 1952, the San Francisco Peace Treaty effectively placed Japan among the league of 48 nations for peaceful coexistence, which ironically rearmed Japan. The Maritime Security Force was created concurrently, and it evolved into the Japan Maritime Self-Defense Force under the newly formed Japan Defense Agency. The end of the Allied occupation in April 1952 lifted the official ban on the aircraft industry. In this context, the revived navy needed wartime expertise such as Miki's. Yet he resolutely rejected solicitations from his wartime colleagues asking him to return to research and development for "national defense" in the air.

Any airplanes could be used for war, Miki reasoned, and any automobiles could be modified into army tanks; only railcars could not be converted for war and, thus, could embody the spirit of peacetime and the renunciation of war forever.[69]

In the postwar culture of defeat, he was not alone in his commitment to sustaining the life of a pacifist civilian. The career transitions of many other ex-navy aeronautical engineers could be reviewed in the membership list of Yokuyūkai, the voluntary association they formed shortly after the war. The 1969 roster lists the names, addresses, and places of employment of 129 registrants. This information, albeit a mere snapshot of thousands of wartime aeronautical engineers, suggests the career choices that they had made by that year. According to this information, the vast majority—80 members, or 62 percent—had remained in the engineering departments of various commercial companies that produced, for instance, agricultural machinery, precision tools, glass, or aluminum. Some of them had worked in the construction and milling industries. The second-largest group—17 members, or 13 percent—had resumed their careers at aviation-related commercial companies. Fifteen members had become either university professors or other professional researchers. Only a distinct minority of 8 members, or 6 percent, had returned to conducting military-related research at the Defense Agency.[70] It was understandable for the majority of the ex-navy engineers to remain comfortably settled in the civilian sector after the revival of the Japanese military in 1954. For them, returning to the military sector meant making a career transition for the third time in their lives. Only the committed minority apparently did so.

A Second Opposition by the Former Aeronautical Engineers

In retrospect, the first opposition movement by former military engineers was quiet and rather innocuous. It changed little in the rail industry, and the JNR headquarters continued to dictate the technological development of mainstream railcars in the industry. The second opposition by ex-aircraft engineers, however, was more contentious than the previous one. It involved the Odakyū rail company, which had developed close, trusting relations with the RTRI by the mid-1950s. In the long run, these two players firmly set a positive tone for subsequent projects to develop high-speed railcars. The resulting technology was a remarkable success. This was possible because the RTRI assumed the technological leadership, fully capitalizing on wartime aeronautical technology to develop the Odakyū SE Model 3000, better known as the Romance Car. It was a wingless plane on rails.

This high-speed rail project had a low-key beginning. The first opposition rescued Miki from a laborious life of supplying data for the central headquarters to design railcars. Leaping into designing his own vehicles, almost habitually, Miki had jotted down ideas about technology for a high-speed railcar. In 1953, the technical feasibility of one of his ideas became evident after a series of mathematical computations. He concluded rightly as it turned out that an electric locomotive with nonpowered cars—all lightweight and aerodynamically refined—could run the Tokyo-Osaka Line of 556 kilometers in 4 hours and 30 minutes at an average speed of 120 km/h. This conclusion was far more realistic than that of the Rail Plane two years earlier. He articulated his idea about high-speed rail service in a journal article—the first solid work of its kind written and published after 1945.[71] On October 17, five major national newspapers carried reports on his idea, and *Asahi Shinbun* placed his picture at the center of the article.[72] A popular science journal fantasized about the high-speed railcar a few months later. "No longer a dream," according to the author, the high-speed train would be "designed by aircraft engineers."[73]

These publications were provocative to some in central management. Miki developed the high-speed rail project during his free time after his regular work, but it was a different matter when the RTRI turned to the national newspapers without consulting the central headquarters in advance. Miki and RTRI director, who authorized the publications, did not follow the tacit protocol of the JNR organization. The unwelcome newspaper article was unexpected for the headquarters of the Department of Engineering, which subsequently lost face.[74] The appearance of the article was untimely, awkward, and embarrassing to central management. At that time, the department was leading a project to electrify the Tokyo-Osaka Line west of the city of Hamamatsu for the high-speed train Tsubame; it was to run the Osaka-Tokyo distance in eight hours at an average speed of 72 km/h. In this setting, the peripheral research establishment offered a faster train service than did the central headquarters. Remaining unfazed, the central headquarters was fully determined to pursue its own course of technological development.

The RTRI was equally adamant. Its defiance stemmed mainly from its confidence and competence. This center-periphery tension took a curious turn when the private railway company Odakyū—which had supported Miki's previous railcar projects—co-opted this task. Miki and the company director, Yamamoto Risaburō, shared a vision of the high-speed rail service of the future. Yamamoto was an electrical engineer by training and had worked at the JNR until 1945. An advocate for high-speed transportation, he wanted to reduce the travel time between Shinjuku in Tokyo to the city of Odawara across 83 kilometers from 130 minutes to 60 minutes, or by 54 percent. Miki's idea came at an opportune moment for Odakyū; it had hosted a series of technical meetings for high-speed ground transportation around the same time to increase the passenger traffic from Tokyo to the resort town of Hakone.[75]

This town remained a popular tourist center due to its location and history. The mountain resort is located 90 kilometers southwest of Tokyo, near Lake Ashi, at the foot of Mount Fuji. The Hakone toll station, constructed in 1618, was a famous checkpoint on the Tōkaidō highway, connecting Edo/Tokyo and Kyoto under the Tokugawa regime that ended in 1868. Especially after the 1920s, hot springs, temples, museums, and golf courses brought travelers to this resort area well outfitted with hotels, sightseeing boats, sport facilities, and vacation cottages. With Japan's fifth-largest daily yield of hot spring water (25,000 tons), Hakone is the closest hot spring town to Tokyo.[76]

Before 1945, a private railway company tried unsuccessfully to directly connect Hakone and Tokyo. On April 1, 1927, Odawara Kyūkō Dentetsu, founded four years earlier, commenced its commercial rail operation between Shinjuku in Tokyo and Hakone's neighboring city, Odawara. Nonstop weekend rail service began on June 1, 1935, to meet the demand of the nation's wealthy seeking a good rest in Hakone. But the outbreak of the Sino-Japanese War in July 1937 put Japan on a war footing. The nation mobilized its resources, such as electric power and overland transportation, merging Odawara Kyūkō Dentetsu into a railway conglomerate in May 1942.[77] Hakone's development was then limited by the wartime mood of voluntary restraint.

Military defeat, however, drastically altered the power relationships in society as well as the demographic composition of those who travelled to the resort town. The US occupation authority seized major hotels and recreational facilities, essentially converting the town into a health resort for high-ranking US officers. With the circulation of Japan travel guides, middle- to low-ranking and petty officers arrived in trucks on weekends, often with Japanese prostitutes. The Korean War from 1950 to 1953 economically revived the town, with soldiers returning from the war zones to enjoy the respite in the hot springs. The historically refined Fujiya Hotel, which had hosted foreign dignitaries after its construction in 1878, grew during the war; at one point, this "OFFICERS HOTEL" retained more than 300 Japanese employees and slot machines for American soldiers. The San Francisco Peace Treaty of 1952 released the hotel and other facilities from the control of the American occupation.[78]

From the occupation era emerged a series of initiatives that democratized Japan and revived Hakone. Big financial combines, such as Mitsubishi, were liquidated under the Anti-Monopoly Law of 1947. On June 1, 1948, the dissolution of one railway conglomerate gave birth to Odakyū Dentetsu.[79] In four months, this private company reopened the express line connecting Shinjuku and Odawara.[80] Its competitor was the JNR, which welcomed the end of World War II. From September 1949, the JNR operated its fleet of high-speed railcars, called "Peace," between Tokyo and Odawara as part of the Tokyo-Osaka operation. On August 1, 1950, Odakyu expanded its rail operations from Odawara to Hakone, which dramatically increased the passenger traffic from 228,000 to 423,000 people in a year.[81] This growth reflected the leisure and economic booms in the early 1950s. Companies organized trips to Hakone for their employees to rest and relax; business managers entertained their important contacts at golf facilities there; and new religious groups took their followers on pilgrimages to Hakone.[82]

In late 1954, the private rail company and the RTRI embarked on their joint high-speed rail project between Tokyo and Hakone, each party with apportioned work responsibility. The plan proceeded with "technical support from the RTRI," and its purpose was to develop a "super express rail service" by "modernizing and maximizing the efficiency of railway technology." Given the "government's tight regulation about the need to stop [any trains] within the distance [of 600 kilometers]," the vehicle's "top speed was set at 125 km per hour," and in the future, it could go up to "147 km per hour."[83] As before, Miki referred to German high-speed rail service for his calculations and prepared a series of design blueprints.[84] Meetings were held regularly, each typically lasting three hours, with 20 to 25 engineers from the RTRI, Odakyū, and several railcar manufacturers. Yamamoto, Miki, and his associates were apparently the only ones among the 890 engineers to attend all 29 meetings, which signified the central role that they played in the project.[85] Given its technological prospects for the future, the Ministry of Transport supported this undertaking.[86]

The result—later known as the Romance Car—was a technological triumph. In September 1957, the vehicle established a world record (145 kilometers) for its high-speed test run on a stretch of JNR's Tōkaidō line. The most important aspect of the success was the technology embodied in the vehicle. Ordinarily, creating a high-speed ground transportation system required improvements in two areas: vehicle and rail infrastructure. Improving the infrastructure was costly. For instance, rail tracks and crossties needed to be strengthened, while railroad crossings and signals needed to be improved. It would have cost the Odakyū company dearly because the roadbed of the tracks, soft in many locations, would need massive civil engineering work. The Romance Car project opted for a cheaper, more viable alternative. Subsequently, the RTRI and Odakyū devoted all their energy to create a high-performance vehicle.[87]

The resulting machinery embodied engineering knowledge and the culture of Miki's former employer, the navy. His approach to developing the high-speed vehicle was as simple as what he had done during his wartime years; that is, he sought to minimize its weight. First, his group reduced the overall car size and employed the monocoque construction of aircraft. To reduced weight, Miki's group used the thinnest metal sheets—1.2 millimeter—for the body's outer shell, whereas the traditional shell plate was 2.3 millimeters thick.[88] After a series of stress analyses, the engineers drilled as many holes as possible in the steel frames to reduce the "fat."[89] This strategy saved roughly 10 percent of the weight from each stringpiece, reducing its weight from 327 kilograms to 296 kilograms.[90] Moreover, his group eliminated one truck per two railcars to reduce the total weight of the undercarriage.[91] A series of weight reduction efforts bore fruit as a whole. Structurally sound with more rigidity and durability, the new vehicle weighed only 370 kilograms per meter, or 26 percent less than a traditional railcar.[92]

Drawing on his aeronautical wartime training, Miki drastically reduced the nonstructural weight of the vehicle. Since passenger seats were the heaviest item among all the auxiliary equipment, he used aluminum alloys—a light, corrosion-resistant material ideal for car construction—and developed a new set of seats. The result of this weight engineering was phenomenal. A similar seat unit widely used in the industry at the time weighed 114 kilograms, and even the lightest reclining seats used on buses weighed 39.4 kilograms. In contrast, Miki's product weighed only 33 kilograms, 3 kilograms lighter than the one used in the world's first commercial jet aircraft, the British de Havilland DH 106 Comet.[93] These efforts to minimize the weight did not translate into a comfortable ride in the summer. Because air conditioning equipment was heavy, Miki did not initially install this feature in the vehicle. Drawing from its wartime experiences, the design team may have expected civilian passengers to display the austerity and discipline of military aviators. Later, as a compromise, the group thought of mounting ice. When that idea failed, the group tried dry ice. When that also failed, the group adopted the weightier option of installing fans in the railcars.[94]

Miki's weight reduction program borrowed techniques from the wartime aircraft industry. One initiative involved craftsmanship, so-called spot welding, which had been a prevailing practice for manufacturing aircraft during the war. This more sophisticated technique aside, the railway industry continued to adhere to house-building techniques even after the war, whereby rivets, or often nails, supported the structural integrity of the passenger cars. Once introduced into the postwar railway industry, the new spot-welding technique removed the

Load test on a steel frame of the vehicle. After careful stress analyses, engineers drilled holes in each frame to minimize the structural weight of each railcar. Photo courtesy of Railway Technical Research Institute, Tokyo, Japan.

redundant crossing of steel plates and unnecessary heads. Consequently, Miki's group minimized the vehicle's weight without sacrificing its structural integrity.[95] Furthermore, the ex-aircraft engineers used a corrugated metal sheet on the floor of the railcar. This technique had been employed in wartime aircraft. This type of sheet covered the surface of the main wings of aircraft, until the surface roughness was shown to cause drag. As the speed of flight increased and required care as meticulous as welding, the corrugated metal sheet no longer appeared on the surface but supported the inner structure of the wings.[96] This wartime technique resurfaced in Miki's group to create the Romance Car for high-speed rail service. A corrugated, thin metal sheet provided enough strength to support the car's floor from underneath.

Miki's creation was successful not only because he minimized its weight, but because he streamlined the railcar according to the conventions of wartime science. His laboratory conducted theoretical and empirical analyses of air flowing over the vehicle as it ran at high speed. Because the RTRI lacked a wind tunnel, Miki turned to the country's top expert in the field of high-speed aerodynamics, Professor Tani Ichirō at Tokyo University. Together, they used the institution's wind tunnel, which was the largest one in the country during the war with a 3-meter-diameter test section. They had known each other from the war years.

In the early 1940s, Tani had developed the fastest experimental army aircraft, the Kensanki, and the MXY7 attacker, the fastest aircraft designed by Miki and deployed for the navy's kamikaze missions.

Miki's research group thus drew on its prominent history in wartime. He created a series of car models in various sizes and shapes using one-tenth scale models to observe air resistance against their noses. Aerodynamically more important than the nose shape were the skin friction and air turbulence. To examine them, the engineers tested a one-fortieth scale model of the long fleet in the wind tunnel. During a test run on rail tracks, the research group closely examined the distribution of air pressure on various points of the nose. They also studied a chief source of drag on the vehicle's body, the viscous layer of air that lay adjacent to the surface. Aerodynamically, the new vehicle proved its excellence as compared to traditional railcars. The thickness of the layer on the new vehicle was one-half to one-third of that of traditional ones, a result signifying the scientific superiority of the new railcar.[97] Behind these techniques were Miki's own experiences from developing highly advanced navy aircraft during the war. Since the surface roughness militated against a high-speed run, the engineers installed a hood over each space between the railcars, and they enclosed the underbody for aerodynamic reasons.[98] Among the comparable railcars of the time, Miki's creation kept the lowest centers of gravity to minimize the airflow effects. The resulting vehicle was stylistically and scientifically sound.

The resulting success became a source of great national pride. The vehicles' world records marked a crucial milestone in Japan's endeavor to develop high-speed rail service. Technologically, the new railcar was superior to a prevailing model that ran on the Tōkaidō line; at a high speed, the former weighed less and faced far less drag resistance than the latter. The fleet of new vehicles, that is, the Romance Car, was 108 meters long, but weighed 147 tons, and its resistance was 1,400 kilograms at a speed of 120 km/h; on the other hand, the Shōnan railcar fleet was comparable in length (100 meter long) but weighed 225.2 tons, or 54 percent more, and its resistance was 1,966 kilograms, or 40 percent more than that of the Romance Car at the same high speed.[99] Less weight and less resistance meant good news to the Odakyū rail company that operated the new vehicle because each operation required less power, thus less electricity.[100] The RTRI-led research proved vital in this project. By Miki's own account, roughly 80 percent of the subsequent high-speed train development in the field stemmed from the Romance Car project; from this historical perspective, there was "nothing surprising" about the later, successful design of the Shinkansen railcar.[101] His group of former aeronautical engineers formed the basis for developing aerodynamic, light rolling stock for the Shinkansen in the following decade.

Since the needed technical expertise was already internalized in Miki and other ex-military engineers, outright and costly technology transfer, such as dispatching Japanese engineers to the West, was remarkably absent in this Romance Car project. Yet visible and direct technology transfer showed in the design and production of the running gear, which was based on technical ties between Japanese and Swiss railcar manufacturers.[102] Meanwhile, sources of inspiration were also sought out. In the mid-1950s, for instance, Miki visited Haneda Airport in Tokyo to examine passenger seats as installed in commercial jet aircraft, such as the American Douglas DC4 and British Comet. His wife, who was bilingual, translated the English for him.[103]

In a larger framework, Miki's approach to high-speed rail service placed his research and development establishment at the center of the entire project. The RTRI directed the course of the technological development from the inception in 1953 to the commercial operation in 1957. This was possible because the RTRI, the originator of the technology-based idea, had created and monopolized the requisite knowledge from the wartime years—knowledge that proved vital in the construction of the highly advanced vehicle. Commercial manufacturers aided the RTRI, which led the process. The private railway company Odakyū provided the RTRI with the needed infrastructure for the project, such as an operator, electricity, rail tracks, and operation control. Furthermore, the wartime technology put the RTRI at the center. Technology was power.

By no means monolithic in orientation, the JNR headquarters exhibited mixed responses to the RTRI's collaboration with private railroad companies. For instance, the Department of Engineering that designed railcars did not like Miki working together with Odakyū on the Romance car project. But the Department of Rail Operation supported the initiative. Equally committed to high-speed rail service, JNR chief engineer Shima Hideo straddled the fence; he took a wait-and-see attitude by tacitly approving the collaborative work.

The pursuit of technological excellence was both a motto and practice in the RTRI-driven engineering project, changing the business culture of the railway industry. Technical "advice" from the RTRI essentially dictated the course of the technological development. For instance, the Odakyū rail company no longer chose a railcar manufacturer for mass production solely because of its attachments to them via the exchange of stock or board members. This criterion became least important in the selection process. Far more important were the voices from the RTRI as well as the technological superiority among the manufacturers as exhibited numerically, such as the weight of their product(s).[104] This approach drastically changed the landscape of the manufacturers in the project. For instance, the Tōkyū Railcar Group—from which Odakyū had grown and had become independent in 1948—did not win a single production contract, despite the strong managerial ties between the two.[105] RTRI advice often shocked many experts in the rail industry. The leader of the Rolling Stock Vibration Laboratory, Matsudaira Tadashi (see chapter 5), scrutinized the weight and vibration characteristics of the running gears produced by several manufacturing companies, and on the basis of his individual decision, one relatively minor manufacturer won the production contract against all odds.[106]

Historically, this RTRI-centered approach was reminiscent of the navy's approach during the war. The research and development establishment occupied the center stage in both military and railway industry. In the navy, the locus of the technological development was at the INA, where Miki had developed highly advanced military aircraft. There, groups of able engineers explored ways to develop military aircraft that could fly the fastest, farthest, and highest in the air. Engineers engaged in research and development occupied the center stage. Commercial aircraft companies helped to manufacture the flying machines with production skills, and the navy supported the research and development establishment with the needed infrastructure, such as aviators, fuel, airfields, and flight control.

Miki transferred the navy research approach to the high-speed rail project. Sharing the same style of research and development, the navy and RTRI engineers made painstaking and costly efforts. As the wartime navy had found it difficult to convert its research-based, meticulously designed aircraft to mass production (chapter 3), the RTRI experienced the same issue in the 1950s. Manufacturing was not considered during the design stage of the Romance Car project. For instance, one major technical difficulty had to do with the car's low center of gravity. It was 30 centimeters lower than a traditional railcar, thus leaving little room for technicians to install a controller in the limited space underneath the floor. Another source of trouble was the vehicle's outer shell. Because it was the thinnest among all the traditional railcars to save its structural weight, it was difficult to build the material and test its physical strength.[107] These technical difficulties were solvable in the postwar rail industry, given no dire urgency for war; but the resulting price tag was harder to solve in the commercial rail project. Miki's new vehicle, as it turned out, was lightweight and sophisticated, but awfully expensive. Much of the equipment mounted on the fleet of the new railcars was specially designed to save weight, and because of this characteristic, processing all the fittings alone would cost 5 million yen, or 48 percent more than the vehicle's predecessor.[108] These difficulties, both technical and financial, were legacies from the navy's approach to research projects.

While other laboratory leaders followed Miki's lead in the opposition movements against the JNR headquarters, they embraced different engineering cultures. Unlike many others, for instance, the Metal Research Laboratory featured a distinct academic atmosphere due in part to its strong ties to former imperial universities; professors from Tokyo University, Tōhōku University, and Tokyo Institute of Technology acted in an advisory capacity.[109] Members of the Brake Mechanism Laboratory actively cherished the value of harmony among them. To use the words of the leader, it was "the motto" of his laboratory. Harmonious relationships were important to carry out projects among 10 chief researchers, 6 of whom had accumulated their hands-on experiences out in the field or on the factory floor. This particular laboratory worked closely with the JNR headquarters with apparently little internal and external tension.[110]

Despite some cultural differences, all the laboratories sought technical information from abroad during the 1950s. The Laboratory of Disaster Prevention, for example, actively sent its researchers to France and the United States to collect useful case studies about natural disasters.[111] The leader of the Welding Laboratory observed the state of the field in France, Germany, Switzerland, and the United States for two months.[112] The head of the Brake Mechanism Laboratory studied the high-speed rail operation in the United States for six months in 1953.[113] Later, two of his researchers compiled a series of detailed reports about the latest brake mechanism used for the high-speed rail service in the Soviet Union.[114]

Overall, however, the success of the Romance Car project in 1957 encouraged the RTRI to move forward independently. Designing rolling stock was no longer the prerogative of the Department of Engineering at the headquarters. The joint venture by Odakyū and the RTRI changed the landscape of the railway industry; with research and development, it could play a central role in the high-speed rail project. The center-periphery hierarchy that had existed within the JNR began to crumble by 1957.

Consequences of the Opposition by the Former Aeronautical Engineers

The RTRI's initiatives rejuvenated Japan's efforts to develop high-speed rail service, but unexpectedly, they also exposed an inherent weakness in the JNR's similar effort in the field. Galvanized by the success of the Romance Car project, the JNR's central management aimed for the development of an express train that could run between Tokyo and Osaka in 6 hours and 30 minutes. Its top speed was a modest 120 km/h. In July 1959, its creation, which would later be called Kodama, operated at a top speed of 163 km/h, establishing the world record in a high-speed test run. Compared to the previous Romance Car project, this success was mainly attributed to longer, better straight rail tracks used for the test runs as well as to the more powerful motor that ran on 1,550 kilowatts of electricity.[115] Two years prior to this accomplishment, the JNR headquarters had set up an office specifically devoted to designing modern vehicles; in February 1957, the so-called Temporary Rail-Car Design Office embarked on the design of a new, electrically operated train. This office of 110 engineers and technicians inherited much of its craftsman mentality from the two sources: roughly 50 engineers from the JNR's Department of Engineering and the rest from various train factories. The leader, Hoshi Akira, initially incorporated few technicians from the RTRI.[116] In this project, RTRI laboratories were involved only partly, or only at the testing phase, or remained completely uninvolved in some fields of study such as high-speed aerodynamics.

In a sense, Hoshi's Kodama design fell short of achieving success. Its nose lacked meticulous streamlining, and as a result, its drag resistance against the front of the entire fleet was far more than that of Miki's product. A series of boundary-layer analyses during high-speed runs revealed the same result. The fleet of Kodama railcars had a thinner layer of air flowing on the surface than that of traditional cars, thereby showing a more refined streamlining—but it was thicker and worse than that of Miki's Romance Car.[117] In addition to this science, simple arithmetic exposed the inferiority of Kodama to Miki's lightweight product. Depending on its type, the Kodama railcar weighed between 1.5 and 1.9 tons per meter, which was approximately 30 percent heavier than Miki's vehicle; moreover, the Kodama fleet weighed 272 tons, or 62 percent more than the fleet of Miki's lightweight vehicle.[118]

These results stemmed from two opposing methods of technological development: one by Hoshi, a traditional railroad engineer, and another by Miki, an experienced ex-navy aircraft engineer. Hoshi's office relied more on intuition and substantially less on science than Miki's research group. In contrast to Miki, who had spent 34 months meticulously improving the aerodynamic design of his vehicle, Hoshi lacked education, knowledge, or experience in the field. Nonetheless, it took his entire office only 10 months from the time of the creation of the initial blueprints to the completion of the entire project.[119] Hoshi's design philosophy was simple. It relied on a symmetrical balance and an aesthetically pleasing appearance, rather than a scientifically sound shape. His office completely lacked expertise in high-speed aerodynamics. In need of the knowledge for how to conduct physical experiments, the engineers turned to the College of Engineering at Osaka University, in contrast to Tokyo University from which Miki had received help, a strategy reminder of the longtime cultural rivalry between Tokyo in the East and Osaka in the West. Osaka University had a wind tunnel with a 3.5-meter-diameter test section, which was constructed later and was slightly larger than the one with a 3 meter diameter at Tokyo University.[120] In the end, the legacy of wartime science at Tokyo University prevailed with better test results in the development of high-speed railcars.

The two groups revealed their different approaches not only in aerodynamics but also in the construction of the vehicle's structure. Miki's task force carefully and obsessively reduced as much structural weight as possible, while conducting a series of stress analyses on the body of the vehicle. In contrast, Hoshi's group relied less on science than on an "intuition that drove the development of light-weight vehicle that structurally had many openings" such as windows and doors. "Nothing was certain until load tests were done on the product," according to a design engineer of the Kodama railcar, but "mysteriously," it was "the sight of the vehicle's body [that] gave [him] the courage [to move on]."[121] Such intuitive, unscientific, and relatively nonchalant thinking would be absurd from the point of view of the former aircraft engineers. At the foundation of the difference was the sense of crisis, or safety concern, for the end users. Their task as aircraft engineers was to create machines that could theoretically and empirically withstand complex three-dimensional forces causing enormous stress and strain during a high-speed flight. Dependence on "intuition," or a lack of rigor at any stage of the creation could easily result in a crash. The intuitive thinking, which could result in deaths, was not a risk that aircraft engineers could afford to take.

From 1945 to 1957, Japan's research and development of high-speed rail service advanced with remarkable success. At the center was the RTRI, or more specifically, the Rolling Stock Structure Laboratory led by Miki Tadanao. His experiences, ideas, and activities epitomized the railway industry's and, to an extent, Japan's conversion from wartime to peacetime. In a way, he exemplified the technological and psychological effects of defeat on Japan. Having worked in Yokosuka, he supported the city's conversion from a wartime to a peacetime economy. This former navy engineer with an unforgettable "sinful" past found consolation in creating a series of technologies that signified the importance of peacetime culture. Commercially successful railcars of his design, such as vehicles in amusement parks and the Romance Car SE3000, affirmed his and the nation's commitment to peace. His creations rejuvenated Japan's effort to create high-speed commercial transportation in the peacetime culture of a defeated country.

His means of pursuing technological excellence were not always peaceful. His quiet insurgencies in the 1950s created tense relations in the JNR's rigid hierarchy, where the headquarters presided over the peripheral RTRI. Technologically, his approach to the high-speed rail project bore fruit in the Romance Car's high-speed world records. In leading the postwar project, as in his wartime years, he placed his research and development at the center of the technological development. His wartime expertise was a source of power that he used to influence in and out of his laboratory and to replicate the wartime navy's approach to technological advancement in the postwar railway industry. Yet the fields of expertise among Miki and other former military engineers were in some ways narrow, confined, and separated from a larger canvas of needs. None of their efforts could have achieved success if not for their integration into a much larger rail system and a growing national economy of the 1960s.

CHAPTER SEVEN

Former Military Engineers and the Development of the Shinkansen, 1957–1964

At 6:00 in the morning on October 1, 1964, leaving behind fireworks and the marching song composed for the occasion, the Shinkansen slowly left the Tokyo station to head for Osaka for the first time, a distance of 515 kilometers. Fifty doves symbolizing peace were released into the air. It was a historic moment; it was "the Dawn of the New Age of Rail Service," wrote the *Mainichi Shinbun.* Even a local newspaper in the remote Saga prefecture carried a report.[1] Ishida Reisuke, president of the Japanese National Railways (JNR) cut the tape at the opening ceremony. His presence at center stage surprised those engineers at the Railway Technical Research Institute (RTRI) who were familiar with his view on the high-speed rail service. During his July 1957 visit to observe their experiments for a high-speed run at 200 km/h, he had reportedly said, "Why does a train need to run that fast? I am not going to ride that."[2] Even more surprising were those conspicuously missing from the ceremony. Ishida was new to the rail industry. Originally, he was a Mitsui entrepreneur who headed the JNR organization from May 1963 after Sogō Shinji, known as the father of the Shinkansen project. Neither Sogō nor his right-hand man chief engineer Shima Hideo attended the ceremony, despite their indispensable contributions to the project over the past decades. They and other RTRI engineers watched the ceremony on television.

A reasonable understanding of the technological and political development that led to this outcome needs to consider the independent judgments of politicians, engineers, and local citizens within various institutional and national decision-making contexts. Especially after 1957, RTRI engineers, the JNR leadership, and government officials began to display compatible, yet disparate interests in high-speed rail service. Only after their temporary marriage of convenience in 1957 did the national project move forward. Each player successfully capitalized on what the others could bring to the table within the growing national economy, each creating its own version of the Shinkansen success story.

A Brief History of High-Speed Rail Service, 1918–1955

Before 1955, the JNR had launched a series of engineering initiatives for long-distance, high-speed rail service across the nation but with modest success. Inter-city transportation over short and medium distances was dominant before 1945. Long-distance rail service made slow but steady progress. From 1890 to 1916, the overall train speed increased on the main Tōkaidō line from Tokyo to Osaka, with the average increase of 0.83km/h each year.[3] In 1890, the long-distance rail service ran from Shinbashi, Tokyo, to Kōbe with average speed of 30.1 km/h. The progress was incremental. In 1907, rail travel was possible from Tokyo to Shimonoseki on the western end of the Honshū Island at the average speed of 45.3 km/h. The fastest rail service before 1945 was by Tsubame train that ran from Tokyo to Kōbe at the 67.5 km/h on average; in 1950, it ran at the average speed of 68.6 km/h.[4]

Apart from this gradual progress, the central government launched a high-speed rail project before 1945 that had many political implications.[5] At center stage was the Ministry of Railways. In June 1939, it established the Committee for Constructing a New Trunk Line (Shin senro kensetsu junbi iinkai).[6] A subcommittee (kansen chōsa bunkakai) subsequently inherited the project and explored the technological feasibility of an entirely new line between Tokyo and Shimonoseki. The following month a committee (tetsudō kansen chōsakai) was formed with representatives from various ministries and specialists in the field. Their proposal, arguing the need for a new trunk line, was submitted officially to the Minister of Railways, marking the beginning of the so-called bullet train project. This massive national plan of 1939 supported the construction of a standard track gauge of 1,435 millimeters in width. The initiative capitalized on the prevailing atmosphere of national emergency associated with the war in China. That same year parliament approved the construction budget, allocating 560 million yen for its completion. This was originally expected by 1940.[7]

This project failed partly because its technological blueprint embodied both realistic and unrealistic elements. On the realistic side, engineers planned for an electric locomotive to pull coach cars from Tokyo to Osaka and then substitute a steam locomotive to haul coaches from there to Shimonoseki on a standard gauge track.[8] To that end, parts of the Tokyo-Shimonoseki distance were to be electrified. Consequently, a study group (Shinkansen denki setsubi kenkyūkai) was established in December 1940, which planned for direct current electrification using a voltage of 3,000 volts. An alternate current system was not adopted apparently because the industry, lacking experience in the field, did not find reliable electrical motors overseas.[9] Developing a suitable domestic alternate current motor did not seem possible at the time.

Promoters of the project who were not engineers lacked a realistic sense of how fast a train could cruise safely. In retrospect, parts of the original plan reflected the wishful thinking of a nation at war. The locomotive vehicle was to run the Tokyo-Shimonoseki distance within 9 hours and 50 minutes, and the Tokyo-Osaka distance within 4 hours and 50 minutes.[10] The requisite speed would have been 150 km/h for cruising and 200 km/h for maximum performance. These numbers were determined rather arbitrarily with little technical justification; the top speed seemed possible because European nations had attained that speed on experimental high-speed runs. Engineers believed that all the curtains should remain closed during high-speed runs because fast-moving landscape would cause vertigo to the passengers.[11]

Few in Japan seem to have been aware of the formidable engineering tasks ahead. For instance, the industry's obsession with streamlining was mostly a product of design fashion culture during the 1930s. Streamlined tricycles were popular among children. Streamlined hairstyle was the vogue of the time. There was little solid laboratory research in high-speed aerodynamics on which to base any serious rail engineering project. Tokyo University and the Aeronautical Research Institute (ARI) welcomed the rail industry's requests for wind tunnel experiments to weather a financially difficult time amid the Great Depression, but many assumptions employed in their empirical studies were questionable.[12] The prewar bullet train project seemed promising in some ways, but the Japanese–American war ended the project and left behind little substantive technological legacy for the future.

The prewar high-speed rail project had its own benefit, however. It temporarily settled a major debate over the width of rail gauge, a tension between staunch advocates for the standard gauge and those for the narrow gauge. Until the late 1950s, all rail tracks for commercial service used the narrow gauge of 1,067 millimeters. This cheaper alternative had a British lineage as found in its overseas territories. Japan's first rail track in 1872 adhered to the British style, owing to railway engineer Edmund Morel (1840–1871) who had capitalized on his earlier experience from constructing railways in New Zealand. The narrow gauge formed the basis for extensions thereafter.[13] Forces to counter this failed in the political arena. In the 1880s, military authorities of the first Itō Hirobumi cabinet mounted the first effort to convert to standard gauge of 1,435 millimeters to increase military transportation. Gotō Shinpei (1857–1929), the first director-general of the Railway Agency (Tetsudōin) in 1908, followed suit after the nationalization of the railways in 1907. His proposal for reconstructing railways to standard gauge earned financial support, roughly 230 million yen, from the Railway Congress (Tetsudō kaigi). The bill did not pass amid the intense political debate of the time, but Gotō did not leave the scene empty handed. The Committee for Reconstructing Railways on the Standard Gauge (Kōki tetsudō kaichiku junbi iinkai) was established in 1911 to investigate technological and economic gains from converting to the standard gauge.[14] Shima Yasujirō (1870–1946), appointed chief engineer in 1918 to lead the project, rightly pointed out the prospect savings from conversion to the standard gauge. He argued that this alterative could provide industry with not only faster service, but more transport capability.[15] The engineering issue generated a heated political battle with a ripple effect in the engineering community. Soon, many civil engineers supported the standard gauge, whereas many railcar engineers embraced the other alternative. The government rejected the standard-gauge plan, effectively marginalizing its supporters along the way.[16]

Nonetheless, the construction of the standard gauge won a top priority in the railroad industry after 1955. This was possible because Sogō Shinji (1884–1981), a staunch advocate for the standard gauge, headed the JNR from 1955 to 1964. As a graduate of the Tokyo University Faculty of Law, Sogō had accumulated management experience in various organizations, including the South Manchurian Railway that used the standard gauge. In 1955, Sogō brought back Shima Hideo (1901–1998), a railway engineer, from a civilian train manufacturer. Shima—a graduate of the Tokyo University Department of Mechanical Engineering—had left the JNR after the catastrophic commuter train accident in April 1951. This highest-ranking engineer in the JNR, advised the technical leadership, but he still lacked the political authority to lead rail projects in any given year.[17] Academically trained as a mechanical engineer chief engineer Shima shared Sogō's view toward the standard gauge; Shima inherited this mind-set from his father, Shima Yasujirō, and Sogō from his protégé, Gotō Shinpei.

The new JNR management adhered to the standard-gauge thinking as a way to increase transport capacity. An economic imperative was crucial. Under Shima's leadership, the Committee for Increasing Transport Capacity on the Tōkaidō Line (Tōkaidō kansen yusō zōkyō chōsakai) was established in May 1956. The 23 committee members held a series of meetings, analyzing the future economic prospects of the transport service but came up with a grim picture. Parts of the major trunk line had already attained 90 percent of their maximum carrying capacity; by 1965, passenger service was expected to increase by 40 percent and cargo operation by 32 percent—which meant an overall increase of 30 to 40 percent.[18] Sogō and Shima tried to build consensus for the construction of the standard gauge. From late 1955 to early 1956, as a self-elected sales representative the JNR president stored a stack of thin pamphlets for the standard gauge in his automobile, hand-delivering them whenever he met pertinent officials.[19] Shima found it convenient to revive the prewar high-speed rail service project based on the standard gauge and electrical railcars.[20]

Their attempts failed, however. The committee proposed plausible alternatives, including doubling all the narrow-gauge government lines, converting to the standard gauge, and electrifying all the rail lines.[21] The impending task focused on how to add rail tracks in the most congested areas, or devising ad hoc, convenient solutions in the railways nationwide. Meanwhile, the JNR leadership announced a highly optimistic plan for high-speed rail service on the Tokyo-Osaka line with a journey to be completed in two and half hours.[22] This planning on paper lacked empirical, solid, technical justification. No one in management had envisioned anything like the way the Shinkansen actually developed at the RTRI after May 1957.[23]

The Last Opposition by Former Military Engineers, 1955–1957

JNR management and RTRI engineers often acted independently to maximize their gains, developing different alternatives for technological development. Their projects had evolved along separate tracks at various points in time. The engineers, with considerable autonomy in what they did and how they did it, saw the trees in the forest; the management saw the forest without seeing the trees in it. What bridged the gap between the wishful thinking of the JNR's top leadership and technological reality was the expertise of several senior engineers at the RTRI. Neither Sogō nor Shima, nor RTRI engineers alone determined the path of technological development leading to the Shinkansen, nor did the technology's internal logic define its somewhat ad hoc, haphazard transformation. Until this point in time, former military engineers remained largely ineffective in promoting the needs for high-speed rail service within the industry. In 1957, they found it necessary to reach out beyond the RTRI for effective, public, and powerful expression of their thinking. The engineers believed that the world could change in their favor: for them, the Shinkansen was the transformative agent, a means by which they could push the technological envelope for faster rail service.

This thinking derived partly from Ōtsuka Seishi, the head of the RTRI from 1949 to 1957. He had graduated from the Tokyo University Faculty of Engineering and had accumulated experience mainly on the factory floor. He encouraged new research initiatives, actively seeking ways to fully use relatively young laboratory leaders. Recalling Ōtsuka's efforts, Matsudaira Tadashi credited him with overseeing the basic research needed for a high-speed train.[24] The director had many qualities that made him a leader, including paternalistic concern for younger associates. For instance, he hired many part-time, young female assistants to perform many time-consuming experiments. Occasionally, he channeled small sums of money from outside to the RTRI, purchased snacks for business meetings, and solicited projects from outside. He eliminated much of the bureaucratic paperwork that had delayed research projects heretofore.[25]

Ōtsuka's efforts accentuated serious disagreements between the RTRI and JNR headquarters. For one, he encouraged the publication of articles about the high-speed rail projects in major newspapers without consulting the headquarters ahead of time. The resulting Toshimaen monorail vehicle and the Odakyū Romance Car created thorny issues within the JNR organization (see Chapter 6). The two discordant parties also remained at loggerheads over how to renew the RTRI. In May 1956, they discussed three alternatives—centralize all the dispersed RTRI facilities and relocate them to Hamamatsuchō or Kunitachi, or decentralize them in three different locations. Any of these choices was considered a way to fortify the research foundation. JNR chief engineer Shima promoted relocation to Kunitachi as the main center for research. But his renewal plan, formulated in 1956 on the assumption of a two-year period of transition, met strong opposition from some RTRI engineers and Ōtsuka who considered Kunitachi an inconvenient venue.[26]

In January 1957, a new RTRI director succeeded Ōtsuka who had retired earlier. Shinohara Takeshi had graduated from the Department of Civil Engineering at Tokyo University, but he was, in his words, "a complete novice in research and development." The JNR headquarters appointed him to the position in part because he had supported the Kunitachi relocation plan. At the time of appointment, he did not know where the RTRI was located. During his first visit, Shinohara was flabbergasted by "the utterly dismal, disorderly state of research and development." One machine to test the strength of a material was 40 years old. "Principal laboratories," his memoir reads, "were small, dark, depressing." He found no researchers in the building but soon learned that they were behind the walls of stacked dusty books, busy conducting experiments. Each laboratory pursued its own research separately. The experience made him think seriously about quitting his directorship altogether. With managerial experience he had accumulated, he proposed a concrete, overarching goal to use able engineers across laboratories more fully. After the arrival of this new director, research blossomed at the RTRI, which had remained as a subsidiary research institute for the JNR headquarters.[27]

Finally, in 1957, real momentum developed at the RTRI for high-speed rail service. The fiftieth anniversary of the RTRI (May 30) provided a timely opportunity to assess social expectations in the matter; the occasion provided the RTRI with "an excuse" to explore the possibilities independently without any prior consent from the JNR headquarters. Shinohara organized a forum that integrated various, separately conducted research projects involving high-speed rail service. In March 1957, 10 senior engineers began to plan for the occasion.[28] By May, the four laboratory leaders—Miki Tadanao, Matsudaira Tadashi, Kawanabe Hajime, and Hoshino Yōichi—concluded that high-speed rail service of less than three hours over the Tokyo-Osaka distance was technologically feasible, regardless of the track width. His task force originally aimed for high-speed rail operations throughout the country, but to avoid a quixotic image, they restricted their plan to intercity service.[29]

This strategy proved effective. On May 30, each of the four senior engineers covered his own domain of technical expertise, proposing in turn an optimistic yet feasible prospect. RTRI director Shinohara Takeshi's opening statement was followed by Miki Tadanao's presentation. The former aeronautical designer rightly emphasized the crucial role of aerodynamically refined, light vehicles for the project. Hoshino Yōichi followed suit, arguing that newly developed steel rails could effectively sustain high-speed travel. Kawanabe Hajime, a former army engineer, proposed a sophisticated signaling system for providing unprecedented safety. Last, Matsudaira Tadashi, a former navy aeronautical engineer, maintained that train vibrations were reducible for the high-speed operation. At the end of the forum, a short documentary film showed an experimental high-speed run in France that established the world record of 331 km/h. Not only the compelling Western example but visual aids, such as diagrams and tables, helped to persuade the lay audience that the proposals were realistic. The two-hour event was a phenomenal success, attracting an audience of more than 500 with some people standing on a rainy Thursday afternoon. Their ovation at the end attested to a social expectation of more convenient, faster three-hour overland travel between Tokyo and Osaka.[30]

This success yielded mixed results within the JNR. First, it generated a backlash from engineering communities outside of the RTRI. In February 1957, the reorganization of the JNR headquarters gave birth to the Temporary Rail-Car Design Office, a subsidiary establishment akin to the RTRI but with the specific task of designing modern railcars. The engineers of the newly created office explored the possibility of high-speed rail service between Tokyo and Osaka, whereby passengers could go and return in one day. Their idea of a "Business Super Express

Train," or Kodama, eventually materialized in November 1958 (chapter 6). But on May 4, 1957, they publicized their idea of a six-hour train ride each way in the national *Asahi Shinbun*.[31] The RTRI's forum on May 30 posed a direct challenge to this concurrent project. About 70 percent of JNR engineers and administrators reportedly opposed the RTRI's proposal.[32]

Meanwhile, the success of the forum created a windfall opportunity for the RTRI to sell its project directly to the JNR top leadership. By invitation, the engineers made the same presentation, this time specifically for JNR president Sogō who could not attend the forum earlier. By that time, Shinohara leaned toward the idea that the internationally standard gauge would help the industry's exports in years to come.[33] This thinking proved convenient for selling the project to Sogō and others who had historically believed in the importance of the standard gauge. During the presentations, his interest lay in the width of the gauge for the project; in fact, he made this clear through direct questioning.[34] At least for the president, the construction of the standard gauge was apparently the end, rather than the means, of the high-speed rail service.

To sell the project, Miki Tadanao, the most outspoken and senior among the four engineers, openly shared his concern about the increased competition from other modes of transportation. Commercial air travel across Japan began in July 1951 with the establishment of Japan Air Lines. The automotive industry was also displaying signs of growth in a revitalized postwar economy. Miki's calculations showed that in light of horsepower divided by weight and speed, high-speed rail service was more efficient than motor transport and air service. Echoing the concern of the railway industry, he directly appealed to Sogō, arguing that newer technology as embodied in the Shinkansen project was the only way to combat this growing competition, salvage the railway industry, and keep it from further decline.[35]

Sogō endorsed the proposal, which had a curious result for the JNR. Soon he successfully garnered political and economic support. The RTRI engineers and the JNR leadership, who wanted to promote technological modernization, had completely agreed on the best means to do so. Meanwhile, the original center-peripheral relationship between the RTRI and the JNR headquarters was maintained officially, but it was unofficially reversed. The RTRI became the center and the leader of the technological transformation, whereas the headquarters provided requisite political and engineering support. In 1961, the JNR established a Shinkansen general office (Shinkansen sōkyoku) with 180 employees. Its bureaus in Tokyo, Shizuoka, Nagoya, and Osaka deployed electrical, civil, and railcar engineers to manage the massive national project at the local level.[36] Many of the technological blueprints, as presented in May 1957, formed the basis of the high-speed rail development in the years that followed. On the basis of a series of meticulous calculations, the original RTRI proposal of 1957 articulated detailed features of high-speed railcars, such as the motor system, the control mechanism, the brake system, the average axle weight, the material used in constructing railcars, car body size, the maximum speed of operation, wheel diameter, the distance between axles, the running gear mechanism, and the signaling system. "Some ideas embodied in the proposal," according to Matsudaira, were "a matter of course in the eyes of aircraft engineers." With some modifications, almost all of the mechanics had remained essentially the same for the next 40 years across the next three generations of the high-speed railcar, including the model 300 that had remained in service until 2012.[37]

As a result of their 1957 proposal, the RTRI engineers gained what they needed: state-of-the-art research and development facilities. Shinohara argued rightly that the world's best rail service would be possible only with the world's best research institute. His original plan included the construction of 10-story high-rise buildings to provide housing for more than 200 families and a recreation center.[38] In October 1958, Sogō and Shima joined the groundreaking ceremony for the new, modern institute in Kunitachi, Tokyo. By September 1959, the RTRI hosted a total of 803 employees, including 18 researchers with doctoral degrees in engineering, 305 employees with bachelor's degrees, and 228 technicians with degrees from technical colleges.[39] The RTRI headquarters, formerly located in Hamamatsuchō, and its three branch offices across Tokyo were all relocated to Kunitachi for more efficient communication among researchers.[40] For the Shinkansen project, the RTRI obtained a new rolling-stock testing plant, which finally replaced the old plant, a 1914 device for testing the performance of steam and electric locomotives. The enormous construction cost of the plant, 300 million yen, attested to the RTRI's commitment to the national project.[41] Soon Japan's first giant electronic computer, Bendix G20, was also installed at the RTRI, rendering its Tiger arithmometers obsolete, and made complex calculations easier and more accurate than before.[42]

Marriage of Convenience: The JNR and the Central Government, 1957–1961

The national project owed less to central planning and policy by technocrats than is generally assumed. The promising engineering project by the RTRI drew together disparate but compatible actors—the engineers, the JNR management, and government officials. Only after a genuine convergence of their interests did the three parties deploy their expertise in developing the mutually beneficial proposal. A national project as massive as the high-speed rail service inherently required financial, political, and moral support from outside of the RTRI engineering communities. The construction was expensive, projected to cost 172.5 million yen (roughly $4.8 million), more than one-tenth of the general account budget for 1957. This enormous expenditure, some claimed, would be a waste in the modern age of highway and airborne transportation. Critics called the Shinkansen project quixotic, equating it to the Great Wall of China, the Pyramids of Egypt, and the giant battleship *Yamato*—all grandiose, but costly, obsolete, and practically meaningless.[43]

One immediate cause that pushed the JNR to rush the high-speed rail project was a lingering court case. The legal dispute began over land use, the genesis of which lay in the Special Measure for Independent Farming (Jisakunō tokubetsu sochihō), a part of the 1946 land reform launched during the occupation era. The legislation was an effort after defeat to democratize landownership. It was designed to help dismantle the pre-1945 landlord system and encourage owner-occupied farming. The legal case during 1957–1958 involved the farmlands that the JNR had obtained from local landowners initially through a loan; the JNR used the property for food production and fed its employees. Under the new land reform law, local landowners sold their property to the Ministry of Agriculture and Forestry in 1953. The JNR, however, continued to plot the lands. This usage, in the eyes of the former landowners, constituted a violation of the land reform law. Subsequently, they filed a civil lawsuit, arguing that the central government had misused the law for its own benefit, and thus, the purchase agreement between the two parties was canceled. In July 1958, the judges in the Supreme Court unanimously supported the plaintiffs. They concluded that the JNR must return all the portions of the landed property to the plaintiff unless it began to use them for "public good."[44]

A more important underlying cause for the JNR to launch the national project was economic: competition from other modes of transportation. Japan Air Lines provided passenger air service between Tokyo and Osaka, and in April 1959, All Nippon Airways gradually moved to enter the market. The planned construction of major highways, such as the one over the Tokyo-Kōbe distance, posed another challenge; once completed, the highway was projected to take away 10 to 19 percent of passenger traffic, and 4 to 5 percent of cargo traffic, from the JNR.[45] Meanwhile, the railway industry was steadily losing its economic vitality. The JNR had kept more than 50 percent of market share both in cargo traffic until 1955 and in passenger traffic until 1960. But it then began to lose share in both markets to the booming auto industry.[46] As a solution, the JNR could introduce faster and reliable trains—not necessarily the fastest one in the world—into commercial services.

From the late 1950s, the JNR management launched a series of ad hoc initiatives to improve the rail structure of the Tōkaidō line. Transport capacity had almost reached its limit. The route length of this line was only 590 kilometers, or about 2.9 percent of the JNR's total rail length, but it was the main artery of the nation's inland traffic, carrying more passengers and freight every year. The line handled 24 percent of the nation's passenger traffic and about 23 percent of freight traffic. Moreover, the districts served by this Tōkaidō line accommodated roughly 40 percent of Japan's entire population and helped to produce more than 60 percent of the nation's total industrial output. The rate of annual increase, both in population and industrial production, was higher in these districts than the nationwide average.[47] The then available railway infrastructure was unable to accommodate projected growth. For instance, when compared to the 1957 figures, the volume of passenger and freight traffic on the Tōkaidō line in 1975 was expected to increase by 200 percent.[48]

The range of plausible solutions remained severely limited. The JNR launched its First Five-Year Plan in 1957, with the estimated expenditure of 598,600 million yen ($1.6 billion) for three tasks: (1) replacing worn-out facilities and rolling stock; (2) providing more comfortable, safer, and faster service; and (3) increasing the transport capacity of the time to 139 percent of passenger and 134 percent of freight traffic.[49] The Shinkansen was not part of the picture at this time. This First Five-Year Plan was aborted in 1960 because it failed to accommodate the accelerated pace of rail traffic growth on the existing railways.[50] The Second Five-Year Plan (1961–66) incorporated the Shinkansen project to fix the issue. Sogō reported in a diet session that the JNR was facing "an urgent need to increase transport capacity in response to [the recent] unexpected rise of demand [in our society] for transportation." Delayed completion of the national project could incapacitate the trunk line as a whole; it would likely paralyze national transport with a serious effect on the national economy.[51]

Within this context, the central government finally embarked on the Shinkansen project. Sogō's nationalistic rhetoric prevailed. On August 30, 1957, the cabinet under Kishi Nobosuke decided to set up a consultative body within the Ministry of Transport to "strengthen [the JNR's] transport capacity and modernize a means of transportation."[52] The resulting Committee for Investigating the Trunk Line (Nihon kokuyū tetsudō kansen chōsakai) consisted of 10 secretaries and 35 experienced members, including Shima Hideo, who emphasized the economic importance of constructing a new standard gauge rail track on the Tōkaidō line. The participants in the January 21 meeting agreed with Shima and discussed how they could sell their idea in the diet more effectively. A representative of the Ministry of Transport asked the committee to "consider how Japanese trains can look modernized in the eyes of foreign visitors." In selling the proposal, appearance was less important than the economic concern over the transport capacity of the Tōkaidō line. The resulting language reflected "careful thinking of the Committee" that factored in both pros and cons of using the preexisting narrow gauge rail track and constructing the standard gauge track.[53] On July 7, the committee submitted a final report to the minister, urging the construction of the high-speed rail service running on a new standard-gauge track with the alternating current electrical system.[54] The committee was dissolved on February 28, 1958.[55]

Meanwhile, Prime Minister Kishi set up a council among the ministers of finance, agriculture and forestry, international trade and industry, transportation, and construction; also included were directors-general of the Board of Hokkaidō Development and the Economic Planning Agency, and the secretary-general.[56] On December 19, 1958, the cabinet approved the council's plan, the "System of Land Transportation between Tokyo and Osaka."[57] Subsequently, the 31st session of the diet on March 31, 1959, approved the budget for the Shinkansen project, appropriating 3 billion yen.[58] Probably no one was more elated by the approval than president Sogō. At the solemn Shintō groundbreaking ceremony at a tunnel the following month, he had the honor to strike the ground with a riding hoe before 80 attendees. At his first powerful swing, a chrysanthemum fell from his chest pocket; then at his third swing, the hoe's head flew off. Shortly thereafter, the participants, including Shima Hideo and the Minister of Transport, shared their joy with beer.[59]

Constructing the new Tōkaidō line would have been impossible had the JNR leadership depended entirely on domestic funds. They strategically sought financial support from the World Bank. The World Bank required that the national government act as guarantor and would be obligated to finance the project continuously until its completion. This financial strategy was based on a proposal from the minister of finance, Satō Eisaku, a former director of the Railway Bureau and later prime minister of Japan (1964–72).[60] In May 1960, the JNR hosted a team from the World Bank that investigated the future prospects of the project. At first, team members were openly dubious about the industry's technical capabilities. Shima then took them on a tour of the RTRI, where they closely observed the state-of-the-art research equipment for relevant engineering fields. This tour

helped the JNR to earn the confidence of the World Bank team.[61] Another chica-
nery worked equally well. Three years earlier, four RTRI engineers had publicly
proposed a 250 km/h high-speed rail operation; subsequently, the RTRI had become the locus of experiments related to that speed. The World Bank, however,
would not invest in any experimental technology projects. To secure the loan,
therefore, the JNR's proposal assumed the modest "maximum speed" of "200
km/h."[62] Eventually, Sogo and the president of the World Bank signed on to the
loan agreement in May 1961 for more than $80 million (28.8 billion yen), the
largest amount ever awarded to Japan.[63]

Reconciling Speed, Safety, and Reliability: Engineers as Problem Solvers, 1957–1963

On March 31, 1963, the Shinkansen successfully established a world speed record
of 256 km/h in a test run. During the years 1957–1963, the Shinkansen remained
experimental, and even the experts had only a vague notion of the physical design
shape it would have. Much of the requisite technology for the high-speed rail service derived from eight research groups at the RTRI. They amply demonstrated
their skills in solving 173 major technical challenges across eight research topics:
25 projects in rails, 29 in vehicle, 17 in operation, 18 in brake system, 25 in overhead wire mechanism, 18 in alternate current system, 19 in signaling system, and
22 in automatic operation system.[64]

The process of solving these problems at the RTRI reflected the engineers'
military lineage in considerable degree (Table 7-1).[65] Eight project leaders—five
of whom were former army engineers and four of whom were former navy engineers—shared specific beliefs about what technology ought to be. Those who
wrestled with this puzzle thought of the vehicle as performing their duty in a
familiar domain of expertise. In the national project, military engineering played
a crucial role in reconciling often competing logics of experience, while crystallizing a set of fundamental relating to speed, safety, and reliability.

Experts looked forward to a new world of speed by enhancing an old one.
Former aeronautical engineers, led by Miki, drew freely on the intellectual capital they had accumulated before and during the war. These professionals faced
major technical challenges in developing a railcar that could run faster than before, even if not the fastest in the world. Nationalistic rhetoric about the "world's
fastest train" was absent in their writings and practice. Their scientific inquiry
underwent no radical change. At the speed of 200 km/h, roughly 70 to 80 percent of the horsepower would be wasted in drag resistance even in a stream-lined
vehicle.[66] A series of wind tunnel experiments helped determine the shape of the

TABLE 7-1
List of the Project Leaders as of 1962

Project Leader	Birth	Wartime Affiliation	RTRI Entry	Education	Research Project
Miki Tadanao	1909	Institute for Navy Aeronautics (1933–45)	1945	Tokyo Univ. Navy Eng.	Railcar structure
Matsudaira Tadashi	1910	Institute for Navy Aeronautics (1934–45)	1945	Tokyo Univ. Navy Eng.	Running gear
Shinohara Yasushi	1916	Institute for Navy Aeronautics (1941–45)	1945	Kyoto Univ. Physics	Automated operation
Ogata Hideto	1915	Navy Technical Research Institute (1938–45)	1945	Osaka Univ. Electrical Eng.	Electricity
Kawanabe Hajime	1914	Army Technical Research Institute (1941–45)	1945	Kyoto Univ. Electrical Eng.	Signaling system
Kumezawa Ikurō	1917	Japanese National Railways (1940–45)	1945	Tokyo Univ. Electrical Eng.	Contact line structure
Kanō Masaru	1910	Japanese National Railways (1935–45)	1945	Tokyo Univ. Mechanical Eng.	Vehicle control
Hirakawa Tomoyuki		Japanese National Railways (1939–45)	1935	Tokyo Univ. Civil Eng.	Track structure

front and rear ends of the high-speed train. The result was phenomenally successful. Following the pattern of development for the Odakyū SE3000 train, as in their wartime years, Professor Tani cooperated with Miki in the theoretical and empirical studies conducted at the ARI. Its wind tunnel provided a controlled environment in which researchers measured air resistance to various train heads, the thickness of the boundary layer, and the distribution of air.[67]

If one considers the importance of wood in Japanese aesthetics and artisanship, one could say that this modern scientific inquiry was surprisingly traditional. For wind tunnel experiments, former military engineers made various clay figures by hand, one of which closely resembled the US passenger airplane Douglas DC 8. At one point, Miki sent one junior engineer to a distant wood-processing plant near the Institute for Navy Aeronautics (INA) for malleable wood; the material proved useful for creating wartime aircraft propellers as well as scale models of high-speed vehicle (Figure 7-1).[68] The result was a major triumph. The drag coefficient of the Odakyu Romance Car was 0.25, better than 0.34 of the Business Express Train Kodama that was developed concurrently at the JNR headquarters from 1957 to 1959. In contrast, the archetype of the Shinkansen train model 0 marked the notable drag coefficient of 0.22.[69] The aerodynamically cleanest and most energy-efficient design at the time capitalized freely on the wealth of wartime expertise in the field, an asset readily available to Miki's laboratory at the RTRI and Professor Tani's research group at Tokyo University.

Expert knowledge in high-speed aerodynamics proved indispensable for rail

Wooden train heads made for wind tunnel experiments. Engineers tested a variety of head shapes in wind tunnels to determine the most aerodynamically clean. Photo courtesy of Railway Technical Research Institute, Tokyo, Japan.

engineers who had known little about this field. A case in point involved the development of a sophisticated structure for the overhead trolley wire. The problem before Kumezawa Ikurō, who was initially unprepared for the task, was how to maintain the contact wire over the entire 515 kilometers distance between Tokyo and Osaka at a fixed height without any undulations.[70] The high-speed rail project goal of 250 km/h surprised him in 1957. It was, to use his words, "a serious matter." The pantograph was to run with the velocity of 250 km/h, or 70 meters per second (m/sec), meaning that the device could jump up and down, through each gantry support set 70 meters apart, at each and every second.[71] Temporary separation of the pantograph from the wire could easily cause an electrical spark, making noise and damaging the wire and contact strip with tremendous heat. The electrical circuits of the train could face power suspensions with damage to the motor's rectifier.[72] In 1957, Kumezawa had just determined how to increase the speed of regular rail operation from 95 km/h to 120 km/h, but the speed of 250 km/h seemed too big a leap.[73]

The research team, led by rail engineers from wartime years such as Kumezawa, turned to experts in high-speed aerodynamics to devise a highly sophisticated contact line structure. For reliable power transmission at high speed, a series of wind tunnel experiments exposed the pantograph to a wind velocity of 100 m/sec. The device employed an attack angle as a way of generating lift at high speed for constant contact with the overhead trolley wire. The pantograph was made small and relatively light,[74] and the resulting mechanism enabled the system to minimize the number of pantographs across the fleet. In the end, only one of the two cars was equipped with this power-collecting device. Also, as a result of wind tunnel experiments, a pantograph was mounted on the car following a thick boundary layer, rather than on the leading car.[75]

Experts in the field of aerodynamics were needed to pursue civil engineering projects, which had remained in the hands of rail engineers. The theoretical study of airflow was indispensable for initiating tunnel construction. Air resistance proved a troublesome issue when a train entered a tunnel and passed alongside another train in that limited space at the speed of 200 km/h. While some useful data were acquired from the Odakyū Romance Car project,[76] many unexplored issues remained. A chief concern was the thickness of the layer of viscous air—called the boundary layer—adjacent to the surface of the running train. This variable determined the space between rail tracks, cross sections of tunnels, and the width of land strips the railways needed to buy from local residents; the sum total of this data dictated the overall construction cost.[77] Because a rough surface militated against high-speed operation, all the tunnels featured a smooth internal surface area.[78]

In pursuit of speed, former military engineers supported civil engineering projects. For instance, former navy engineer, Satō Hiroshi, developed the durable ballast structure after theoretical and empirical investigations.[79] He formulated a highly useful theory about the vibration of rails; using it, he computed the optimum hardness for the rubber pad underneath the rails as an effective shock absorber. The resulting rail structure could support high-speed operation as fast as 300 km/h.[80] Ōta Seisui, a former materials engineer at the INA, assisted the development. During the war, for military aircraft, he had designed organic bulletproof glass—a solid, glasslike material made out of transparent plastic. His accumulated knowledge and experience in polymer science proved valuable in developing the rubber shock absorber.[81] Another important engineer in the field was Yamana Naruo, a former expert in materials engineering at the INA who, after the war, examined the strength and durability of different types of wood for optimum sleepers.[82] The vehicle ran chiefly on steel rail mounted on concrete sleepers; but his product, durable for an average of 20 years, occupied roughly 10 percent of all sleepers on the track, such as bridges and concrete ballast.[83] To improve strength and antiwear characteristics, all steel rails for the track were heat treated, a technique used for developing weapons during the war.[84]

Not surprisingly, the significance and value of speed generated a sharp debate. The first test run over the 37 kilometers distance achieved 70 km/h in June 1962; and, eventually, in March 1963, the vehicle established its high-speed world record at 256 km/h (Figure 7-2). Subsequently, speed was set at a continuous rat-

A high-speed test run of the Shinkansen model o in 1962. Photo courtesy of Railway Technical Research Institute, Tokyo, Japan.

ing of 168 km/h and a maximum at 250 km/h for daily operations.[85] This was not terribly fast, but there was a cultural rejection of speed outside the industry. One month before the commercial service began, a newspaper surveyed 6,000 responses and revealed that 2 percent feared high-speed runs at 210 km/h; they replied that probably they would not ride on the Shinkansen.[86]

In their pursuit of speed at all costs, former military engineers tended to push the technological envelope, resulting in extremes. For the emergency brakes during high-speed run, a team of engineers in Miki's laboratory calculated the effectiveness of using a parachute at the rear end of the fleet;[87] some proposed a jet propulsion of the train against the running direction.[88] Miki himself proposed to use air resistance brakes. When properly positioned, four panels on the roof and two panels on both sides would effectively increase air drag and reduce the velocity of the high-speed train (Figure 7-3). This idea derived from the air brake mechanism he had installed in his wartime bombers. A series of wind tunnel experiments revealed that the drag increased in a fleet of three railcars by 2.3 times. Studies showed that the lighter the vehicle, the more effective the air brake was,

Air resistance brake installed on the scaled model of an experimental train head. A total of six panels—all erected in case of emergency, as shown here—would increase air drag and thereby effectively reduce the speed of the train. Photo courtesy of Railway Technical Research Institute, Tokyo, Japan.

particularly at speed greater than 200 km/h.[89] But the device was useless during a high-speed run in tunnels.

The pursuit of speed was effectively subordinated to safety concerns of the JNR leadership. President Sogō once declared in a magazine for a popular audience that the industry's mission was primarily about speed. Wasting time and space would be minimized as a result and efficiency would improve. This thinking, wrote Sogō, was part of the needed modernization of ground transportation.[90] In a 1959 diet committee meeting, however, he shared a more reflective view, envisioning that the "super-express train . . . [would] stop only at Nagoya, and run the Tokyo-Osaka distance in about 3 hours . . . [and] its [subdued top] speed of about 200 km/h would enable the most efficient and safest 3-hour travel."[91] Credibility was now at the heart of the matter for Sogō. In a 1961 diet session, he said he would "pay particular attention" to the Shinkansen project because "any problems could have international repercussion" and jeopardize the world's "trust in the nation."[92] JNR chief engineer Shima shared this view in part because of his painful personal experience. He had once left the organization under intense social pressure, held responsible for the horrific commuter train accident in April 1951 that left 92 passengers injured, 106 dead. He well understood the fragility of JNR's reputation.

Moreover, the emphasis on speed was compromised by cost considerations, a matter of little concern to the RTRI engineers but weighed heavily on JNR chief engineer Shima's mind.[93] For instance, the original proposal of 1957 had four-seat cross sections, a 3,000 millimeters width for the vehicle to minimize surface area and drag for speed. But eventually for financial reasons, the vehicle was designed for five seats with a width of 3,400 millimeters.[94] Because of this, the floor space of the type o railcar was 40 percent larger than a regular railcar. Consequently, vigorous weight reduction became paramount.[95] The normal weight was not to exceed 60 tons with passengers or approximately 54 tons when empty. The train's front end reflected cost concerns. The more spear shaped the train, the fewer seats could occupy the front car, thus reducing profits from rail service. Originally, a prototype of the type o railcar had a sharper front end.[96] A decade or so after successful operations began in October 1964, the rail industry, now with more financial resources, incorporated light hollow axles and a more needle-sharp front car for velocity.

A combination of wartime Japanese studies in aerodynamics and fragmentary knowledge about American suspension system helped the RTRI to resolve competing issues of speed, safety, and cost. A leading example is the contribution of the wartime experts to solving the flutter phenomenon, self-induced vibrations of the high-speed train. The fastest trains at the time lacked an adequate, flexible device for resisting truck rotation and preventing railcars from wiggling and even derailing. In this situation, Matsudaira's laboratory successfully developed two-axle rolling stock equipped with air-filled bellows and leveling valves. But this air-suspension system was neither new nor original. German engineers had conducted research on it but were unable to reach the stage of practical application. Matsudaira derived an effective suspension system from a mechanism used in American Greyhound buses. He had read an article about the system but was unable to obtain the necessary technical information from the United States. Undaunted, he developed his own mechanism largely through trial and error, applying it to the express trains of the time, Asakaze and Kodama.[97] For the high-speed rail project, his laboratory built a testing stand and empirically observed how a full-scale car truck behaved at high speed.[98] The resulting air spring mechanism proved effective for curtailing both lateral and vertical vibrations of the vehicle, enabling the train to run safely at a speed of 250 km/h.

Especially among those who had lacked substantive research experience in the military, what mattered most was not technical knowledge but esprit de corps of the wartime navy. The experience of an electrical engineer, Hayashi Masami, offers a revealing example. He joined the navy in 1944 after his shortened, compromised education in electrical engineering at the Tokyo Institute of Technology. As a junior engineer, he had gained some exposure to avionics at the INA but lacked substantive experience in his field. At the RTRI after the war, he developed a power supply system that constantly fed an unprecedentedly large amount of electrical power to the running vehicle. His task from 1959 to 1964 proved daunting because a train operation at the speed of 200 km/h required three times as much power as a regular express train did. In fact, the high-speed project adopted electric traction with 25 kilovolts of alternating current. The faster the speed and the greater the electrical power required for the purpose, the more problematic the issue of reliable power feeding became.[99] As a result, electrically, two cars were organized as one unit. The distribution of electric motors across the entire fleet was a way of providing reliable service in case of technical trouble in one car.[100] Hayashi's wartime activities in the navy and postwar research at the RTRI were not closely related; however, referring to the networks of former navy engineers, which supported him during tough times, Hayashi ascribed his postwar accomplishment to "his pride as a (former) member of the navy" and "a sense of solidarity and support" within and outside the RTRI.[101]

Beyond the field of high-speed aerodynamics, former navy engineers contributed to the reliability and clockwork punctuality of the high-speed rail service. Their creation included an elaborate control mechanism, called the centralized traffic control system. It managed all the Shinkansen traffic without any time lag. Equipped with a bird's-eye view, a dispatcher at the central office in Tokyo remotely choreographed rail transit via communication with each train's operator. This automated mechanism placed as many trains as possible on the track, enabling the central command to maximize transport capacity and reduce delays.

Daily rail service was both punctual and reliable and owed its success partly to the laboratory leaders. One was Shinohara Yasushi, a former navy expert in aircraft instrumentation at the INA, the other was Hobara Mitsuo, a former navy electrical engineer. The system they implemented was not entirely the product of their wartime years; it derived initially from the single-track railway community in the United States. The system had effectively freed American train operators from a need to understand complicated timetables and to exchange written messages in the field. It reduced management costs and increased transport capacity, thus appealing to the railway industries in Europe and Japan.[102] From 1964 to 2011, the Shinkansen made approximately 362,000 runs each year, and the average delay per trip was less than one minute, a statistic that includes weather and earthquake-related delays.[103]

While former navy engineers dominated the research and development scene, former army engineers often played a crucial role in improving safety. RTRI research groups relied more on technology, less on people. They rightly concluded that conventional signaling equipment was inadequate for high-speed operation at 200 km/h. The operator's range of vision was compromised at night. It depended dangerously on topography and such weather conditions as fog and rain. Greater velocity would invariably result in less time available for the operator to recognize a distant signal by sight and react appropriately.[104] Naked eyes could not detect any object on rails beyond 1 kilometer.[105] In addition, faster operation inherently required more distance for a train to come to a complete stop. For instance, a fully loaded train running at 200 km/h would require roughly 2 to 3 kilometers to stop.[106]

While speed and safety seemed hard to reconcile at first, Kawanabe Hajime, a wartime expert on physiology audio-signaling, ultimately developed a reliable, electronics-based brake system. In 1957, he successfully combined audible frequency electric wave and warning signals to develop sophisticated track circuits.[107] The resulting automatic train control system was a technological triumph. The speed meter indicated the top speed for each section of the Tokyo-Osaka distance through the track circuits. For the operator, electronic equipment automatically compared the operating speed as well as the speed specified by the signal, immediately bringing the train at a fixed rate of deceleration down to or below the specified velocity.[108] This technological contribution still continues to protect passengers' safety. In 2004, the automated train control system successfully replaced an inattentive operator during the train's 270 km/h operation. When the train came to a complete halt after a safe and automatic deceleration, the chronically narcoleptic operator was fast asleep.[109] Since 1964, approximately 9.2 billion passengers have used the Shinkansen with no fatalities.[110]

Not all the wartime studies proved useful in the high-speed rail project. Amamiya Yoshifumi was a former avionics expert at the INA. At the RTRI, he developed a radar detection system on the basis of Miki's concern for the safety of high-speed runs. He obtained a patent in 1958; his idea, supported by Kawanabe Hajime all along, seemed promising at the time for the high-speed train.[111] The front car transmitted electric waves 20 centimeters above the ground along the rails. Subsequently, the radar detected an object of obstruction on the way, automatically computed the distance between that and the approaching train, and showed the resulting graph on a television screen in the operating room. The problem was the regular electric waves could not handle curves and tended to spread outside the 3 kilometers range.[112] Radar successfully detected an object as small as a soccer ball, but it also detected crows that rested their wings on rails in the midst of rice paddy fields. The radar could not distinguish a rock from a crow, thus failing to provide needed safety for high-speed operation.[113]

Not only military studies of wartime but postwar research in civilian industry proved useful for the project. Postwar aeronautical technology helped reconcile speed and safety by effectively alleviating the impact of a crash. Because the vehicle ran twice as fast as a regular train, the impact of a collision would increase inherently by some 400 percent.[114] For safety, the train track lacked all grade crossings and ran on underpasses or overpasses, contrasting markedly to over a thousand crossings on the regular Tokyo-Osaka line.[115] But this precaution was insufficient. The safety issue became particularly alarming even in the high-speed test section. Some miscreant left rocks on the rails; someone jumped into the path of a running train to commit suicide. Subsequently, the rail guard of the front car, in part designed to prevent lift from generating during a high-speed run, gained more thickness and strength for protection.[116] But another threat came from birds. In 1963, engineers borrowed a giant air gun that proved vital for developing the windshield of Japan's first domestically designed commercial airplane, YS-11. They obtained dead birds, stuffed them with cheap whale meat, and weighed each creation carefully. They simulated the two most common types of bird in the country, a crow (0.8 kilograms) and a black kite (1.2 kilograms). The air gun projected the bird bullets onto tempered glass, providing the engineers with empirical data to decide the optimum thickness of the railcar's windshield.[117]

Sometimes the work of former military engineers required further modifications even after the Shinkansen became commercially operational in October 1964. As Japanese train engineers and some French noted, staff and passengers experienced "a very unpleasant sensation" in their ears when entering tunnels. This was because air pressure changed within any vehicle that lacked an airtight structure.[118] Suggestions to solve the problem included distributing chewing gum to passengers at each tunnel. This idea was soon rejected given the sheer number of tunnels on the Tokyo-Osaka line, 67 in total.[119] The cars subsequently became airtight as in commercial aircraft to protect the passengers from the noxious phenomenon. Useful data for the purpose derived partly from wartime studies on aeronautical medicine.[120]

The resulting vehicle did not fully meet the engineering specifications, and several passengers paid dearly as a result. Because the initially mass-produced railcars lacked the airtight structure in restrooms, some passengers were trapped behind the door that remained shut under a particular air pressure. When the train entered a tunnel and changed the internal air pressure of the confined space, some passengers received an unwanted shower of human waste from the toilet.[121] Only after serious research did the railcars adopt the airtight structure in restrooms. Human waste was stored in the tank underneath the floor and disposed of at the destination station.[122] The result was obviously more pleasant rail service and greater social acceptance of the vehicle.

While the Shinkansen's technology derived heavily from wartime studies, their contribution should not be overstated. Hoshino Yōichi, an expert in rail track structure from the prewar era, amply demonstrated his particular—non-military—skill in the high-speed rail project. The faster the vehicle, the larger the amplitude of rail vibration tended to become, producing critical damage to the rails. Hoshino concluded that the rails for the high-speed operation would require 5 to 10 times more maintenance than did conventional rail sections of 10 meters in length.[123] For the Shinkansen, he developed a new theory according to which each rail would be welded to measure about 1.6 kilometers in length. These sections would then be linked together by expansion joints with double elastic fastenings on prestressed concrete ties. The resulting steel rail was heavy but durable, weighing 53.3 kg/m as compared to 50 kg/m for a conventional rail section. Similar research projects moved on concurrently at the JNR headquarters, which finally supported Hoshino's effort at the RTRI.[124]

The high-speed rail service was possible also because civil engineering and the political cunning of proponents blazed the trail. Many newly constructed tunnels made it possible to connect Tokyo and Osaka via the shortest distance. By design, the curves of the track were gentle for high-speed operation. The smallest radius of curve before then had been 400 meters but that for the Shinkansen was set at 2,500 meters. The standard-gauge line between the two cities was 40 kilometers shorter than the narrow-gauge line; raised ground occupied 44 percent of the distance, viaducts 22 percent, bridges 11 percent, tunnels 13 percent, and cutting sections 9 percent.[125] The speedy construction of this rail line was possible in part because the JNR had already purchased 95 kilometers of the 515 kilometers, or roughly 20 percent of the land for the Tokyo-Osaka line during World War II. Excluding the areas for the construction of bridges and tunnels, acquiring the lands for the remaining line of 325 kilometers was among the most arduous part of the entire project. Probably the most tenacious local landholder among approximately 50,000 in total was a retired professor.[126] As late as in January 1964, he adamantly refused to give up his land at the foot of Mount Fuji, given his personal attachment to his countless tulip bulbs under the ground.[127]

This extreme case signified a larger cultural issue. Shrines were strongholds of local communities that worshiped indigenous Shintō gods. Graveyards at Buddhist temples were the eternal resting place of the ancestral spirits. For this reason, the JNR management perforce took into account not only topography but also the locations of schools, hospitals, shrines, and temples in their planning. Severely limited flat areas in the mountainous nation constrained Shinkansen's route system.

To date, RTRI engineers and JNR management have been unable to solve all geographical and environmental concerns that arose unexpectedly from the high-speed rail service. Aerodynamics provides a salient example. On entering a tunnel, air pressure would increase before the train head, and the resulting air-wave front exited the tunnel as fast as the speed of sound. The RTRI had lacked useful empirical data, and the former aeronautical engineers had examined this phenomenon purely from a theoretical point of view.[128] After commercial operation began in October 1964, the engineers learned that the wave front caused a large, obnoxious popping noise on exiting the tunnel. Local residents expressed their concerns in their protests. Environmental issues, which had not much occupied the minds of the engineers or management, proved serious in the years that followed.

Soon after its introduction, the Shinkansen became a builder of national enthusiasm for modern technology at the expense of its key architects. The public reaped the benefits from the fast, safe, and reliable rail service that had not existed before. Meanwhile, politics obscured their technological accomplishments. JNR president Sogō was held responsible for the fact that total expenditure had far exceeded the original estimate. The projected figure for the project had been originally 30 billion yen, but he intentionally reduced the amount by half to win approval from the diet.[129] This machination backfired with ironic, unintended consequences as the total construction cost increased steadily. The budget estimate was 172.5 billion yen, but the actual cost increased from 197.2 billion yen to 380 billion yen; and in 1963 alone, the project fell short by 90 billion yen. The remarkable economic growth of the time inflated the costs for materials, labor, and land acquisition.[130] When Sogō resigned in 1964, chief engineer Shima and other senior management also left the JNR in protest. By this time, key senior engineers who had gained substantive research experience during World War II, such as Miki, reached their retirement age, left the RTRI, and moved on. Thus, none of them attended the opening ceremony for the Shinkansen on October 1, 1964. The national project, after all, embodied a temporary but successful marriage of convenience among parties on the developmental side, a marriage with longlasting political, economic, and cultural results.

International Competition and High-Speed Rail Services

After its successful commercial operation, the Shinkansen became a cultural symbol of Japan's technological modernity. Children rhythmically praised the high-speed train as they sang the popular song, "The Dream Super Express Train." The Shinkansen not only occupied the minds of the young generation but also appeared in a 1964 commemorative postage stamp. The Shinkansen's economic importance was reinforced by the 36,128 passengers on its first day of operation. Ridership continued to grow thereafter, reaching 10 million after 170 days and 50 million on the 619th day.[131] By the third year, the rail operation had generated a net profit. The success of the high-speed rail operation between Tokyo and Osaka underlay the subsequent project to extend the line farther west. In 1975, the newly constructed line reached Hakata station, 554 kilometers away, in the city of Fukuoka on north Kyūshū.[132]

The train began to carry symbolic, nationalistic meanings outside of Japan after the 1964 Summer Olympics. Between October 10 and 24, 5,152 athletes arrived from 93 countries to compete against one another in Tokyo.[133] This 15-day athletic festival was televised internationally. As Emperor Hirohito announced the opening of the event, the ceremony culturally symbolized the full reintegration of Japan into the postwar international community.[134] Within a year, France developed "a legitimate curiosity about [the Shinkansen] especially after the Tokyo Olympic Games."[135] Soon thereafter, developing nations began to praise the train as a symbol of non-Western technological modernity. From 1972 to 2001, the Shinkansen was featured on postage stamps in nations throughout the world—7 in Asia, 8 in Latin America, and 17 in Africa. In 1972, Umm al-Quwain in the United Arab Emirates and Paraguay featured the train in their postage stamps, followed by Mali (1973), Comoros (1977), and Mongolia (1979). The only European country to follow suit was Hungary (1979). All of the other 32 nations were non-Western developing nations, and most—including North Korea (1981)—shared a colonial past.[136] The Shinkansen came to represent a national, technological marvel outside of Japan.

Technology and national pride seemed more closely tied in France, which had developed its own high-speed rail technology. The National Corporation of French Railways (Société Nationale des chemins de fer français, or SNCF) was the country's state-owned railway company that today operates its high-speed rail network, including Train à Grande Vitesse, TGV. After Great Britain, France had the second-longest history of railway operations in the world, which began in 1832. Mirroring its history, in March 1955, the SNCF established the high-speed world record of 331 km/h in a test run on the standard gauge, using the tractive effort of two electric locomotives.[137] For the next several years, however, the SNCF engineers pursued no further research into high speed trains, until the news about the Shinkansen reached France. In December 1961, a team of SNCF engineers visited the RTRI but initiated no immediate action.[138]

The successful establishment of the Shinkansen in 1964 sent shock waves to the communities of French rail engineers. Among them was Philippe Roumeguère, a chief engineer at the SNCF and later director general of the International Union of Railways. Spurred on by the Shinkansen's success, he traveled to Japan on his honeymoon; what impressed him were Shima Hideo's leadership and the roles of the former military engineers in the bullet-train project.[139] "Undoubtedly," reported a leading French railway magazine, the train set "a landmark in the history of railway operation." Seeing Japan through the lens of techno-nationalism, French observers "felt that every citizen [of Japan] has done his job better in the mood of strong support for the project."[140] Japan's "technically quite extraordinary" train raised the question: "to what extent is the creation of a high-speed train possible in Europe?"[141] Blinded in part by their cultural pride, the SNCF had remained accustomed to seeing itself as the world's leader in rail technology. France had not considered constructing a high-speed rail line, but that changed radically after October 1964. In the eyes of the SNCF, the top speed of the Shinkansen, 250 km/h, was not terribly fast. French trains could reach faster speeds on certain straighter, newer lines. But France lacked a specialized track for a continuous, regular, and reliable high-speed service that could run like a metro commuter train leaving and arriving at very short intervals.[142]

Having lost its prestige in the world, the SNCF responded swiftly. In 1966, it established a modern research department for faster rail service via limited stops between major French cities. In December, the SNCF publicized the national TGV project, C03, which commenced the following year. On May 28, 1967, the press reported on the long-distance, high-speed run project between Paris and Toulouse. Meanwhile, the government laid the path toward the world's fastest commercial train by increasing the maximum speed between Les Aubrais and Vierzon.[143] In France and elsewhere, nationalism thrived in speeches more so than in practice. Jean-Pierre Audoux, director-general of the French Railway Supply Industry Association, spoke before French rail engineers, evoking their sense of nationalistic pride vis-à-vis Japan's high-speed rail service. He said that he had received Christmas cards every year from Japan, half of which featured the Shinkansen—and that the French engineers ought to make their own high-speed train to compete with it. His speech was greeted by applause.[144] In response to the train's success, high-speed rail service such as TGV gradually became a cultural, national icon symbolizing the technical prowess of modern France.

The importance of techno-nationalism in France should not be exaggerated, however. It was a mere driving force among multiple intervening factors that operated for the growth of the nation's high-speed rail project. A more crucial factor than national pride was competition from other modes of transportation. In 1959, Air France launched a commercial jet airliner to revive the French aircraft industry and increase the volume of domestic and international air traffic. An equally serious threat came from the auto industry. Its output doubled from 1952 to 1958, and the total number of private vehicles surged from 1.7 million to 4 million during 1951–1958. The success of the Tōkaidō Shinkansen, however, loomed large in the SNCF. It was partly responsible for the SNCF's decision to create its own high-speed rail line from the mid-1960s.[145]

Japan's pioneering example in 1964 sparked national projects to develop high-speed rail service in Europe. French delegation teams observed the Shinkansen technology and operation in Japan.[146] In June 1968, 349 rail engineers from around the world discussed the prospect of high-speed rail service in the six-day symposium, which recognized Japan's technological success. This event in Vienna was hosted jointly by the International Railway Congress Association and the International Union of Railways. At a high point of the Cold War, the 29 participating nations included both Western democratic states (such as the United States, Canada, Great Britain, France, Italy, and West Germany) and Communist regimes (such as the Soviet Union, East Germany, and Yugoslavia). As a chief architect of the Shinkansen, Matsudaira Tadashi chaired the first of the five sections of the symposium: traction and rolling-stock problems.[147] From the mid-1960s, European nations other than France initiated their own high-speed rail projects. Great Britain, for instance, moved forward with its ill-fated advanced passenger train project. The top speed of 240 km/h, the cruising speed of 160 km/h, seemed technologically possible at the time with the new power source: gas turbines. In early 1969, Italy embarked on its own high-speed rail venture to connect the cities of Florence and Rome across 120 kilometers. This ambitious plan set the train's maximum speed at 250 km/h.[148] An era of truly international competition for the world's fastest rail service began in Europe and then in Asia.

Legacy of War and Defeat

For about a century following the Meiji Restoration of 1868, Japan had to acquire or develop appropriate technology in a very challenging environment. The nation capitalized on both the intended and unintended consequences of war, and especially after World War II, the latter carrying a greater impact on technological development than the former. As war and peace succeeded one another over the decades, the power dynamics changed at multiple levels, ranging from international to national and in regional, local, industrial, institutional, and laboratory settings. In roundabout ways, these changes after defeat prompted various engineering communities to promote the nation's postwar reconstruction, carrying out state-sponsored projects along the way. The result was a haphazard, unsystematic, nonlinear progression of technological development. The conversion of military technology after the defeat was a value-laden, internally conflicting, and contingent process. It did not occur automatically in a preordained, orchestrated manner.

This is not to suggest that Japan's effort to build modern technology was unsystematic or chaotic. From 1868 to 1945, the nation rather systematically trained and recruited a modern engineering workforce for war by using the geographical concentration of resources in key regions. Before, during, and after World War I, a national security imperative underlay the expansion and changing content of modern engineering. The educational and research infrastructures, which remained unscathed from conflicts abroad, supported the country's imperial ambition. The nation's new capital, Tokyo, was an ideal location for all forms of top-down modernization because it had amassed the requisite financial, political, and human resources from the earlier Tokugawa era. One illustrative case was the creation of Tokyo University, from which institutionalized engineering education spread to the periphery.

Engineering education and war reinforced each other. Japan developed its engineering education programs through four successive phases—(1) 1895–1897, (2) 1905–1911, (3) 1918–1924, and (4) 1938–1942—each roughly corresponding to an external conflict that Japan waged. After Tokyo University came Kyoto University (1897), which marked the first phase of engineering education expansion immediately after the Sino-Japanese War (1894–1895). The second phase of expansion followed the Russo-Japanese War (1904–1905), as exemplified by Kyūshū University and technical schools in urban centers across the country. World War I (1914–1918) yielded a third wave of expansion between 1918 and 1924. The national security imperative strengthened engineering education at imperial universities in Tokyo, Kyoto, and Kyūshū. Meanwhile, Hokkaidō University and Tōhoku University built solid, modern engineering programs. Incrementally, systematic education also became available at technical schools across the country.

In respect to scientific and technological development, each international conflict until 1937 was, to use the words of John Dower, a "useful war" for Japan.[1] As the nation moved from war to war, its engineering education was severely tested by conflicts abroad. For the Asian latecomer to the age of imperialism, each war conveniently meant the rationale to expand engineering education across the country. World War I in particular reduced the nation's financial constraints on engineering education, legitimizing aeronautics as an academic field of study for national security. The war, which had left Japan's homeland unscathed, had been the most "useful" in Dower's sense. It had stimulated the research interests of Japanese civilian and military officials to build research and development institutions, especially for air power. Driven by a national security imperative, the military turned the destructive European war to constructive use; it choreographed technology transfer from overseas, devising a system in which foreign engineers competed with one another for Japanese civilian aircraft companies to advance the field of aeronautics.

What substantially affected Japan's technological development before, during, and even after 1945 was interservice rivalry in the imperial military, which ironically undermined the nation's capability to wage war abroad. Despite, or because of, threats from the West, the army and navy contended with each other in a quiet, subterranean war at home that surfaced by the 1930s. In the end, the navy prevailed over the army by recruiting elite students through maneuvering; it often "stole" promising engineers directly from the army, while Japan had chronically suffered from a shortage of such manpower in support of imperialism. The two services had erected barriers between them, hindering the spread of engineering knowledge across the board.

The fast rise of the navy air power stemmed in part from the navy's victory over the army in the war at home as much as from its engineering culture, as epitomized at the Institute for Navy Aeronautics (INA). The organization's most remarkable asset was its engineers. The navy trained its own engineering workforce for war, whereas the army outsourced its research and development projects and neglected its human resources. Before 1942, navy engineers could afford ample time, energy, and material resources for their relentless pursuit of excellence at all costs, advancing basic research with relatively few financial or time constraints. The navy's engineering culture was sustainable in the research atmosphere of optimism.

Japan's military engineering revealed its inherent weakness after 1943, the point of an irreversible shift from optimism to pessimism in the research and development environment. One key cause was the perennial shortage of engineers in society. As exemplified in the Central Aeronautical Research Institute (CARI), the nation's effort to mass-produce engineers hurriedly proved counterproductive. Engineering education was severely compromised at all levels across the country, from the top imperial universities to technical schools in remote areas. Likewise, engineering infrastructure faced dire challenges from late 1944. The Allies' assault from the air, real or imagined, successfully crippled the Japanese strategy of resource concentration in Tokyo for research and development, dispersing key military facilities from the capital to remote areas.

No technology other than the kamikaze attacker, MXY7 evidenced the degradation of engineering culture as well as the pursuit of excellence within engineering communities. The shift in the research atmosphere from pessimism to desperation gave birth to the wonder weapon that signified a paradox: it embodied the engineers' nationalistic support for the unethical military operation as much as their ethical, humanistic drive to maximize the aviator's chance of survival. This tragic weapon was a product of desperation, professionalism, secrecy, and autonomy within the engineering communities—all present at the INA. The degrees of autonomy and professionalism that the military engineers displayed would resurface more successfully in the postwar years.

In the summer of 1945, defeat set the international, national, regional, and local contexts for the subsequent, somewhat idiosyncratic, situational transformation of wartime technology, the use and value of which reflected the new peacetime order. The emergence of technology unrelated to weapons but with a

military legacy stemmed in part from the drastically reconfigured power relationships and cultures after defeat. From 1945 to 1952, the Allies' occupation policy
ended the old regime, demilitarizing Japan and its engineers, acting on genuine
aspiration for peace among millions of ordinary people. The nation's first experience of defeat discredited technology and engineers for war; it brought about
metamorphosis that constructively turned them to something useful and meaningful for peace. Defeat had a far-reaching, unintended yet overall positive impact
on the patterns of the diffusion of engineering knowledge as well as engineers'
migration in both the domestic and international settings.

The Cold War was a blessing in disguise for Japan's post-1945 reconstruction.
A few counterexamples notwithstanding, the geopolitics in East Asia during the
early phase of the war pressured Japanese engineers to remain at home. Economic and legal barriers alike barred their departure. Sociocultural norms also
tended strongly to keep them at home, unlike their German counterparts. With
neither the means nor the opportunities to pursue an exodus, Japan's former
military engineers were effectively contained at home and only migrated domestically into the civilian sector. Their lack of migration abroad in any significant
numbers during the occupation era ensured that their wartime expertise was
passed on exclusively within Japan.

In the domestic sphere, defeat allowed ordinary citizens, such as former military engineers, to inaugurate an era of change in the common technologies of
daily life. Among them was the rail system operated by the Japanese National
Railways (JNR). It embodied neither the pattern of a linear progression of technology, nor the wartime techno-nationalism that had pervaded state-sponsored
technological projects. From 1945 to 1952, Japanese society—to be more specific,
the occupation authority, thousands of ordinary rail passengers, and railway engineering communities—shared a mundane, collective, and peace-oriented vision:
safer, more reliable, and comfortable daily rail service. The rail infrastructure
that had survived the war remained damaged, jeopardized the lives of riders, and
alarmed the occupation authority. Rampant train accidents created a niche in
the JNR for the arrival of former military engineers who Japan's unconditional
surrender had released into the job market. Only during the occupation era did
various engineering groups mingle; consequently, this enabled a successful marriage of engineering cultures between that of former military engineers and that
of preexisting rail engineers for the postwar development of the high-speed rail
service.

The end product Shinkansen represented the impact of war and peace that altered power structures and cultures in the national, regional, local, institutional,

and laboratory settings. Most of the pre-1945 nationalistic aspiration for war and its related technology was discredited. This reorientation of wartime values lay at the heart of the almost voluntary demilitarization of cities across Japan, especially Yokosuka, that had hosted military facilities during the war. The local railway and tourist industries welcomed the social conversion because it revived the regional economy. Meanwhile, defeat reconfigured the engineering communities in the rail industry. One result was cultural tensions that erupted between preexisting rail engineers and former aircraft engineers, as typified at the Railway Technical Research Institute (RTRI). The former military engineers launched a series of opposition movements against the preexisting hierarchical structure between the peripheral RTRI and the JNR headquarters. They had solved technical problems, gained autonomy, and established themselves firmly in the JNR—and their tool was the engineering knowledge that they had accumulated during the war. These engineers began to reshape their fields of study, power relationships in the JNR organization and in the rail industry, and the national project of the high-speed rail service.

While the Shinkansen was a prominent consequence from the successive experiences of war and peace, its success owed less to central planning and management than is generally assumed. The project eventually bore fruit after the RTRI engineers, the JNR management, and government officials began to put different, yet reconcilable, interests on the table. What unified the parties on the development side was something concrete, practical, and beneficial in the peace-loving era. Redundant projects and the cultural tensions that had existed in the industry subsided with this development. The Shinkansen embodied the independent judgments of the rail engineers and leadership, the reconfigured power structure in the railway industry, and the political economy within which the engineers, management, and politicians made their choices. The Shinkansen was the result of the extensive, unintended impact of military defeat on value-laden, nonweapon technological development now suited for peacetime use. It was a technological and cultural product of defeat.

Trans-war History of Technology: Continuity and Discontinuity

This sociocultural history of Japanese technology presented the key legacies of war bequeathed to the defeated nation. In many ways, the summer of 1945 drew the line of demarcation between war-oriented, militarized technology and society on the one hand and its peace-oriented, demilitarized counterparts on the other. On one level, the experiences of defeat transformed the thinking, lifestyles, and communities of thousands of ordinary citizens, such as engineers, in postwar

Japan. On another level, however, people resumed how and what they had done with some brief interludes. The technological transformation from 1868 to 1964 drew on the tension between the continuity and discontinuity.

Compared to the effects of Japan's military victories before 1941, its defeat had a more far-reaching, often ironic, impact on the transformation of culturally coded technology. World War II lasted longer than any of the previous external conflicts, and the war engulfed the motherland for the first time in the nation's history. The unconditional surrender in the summer of 1945 drastically altered its geopolitical, economic, and sociotechnical landscape. Defeat in fact reconfigured the power relations of East Asia, Japan as a whole, Tokyo and its adjacent regions, the railway industry, and research laboratories. Meanwhile, ordinary citizens—including former military engineers—developed new power relationships from the bottom up. Japan's defeat fortuitously empowered the communities of former military engineers for postwar buildup. In a roundabout, ironic way, the defeat rejuvenated the former wartime technology, aeronautical engineering, for the Shinkansen.

Defeat also functioned as a catalyst, giving birth to a new hybrid engineering culture for peacetime society. It merged the theory-driven engineering style of the former aeronautical engineers with the empirically based style of automotive and, above all, rail engineering. This transformation was possible because of Japan's unconditional surrender, a circumstance that dissolved the entire military structure and released wartime engineers into the job market. The two engineering communities experienced cultural tensions, which were healthy enough to generate useful competition in the national high-speed rail service project. These tensions signified a certain cultural discontinuity that supported Japan's postwar reconstruction.

One striking element of continuity was Japan's technology transfer from abroad. Building modern technology for war and peace required the nation to borrow effectively from the West. In the first half of the twentieth century, Japan had needed various foreign models to jump-start the field of aircraft engineering and build air power for war. Technology transfer from abroad, however, was chancy and expensive. Only the state, or big business, could afford the expense of retaining foreign engineers and purchasing license agreements. While not all foreign models yielded concrete benefit, the top-down planning and execution of technology transfer from abroad proved successful for the most part. Systematically managing the flows and volumes of foreign technology was critical for a country that lacked the option of incidentally diffusing engineering knowledge from abroad given Japan's geographical isolation.

World War II in particular exposed the nation's vulnerable geographical configuration. The Japanese-American war almost completely separated Japanese engineers from their counterparts in the West, interrupting much of the flow of technical information and engineers from overseas. Defeat ended this isolation, but little else changed. Japan still shared no land borders with other countries, and its geography tended to restrict the entry of foreign expertise. Before and after 1945, Japan continued to devise some kind of artificial, institutional framework to facilitate the technology transfer for its state-sponsored projects. Foreign experience retained its importance in the aircraft and railway industries as much as elsewhere, within and outside of Japan.

In retrospect, the participants in the technology transfer were partly responsible for the potent, lingering image of Japan as a copier of foreign technology. They documented their paid visits abroad by keeping written records of on-site observations, meetings, budgets, and conversations. Vigilant over the flows of foreign expertise and scarce hard currency, government offices produced other quantifiable, widely used data, such as the number of approved technology contracts. Meanwhile, the informal, contingent, and unplanned diffusion of engineering knowledge in the country had less chance to be documented and, thus, studied by later historians. This is unfortunate because the diffusion of technology and any explicitly planned technology transfer were equally important for Japan as they were elsewhere.

Such was the case in the Japanese railway industry after 1945. The end of the war meant the end of many artificial barriers that had impeded the unplanned diffusion of engineering knowledge across the country. Compared to explicit technology transfer, the course and impact of the technology diffusion was less predictable, visible, or traceable. Yet, defeat offered a cheaper, more readily available option than technology transfer from abroad. While the nation's geographic isolation hindered the transfer of foreign technology, it aided technology diffusion by containing much of the nation's expertise within the country. In the national high-speed rail service project, the most successful technology had less to do with any deliberate, intended, top-down attempt than an unplanned, informal, idiosyncratic diffusion of engineering knowledge at the grassroots level.

In trans-war Japan, there was a more obvious linkage at the organizational level: key institutions remained with different labels or remained in less visible ways. For instance pre-1945 imperial universities, especially Tokyo University, maintained their prestige as the nation's top academic institutions. Their old-boy networks remained strong within and outside of the respective regions. Postwar research institutions in Japan carried less visible prewar legacy unlike compa-rable cases in the United States. For instance National Advisory Committee for Aeronautics, created in 1915 for America's aeronautical research, grew massively into NASA (1958–present) in the midst of the Cold War. In Japan, except for the RTRI that had existed after 1907, many prewar research institutions underwent phases of more radical reorganization, contraction, and expansion under dissim-ilar headings. The Aeronautical Research Institute, for instance, was dissolved in 1945 during the Allied occupation; by 1964, however, it had been reestablished as the nation's research center for aerospace technology, and today, it functions as the Research Center for Advanced Science and Technology (1987–present). The INA ceased to exist in late 1945, but its material assets, such as land, plant, and equipment proved useful and productive for the growth of heavy industries in the area. The wartime CARI formed the basis of the Transportation Technology Research Center (1945–1963), the Research Center for Marine Technology (1963–2001), and the National Maritime Research Institute (2001–present). Likewise, in Germany, the Kaiser Wilhelm Institute for Flow Research (1924–1948), originally created for studying aerodynamics and hydrodynamics, formed the basis of the present Max Planck Institute for Dynamics and Self-Organization.

The continuity of human resources is another obvious linkage, though it remains insufficiently understood. Historians have noted the visible continua-tion of personnel in the government bureaucracy who contributed to postwar reconstruction.[2] The nominal existence of particular individuals in pre-1945 Ja-pan, however, does not necessarily signify their substantive contributions after the war. For instance, in the field of engineering, many young associates moved from the CARI to the RTRI after the summer of 1945 (chapter 6), but as ju-nior researchers, they had accumulated little concrete research experience in the fledging research facility, consequently making few contributions to postwar re-construction. The army offers a starker example. At least a few hundred former army aeronautical engineers survived into the postwar era (chapters 2 and 6), but they had lacked substantive research experience before 1945, and correspond-ingly, their presence remained negligible in the postwar technological landscape. Many cases of navy aeronautical engineers suggest an underlying cause. Those who had undertaken research and development as senior researchers *before* 1942 were the most educated and best trained, and thus, they had a visible and pro-found impact on the postwar technological transformation. This so-called pre-1942 generation of navy engineers exhibited its skills by minimizing the drag resistance, structural weight, and the vibrations of the parts in wartime aircraft and the postwar railway industries. A closer look into the continuity of human resources reveals that age and the quality of experience mattered. To use Dower's word, the war before 1942 was more "useful" than that after the year.

One other key continuity in trans-war society was how the engineering communities responded to the call for their service in national projects. Their technical knowledge proved indispensable for building Japan's air power before 1945 *and* the nation's high-speed rail service after the war. Minor institutional differences aside, the engineering communities had consistently displayed their loyalty to the professional pursuit of excellence, not necessarily for the sake of a collective vision of state-sponsored project, or ideology of the nation as a whole, if any. Their autonomy in the national projects was partly a product of their lifetime employment in their state-sponsored research establishments; lacking a fear of losing their jobs, they freely, stubbornly, and continuously exercised their autonomy as part of—and often against—the ongoing national projects in wartime and peacetime. As technical problem solvers in Japan as elsewhere, they were by no means faceless, voiceless, or powerless. The engineering communities were far more pragmatic and far less ideological, or even nationalistic than previously portrayed in the literature; in this sense, their rhetoric and action lacked anything inherently exotic and stereotypically quintessential to Japan, such as "samurai values" or the Japanese *wakon* spirit.[3] The transformation of engineering knowledge was more situational, fluid, and contingent and less "cultural" than what essentialism claims.

In part because of the engineering communities' loyalty to their profession, not necessarily to their nation, internal cultural conflicts were the norm in the national projects of trans-war Japan. The engineering communities of the army and navy quietly fought against each other until Japan's defeat. After 1945, somewhat similar tensions gave birth to different high-speed rail projects—three coevolving in parallel—within the national rail industry. The Shinkansen, after all, was a national project by heterogeneous engineering communities with little, if any, unifying sentiment of techno-nationalism. Ironically, however, only after the success of the project did nationalistic fervor emerge, first, outside of Japan after 1964, which then echoed within the nation.

The year 1964 marked a new phase in the global competition to set a record for the fastest speed in commercial rail service. Soon after the success of the Shinkansen project, Japanese engineers employed the radically new method of magnetic levitation in rail transport. Motor engines were no longer the power source.

Superconducting electromagnets would levitate and, with a linear motor, propel the train with neither sound nor friction between physical properties. This system was expected to solve the technical difficulties inherent in high-speed rail runs around the world, an issue resulting from the physical contact between the rails and wheels, and between the overhead trolley wire and pantographs.[4] In March 1955, for instance, French engineers successfully established the world's high-speed record of 331 km/h in a test run, but the mechanical contact ruined the rails and overhead wire.[5]

Research and development of the magnetic levitation system had a quiet beginning at the RTRI in the early 1960s. Hara Tomoshige, a former navy engineer, began a series of tests for the high-speed run in his laboratory. Born in 1910, he had earned his degree in physics at Tokyo University (1936) and had examined the airflows of the navy dive-bomber D4Y Judy at the INA during World War II. Three other researchers, all of whom earned a degree in physics before 1945, supported the research in his laboratory.[6] Together, they built small-scale models and studied the physical characteristics of high-speed runs.[7] This research on aerodynamics merged with the theoretical studies about superconductivity that flourished in Europe at the time. Once introduced in Japan, the idea of a magnetically levitated train seemed useful to solve the environmental problems that arose from the operation of the Shinkansen, namely, the noise and vibrations felt among local residents. Soon the RTRI engineers explored technical possibilities in their study groups.[8] In 1965, the so-called linear motor car project became "the most important research topic" at the RTRI.[9]

The JNR headquarters supported the research effort with an increasingly nationalistic overtone. In 1966, the "High-Speed Rail Study Club" was established in the organization. The two separate study groups, one at the RTRI and the other at the JNR headquarters, theretofore moving along in parallel, together endorsed the idea of a magnetic levitated train in 1969. The national project officially began the following year. The JNR built a test track across 7 kilometers in Miyazaki prefecture on southern Kyūshū. In December 1979, an unmanned test train ran successfully at the speed of 517 km/h.[10] Observing the project, the minister of state, Fukunaga Kenji, declared during a diet session that "Japan is advancing far ahead of Western nations, which keep watch on our linear motor [technology]." He strongly argued that "Japan should put this [experimental] technology to practical use." He then went on to say, "once Japan beats foreign countries [on the ground of technology], they shower abuse [on Japan] . . . I believe Japan will lead the technology as absolutely the best in the field, [and] the

Japanese race needs to demonstrate its latent strength as this [national] project grows."[11] Hence, the magnetic levitation technology began to symbolize the ethnic pride of the Japanese.

Any strong political support, such as Fukunaga's, was welcomed by the JNR, which was then operating in the red. The Shinkansen between Tokyo and Osaka was generating revenue, but other rail operations were not. The year 1964 marked the beginning of many successive years of deficit, which worsened year after year.[12] From 1971, the JNR's expenditure (excluding depreciation) exceeded its revenue; without a capital reserve, investment was possible only by borrowing. The JNR raised fares four times in the 1960s and seven times in the 1970s to increase revenue. But this initiative backfired. Passenger traffic peaked in 1974, and cargo traffic in 1970, losing many long- and medium-distance rail passengers to road and air transport.[13] The JNR nonetheless constructed the Shinkansen line from Osaka to Hakata in southwestern Japan, although the density of the passenger traffic on the extended line was only 40 percent of that on the Tokyo-Osaka line. In 1970, the government endorsed the JNR's massive project to extend the high-speed rail networks to the north (Jōetsu Shinkansen) and northeast (Tōhoku Shinkansen). The JNR's hesitation to downsize staff levels and close rural lines aggravated its financial health. By 1975, the JNR had lost its ability to recover.[14]

The JNR's massive deficit jeopardized the ongoing magnetically levitated train project, alarming its architect, Kyōtani Yoshihiro, who presented a new generation of engineers without wartime experience. Born in 1926, he held a degree from Kyoto University, and from 1948, he advanced his career inside and outside of the Ministry of Transport. At the JNR, he carefully compared the high-speed rail projects in France, Great Britain, Spain, and West Germany among those in other nations. He argued that Japan's research and development efforts for high-speed rail service were diffused, inadequate, and poorly funded. "It is a matter of fact," he concluded in his 1971 co-authored book that Japan was "regrettably lagging behind in the world." The rhetoric of "catching up with the West," resurrected from the pre-1945 era, conveniently echoed in support of his ambitious research. Making no reference to competitions from road and air transport within Japan, he asked for massive financial support: 31.5 billion yen for research and development and 1 trillion yen for constructing the infrastructure of the linear motor car system, which would be "the project of the Japanese people."[15]

Kyōtani's subsequent action acquired more "nationalistic" fervor as he directly appealed to the Nippon Foundation for financial support. It was the private, non-profit organization established in 1962 by Sasagawa Ryōichi, a philanthropist and ultranationalist who had advocated Japan's expansion during World War II and consequently served in prison as a Class A war criminal. Kyōtani's voice entered national politics soon thereafter. In 1984, the chairman of the foundation personally relayed the plea for money to the minister of transport, Ishihara Shintarō. Ishihara had founded the Seirankai (Blue Tempest Society) in 1972 and affirmed his commitment to Japan's national defense as the leader of the hawkish, right wing of the ruling Liberal Democratic Party.[16] His nationalistic rhetoric and Kyōtani's engineering project began to reinforce each other.

By this time, Ishihara had good reason to support the JNR's high-speed linear motor car project. In February 1981, French engineers successfully established the high-speed world record of 380 km/h on a stretch of rail south of Paris. Seven months later, the Train à Grande Vitesse (TGV) high-speed rail operation commenced between Paris and Lyon. At the opening ceremony, the newly elected president of France, François Mitterrand, praised France's technological innovation as exemplified in the TGV. He proudly declared that the train was "a sign for the whole world that France intends to remain a great, innovative nation."[17] While the TGV redeemed France's prestige internationally, German engineers challenged its high-speed world record in their own high-speed rail operation, InterCity Express. Meanwhile, they advanced their magnetically levitated train project. The TGV's phenomenal success and the Germans' effort alarmed Japanese engineering communities. Many pundits began to voice their concerns, claiming "the Shinkansen is no longer the best in the world" and "France surpassed Japan in the field of [railway] technology."[18] The nationalistic slogan of historic origin, "Catch up with the West, and surpass the West" resonated in the rail engineering communities amid an international competition for the world's fastest rail operation.

No politician was likely more committed to the nationalistic cause of the international competition than Ishihara. During a 1987 diet meeting, the minister of state argued that Japan needed to develop its linear-motor magnetic levitation technology. "The Germans are catching up," Ishihara said, and Japan "needs to work on it without a moment's delay, so as not be beaten [by them]." The Ministry of Transport would, hence, "devote itself fully to the development of the technology of the next generation." His rhetoric intensified after an observation trip to research facilities abroad. The German research infrastructure used larger areas for testing, but he returned home fully convinced that "technologically, Japan is superior to Germany."[19] Meanwhile, the United States began planning for a high-speed rail project between Las Vegas and Los Angeles, asking a third nation, Canada, to assess the superconduction engineering of Germany and Japan.[20] The result was, to use Ishihara's words, "utterly regrettable." During a 1988 diet session, he proposed that Japan set up a new research facility, put the linear motor car, "the dream of all the Japanese," to commercial use, and promote it as a nonpartisan, national project.[21] Nationalism echoed loudly in speech and action.[22] Setting aside the Shinkansen's lineage of 1945–1964, Japan's high-speed magnetic levitated train project acquired a nationalistic legitimacy that resonates to the present.

The socio-technological transformation in many countries has been treated as the product of some sort of artificial and inorganic system—be it military, political, economic, social, or environmental. With few exceptions, recent scholarship subscribes to some form of top-down verticality, focusing on central civilian/ military leadership, government bureaucrats, and technocrats as the "shapers" of the national and technological developments. While paying careful attention to the top-down system approach, I chose to examine Japan's trans-war history through the lens of engineers. The engineers to whom I refer are neither technocrats, nor intellectual writers, nor business leaders who voiced their opinions broadly in writing and in media. In many ways, the engineers whose names appear in this book are different from renowned, vocal activists in political, scientific and engineering communities who actively constructed the international science or national ideology. They instead are ordinary citizens of Japan or part of the "silent majority." More specifically, my empirical study relies heavily on accounts of former military engineers who experienced the militarization followed by the demilitarization of Japan. Trained to solve technical problems—seeing the details in the bigger picture, seen mistakenly as passive agents of technological transformation—many of these engineers have often left traces of their thoughts in a series of design drafts, diaries, memos, minutes, detailed calculations, and operation manuals. These sources are difficult to obtain and decipher, but they reveal more than the technical challenges that these individuals faced; they reveal the engineers' struggles, remorse, and joy in (re)constructing technology and culture for different needs in society, while forming their own identities and communities for engineering war and peace. Thus, my study employs a bottom-up approach as well as archival, nonarchival, and biographical sources. Engineers in this study seem to loom larger than usual. To be sure, neither technology nor engineers solely drive history, nor should they offer a simple and singular panacea to our understanding of the complex interplay involved in technological change.

Tracking the career paths of the engineers was possible mainly because public institutions kept their directories. My focus is directed at their education as well as research and development projects in public institutions. Private organizations failed to record such information or kept such information, but did not disclose it for the sake of privacy. The directories I gathered from archives and personal networks are rich with information, including engineers' names, dates of birth, addresses, phone numbers, military ranks/positions in the institutions, educational backgrounds (i.e., fields of study and years of graduation), and the year in which they retired. Access to these types of records is normally highly restricted. But they are detailed, reliable, and useful when they become accessible through various means. Most of the engineers discussed in my study are highly educated with baccalaureate degrees from the nation's top universities. Some earned doctoral degrees in engineering. Information about technicians and craftsmen without college degrees is seldom available, or it is not detailed enough to observe career transitions in trans–World War II Japan. Thus, these experts do not occupy the center stage of my study.

My study also relies on personal interviews with former military engineers and/or their relatives. Direct interviewing is inherently problematic. People often craft their own history. They are always right in their own ways. Informants remember what they want to remember and, thus, knowingly or unknowingly, recall what they want the interviewer to remember and record. Memory is inherently fragile; it fades, often conveniently. Remembering and forgetting are equally motivated; each offers clues about who the informants are.

Yet interviews are resourceful and useful when used cautiously. Sources, written or otherwise, can be checked against one another. Different perceptions of the "same" events can be compared, assessed, and put together to grasp a fuller picture of the "same" event. For these reasons, I directly interviewed several key trustworthy individuals at least four times over two years, while consulting other evidence that supported or undermined their stories. First, direct interviewing was always fraught with issues, such as a lack of a rapport between the informants and the interviewer. Some former wartime engineers were more anti-American than nationalistic; they harbored lingering anti-American sentiment, and some were suspicious of my affiliation with American institutions. While the first direct interview usually lacked reliable information, it did function as the gambit, or the first step to build a rapport with the informants for the next two years or more. My background—a Japanese native born in a nearby city and a native speaker of the language—helped build the necessary rapport over the two-year period. In their seventies and eighties, the informants were willing to share once the rapport had been established. By the third follow-up interview, after a year or so, they candidly admitted inconsistency in their recollection of the past when faced with evidence that I had brought from other informants and/or written sources. I always sought records written *before* 1964, the starting year of the Shinkansen's commercial operation, or sought the voices of different informants. In many ways, what some informants told me was less useful than what they had *not* shared with me.

Individual memory is as fragile and malleable as collective memory. Like interviewees, institutions such as private enterprises and government establishments can craft their own history, writing what they want to be remembered and how they want to be remembered. Especially in the case of company history, "success story" appears altogether too often in their presentations of what happened in the past. Rarely, if ever, do we see any "failures" of business undertakings in presentations of company history in Japan. For my study, I cautiously assumed that before 1964 no one had a preordained vision of the eventual outcome of the Shinkansen high-speed rail service. Many readily available sources, such as memoirs by bureaucrats and institutional/company histories, were typically published after the success of the rail operation. Such highly personalized and/ or official histories commonly look back at the technological development of the Shinkansen from a presentist perspective, crediting the "foresight" and "long-term thinking" of the top leadership. A great many studies in Japan subscribe to this notion to some degree; after all, strong, top-down power structure needs to be placed in order to coordinate railcars, passengers, stations, tracks, and ticketing among others on a daily basis. A more reasonable understanding of the technological transformation presents the development as it unfolded *before* that year. To avoid a "retrospective," or "after the fact," reconfiguration of complex realities, I disregarded many official historical accounts prepared by the Japanese National Railways after 1964; I used their information only when it corroborated accounts published before that year.

List of Informants

Names/dates	Means of communication
Akatsuka, Takeo	
December 10, 2002	Letter
Aoki, Yoshirō	
May 10, 2004	Telephone
Enomoto, Shinsuke	
April 5, 2004	Letter
May 10, 2004	Letter
June 2, 2004	Letter
Fujita, Kōmei	
April 7, 2004	Letter
June 25, 2004	Letter
Hirose, Kengo	
April 5, 2004	Letter
Kanō, Yasuo	
March 22, 2004	Oral interview in Tokyo
Kawanabe, Hajime	
April 3, 2004	Letter
Kimura, Shōichi	
May 25, 2004	Letter
June 16, 2004	Letter
July 7, 2004	Letter
Kumezawa, Yukiko	
March 6, 2004	Letter
March 22, 2004	Letter
Miki, Tadanao	
December 22, 2002	Oral interview in Zushi
June 19, 2003	Oral interview in Zushi
August 1, 2003	Oral interview in Zushi
August 8, 2003	Oral interview in Zushi
Nakamura, Hiroshi	
March 14, 2004	Telephone
March 17, 2004	Telephone
March 24, 2004	Telephone
March 28, 2004	Letter
April 29, 2004	Oral interview in Kyoto
June 4, 2004	Telephone
August 9, 2004	Telephone

Nakamura, Kazuo
 November 7, 2002   Oral interview in Tokyo
 January 16, 2003   Letter
 January 21, 2003   Letter
 March 3, 2004   Letter
 March 8, 2004   Letter
Shinohara, Osamu
 March 11, 2004   Letter
Takabayashi, Morihisa
 November 25, 2002   Letter
 December 14, 2002   Letter
 June 18, 2003   Letter
 July 7, 2003   Letter
 July 17, 2003   Letter
 July 22, 2003   Letter
 July 26, 2003   Letter
 August 4, 2003   Letter
 August 14, 2003   Letter
 February 20, 2004   Letter
 April 21, 2004   Letter
 July 23, 2004   Letter
Tanaka, Shin'ichi
 November 28, 2002   Oral interview in Tokyo
Ueda, Toshio
 April 29, 2004   Oral interview in Tokyo
Umeno, Takeyasu
 April 13, 2004   E-mail
 April 14, 2004   E-mail
 April 18, 2004   E-mail
 April 26, 2004   E-mail
 May 1, 2004   E-mail
Watanabe, Saburō
 July 16, 2003   Oral interview in Tokyo

Note: Anonymous informants are excluded.

Introduction • Technology and Culture, War and Peace

1. NHK Project X seisakuhan, *Project X chōsenshatachi* (Tokyo: Nihonhōsō kyōkai, 2000), 2: 14–52; and Rokuda Noboru and NHK Project X seisakuhan, *Project X chōsenshatachi: Shūnen ga unda Shinkansen* (Tokyo: Nihonhōsō kyōkai, 2002).

2. Stuart Leslie, *The Cold War and American Science: The Military-Industrial-Academic Complex at MIT and Stanford* (New York: Columbia Univ. Press, 1994); Atsushi Akera, *Calculating a Natural World: Scientists, Engineers, and Computers during the Rise of U.S. Cold War Research* (Cambridge, MA: MIT Press, 2006); Paul N. Edwards, *The Closed World: Computers and the Politics of Discourse in Cold War America* (Cambridge, MA: MIT Press, 1996); Alex Roland, "Technology and War: A Bibliographical Essay," in *Military Enterprise and Technological Change: Perspectives on the American Experience*, ed. Merritt Roe Smith (Cambridge, MA: MIT Press, 1985), 347–79; Alex Roland, "Technology and War: The Historiographical Revolutions of the 1980s," *Technology and Culture* 34 (1993): 117–34; Gabrielle Hecht, *The Radiance of France: Nuclear Power and National Identity after World War II* (Cambridge, MA: MIT Press, 1998); Edmund Russell, *War and Nature: Fighting Humans and Insects with Chemicals from World War I to Silent Spring* (New York: Cambridge Univ. Press, 2001); and Eric Schatzberg, *Wings of Wood, Wings of Metal* (Princeton, NJ: Princeton Univ. Press, 1998).

3. Thomas P. Hughes presents a masterly comparative work in *Networks of Power: Electrification in Western Society, 1880–1930* (Baltimore: Johns Hopkins Univ. Press, 1983); Eric Schatzberg, "Wooden Airplanes in World War II: National Comparisons and Symbolic Culture," in *Archimedes: New Studies in History and Philosophy of Science and Technology*, vol. 3, *Atmospheric Flight in the Twentieth Century*, ed. Alex Roland and Peter Galison (Dordrecht, the Netherlands: Springer, 2000), 183–205; and Kevin McCormick, *Engineers in Japan and Britain: Education, Training, and Employment* (London: Routledge, 2000).

4. While the meanings of "military technology" could be understood relative to those of "nonweapons technology," the line of demarcation between the two fields was often blurred. In the United States during the twentieth century, as Edmund Russell shows in his study about chemical warfare and pest control, the distinction between civilian and military was a binary, "false" construct amid a series of almost consecutive wars abroad. In some cases, the distinction depends on the users, situations, and purposes. If ex-military passengers discuss a plan on their commercial mobile telephones to retain control over their hijacked commercial airplane, the tool could be seen as a military communication

device. A better example involves state-sponsored, aerospace science and technology, to which one can refer as dual-use technology because it empowers both civilian and military industries. At least in the context of Japan, however, high-speed aerodynamics, that is, both theoretical and empirical studies of airflow as fast as the speed of sound, was strictly of military origin and application until defeat. It was a technology for war. Substantive applications of research and development in the field were confined to the creation of a highly advanced weapon, namely, military aircraft.

Serious scientific inquiries in the field before 1945 qualitatively differed from the popular culture of "streamlining" that permeated Japan and the United States during the 1930s. In Japan before 1945, the trend meant fashionable technological artifacts, including "streamlined" tricycles for playful children, "streamlined" haircuts for the youth, and "streamlined" furniture for those who liked the new style. Furthermore, there are many examples of commercial applications of aerodynamics/hydrodynamics in the twenty-first century, ranging from golf balls, swimsuits, and racing cars. At least in Japan, this type of civilian application was strictly a postwar phenomenon. And Japan's experience of defeat reconfigured the position of the field and their roles of the experts in it.

5. Renowned sociologist Karl Mannheim emphasized the importance of key historical events such as war and revolution that help shape a sociological generation. Not defined in biological terms and by no means monolithic, the sociological generation tends to share a common experience, memory, and view. Karl Mannheim, *Essays on the Sociology of Culture* (New York: Oxford Univ. Press, 1956); Philip Adams, *Historical Sociology* (Ithaca, NY: Cornell Univ. Press, 1982), 227–266; Howard Schuman and Cheryl Rieger, "Historical Analogies, Generational Effects, and Attitudes toward War," *American Sociological Review* 57, no. 3 (1992): 315–326; and Jane Pilcher, "Mannheim's Sociology of Generations: An Undervalued Legacy," *British Journal of Sociology* 45, no. 3 (1994): 481–495; and Eric Ericson, *Identity, Youth, and Crisis* (New York: W. W. Norton, 1968).

6. Edward W. Constant II, "Science in Society: Petroleum Engineers and the Oil Fraternity in Texas, 1925–65," *Social Studies of Science* 9, no. 3 (1989): 439–472; Edwin Layton Jr., *The Revolt of the Engineers: Social Responsibility and the American Engineering Profession* (Baltimore: Johns Hopkins Univ. Press, 1971); David Mindell, *Between Human and Machine: Feedback, Control, and Computing before Cybernetics* (Baltimore: Johns Hopkins Univ. Press, 2000); *Donald Mackenzie, Inventing Accuracy: A Historical Sociology of Nuclear Missile Guidance* (Cambridge, MA: MIT Press, 1990), 11–14; and Satō Yasushi, "Systems Engineering and Contractual Individualism: Linking Engineering Processes to Macro Social Values," *Social Studies of Science* 37, no. 6 (2007): 909–934.

7. Merritt Roe Smith, "Introduction," in *Military Enterprise and Technological Change: Perspectives on the American Experience* (Cambridge, MA: MIT Press, 1985), 4.

Chapter 1 · *Designing Engineering Education for War, 1868–1942*

1. Ōkubo Toshikane et al., ed., *Kindaishi shiryō* (Tokyo: Kikkawa kōbokan, 1965), 97.

2. Takashi Fujitani, *Splendid Monarchy: Power and Pageantry in Modern Japan* (Berkeley: Univ. of California Press, 1998); and Yasuo Masai, "Tokyo: From a National Centre to a Global Supercity," *Asian Journal of Communication* 2, no. 3 (1992): 68–72; and Nicholas J. Entrikin, *The Betweenness of Place: Towards a Geography of Modernity* (Baltimore: Johns Hopkins Univ. Press, 1991), 14–16.

3. Carola Hein, "Shaping Tokyo: Land Development and Planning Practice in the Early Modern Japanese Metropolis," *Journal of Urban History* 36, no. 4 (2010): 447–484.

4. Kagaku gijutsu seisakushi kenkyūkai, *Nihon no kagaku gijutsu seisakushi* (Tokyo: Mitō kagaku gijutsu kyōkai, 1990), 17; and Gregory Clancey, *Earthquake Nation: The Cultural Politics of Japanese Seismicity, 1868–1930* (Berkeley: Univ. of California Press, 2006), 13.

5. Tetsurō Nakaoka, *Nihon kindai gijutsu no keisei: Dentō to kindai no dynamics* (Tokyo: Asahi shinbunsha, 2006), 431.

6. *Tokyo Kogyōdaigaku 90-nen shi* (Tokyo: Zaikai hyōronshinsha, 1975),16; Clancey, *Earthquake Nation*, 13; Nakaoka, *Nihon kindai gijutsu no keisei*, 437; Graeme Gooday and Morris Low, "Technology Transfer and Cultural Exchange: Western Scientists and Engineering Encounter Late Tokugawa and Meiji Japan," *OSIRIS: Beyond Joseph Needham: Science, Technology, and Medicine in East and Southeast Asia* 13 (1998): 109–111.

7. Tokyo daigaku 100-nenshi henshū iinkai, *Tokyo daigaku 100-nenshi bukyokushi* (Tokyo: Tokyo daigaku, 1987), 3: 5–6.

8. James Bartholomew, "Japanese Modernization and the Imperial Universities, 1876–1920," *Journal of Asian Studies* 37, no. 2 (1978): 258.

9. Hiroshige Tetsu, *Kagaku no shakaishi: Sensō to kagaku* (Tokyo: Iwanami shoten, 2002), 1: 29–30.

10. James Bartholomew, *The Formation of Science in Japan: Building a Research Tradition* (New Haven, CT: Yale Univ. Press, 1989), 93.

11. Nakaoka, *Nihon kindai gijutsu no keisei*, 441–442; Eiichi Aoki and others, *A History of Japanese Railways, 1872–1999* (Tokyo: East Japan Railway Culture Foundation, 2000), 12–14; and Tokyo daigaku 100-nenshi henshū iinkai, *Tokyo daigaku 100-nenshi tsūshi* (Tokyo: Tokyo daigaku, 1985), 2: 188.

12. Nakayama Shigeru, "Japanese Science," in Helaine Selin, ed., *The Encyclopedia of the History of Science, Technology, and Medicine in Non-Western Countries* (Dordrecht, the Netherlands: Kluwer Academic Publications, 1997), 469.

13. Tokyo daigaku 100-nenshi henshū iinkai, *Tokyo daigaku 100-nenshi bukyokushi* (Tokyo: Tokyo daigaku, 1987), 3: 21.

14. Ibid., 263.

15. Bartholomew, *Formation of Science in Japan*, 94.

16. Kyoto daigaku 70-nen shi henshū iinkai, *Kyoto daigaku 70-nen shi* (Kyoto: Kyoto daigaku, 1967), 14–15; and Hokkaidō daigaku, *Hokudai 100-nen shi, bukyokushi* (Sapporo: Gyōsei, 1980), 2: 703–704.

17. Bartholomew, "Japanese Modernization and the Imperial Universities," 252–253.

18. Bartholomew, *Formation of Science in Japan*, 106–108.

19. Kyoto daigaku 70-nen shi henshū iinkai, *Kyoto daigaku 70-nen shi*, 16–17.

20. Ibid., 643.

21. Nihon kagakushi gakkai-hen, *Nihon kagaku gijutsushi taikei, dai-2kan tsūshi* (Tokyo: Daiichi hōki shuppan, 1967), 164.

22. Nakaoka, *Nihon kindai gijutsu no keisei*, 445; and *Tokyo Kogyōdaigaku 90-nen shi*, 110–111, 150–154; and *Osaka daigaku 25-nen shi* (Osaka: Osaka daigaku, 1956), 351–352.

23. Ishii Kanji, *Nihon no sangyō kakumei: Nisshin, nichiro sensō kara kangaeru* (Tokyo: Asahi shinbunsha, 1997), 157–158; and Kagaku gijutsu seisakushi kenkyūkai, *Nihon no kagaku gijutsu seisakushi*, 22.

24. Dainihon teikoku gikaishi kankōkai, *Dainihon teikoku gikaishi* (Tokyo: Dainihon teikoku gikaishi kankōkai, 1927), 5, 1151–1154.

25. Kyūshū daigaku sōritsu 50-shūnen kinenkai, *Kyūshū daigaku 50-nen shi, tsūshi* (Fukuoka: Kyūshū daigaku sōritsu 50-shūnen kinenkai, 1967), 1: 90.

26. Ibid, 89, 94, 99, 125.

27. Tomizuka Kiyoshi, *Showa umare no waga oitachi* (Tokyo: Tomizuka Kiyoshi, 1977), 346.

28. Tokyo daigaku 100-nenshi henshū iinkai, *Tokyo daigaku 100-nenshi, tsūshi* (Tokyo: Tokyo daigaku, 1985), 2: 270. The Faculty of Engineering grew marginally after 1924. No chairs were added until 1933. In the midst of the subsequent war in Asia and the Pacific, three chairs were added during 1934–38 and eight during 1939–43.

29. Tokyo daigaku 100-nenshi henshū iinkai, *Tokyo daigaku 100-nenshi bukyokushi*, 3: 34.

30. Kyoto daigaku 70-nen shi henshū iinkai, *Kyoto daigaku 70-nen shi*, 65, 75.

31. Hokkaidō daigaku, *Hokudai 100-nen shi bukyokushi*, 702.

32. Ibid., 711.

33. Ibid., 719.

34. Bartholomew, *Formation of Science in Japan*, 114.

35. *Tokyo Kogyōdaigaku 90-nen shi*, 259; and Bartholomew, "Japanese Modernization and the Imperial Universities," 262.

36. Kagaku gijutsu seisakushi kenkyūkai hen, *Nihon no kagaku gijutsu seisakushi*, 25.

37. Tokyo daigaku 100-nenshi henshū iinkai, *Tokyo daigaku 100-nenshi tsūshi*, 2: 270.

38. Hokkaidō University gained its Faculty of Engineering in 1924. In 1933, Osaka University absorbed Osaka Technical School to create the Faculty of Engineering that had eight programs, including mechanical engineering, applied chemistry, metallurgy, shipbuilding, and electrical engineering. To meet the war demand, the faculty soon added new subjects of study to its curricula, namely, telecommunication in 1940 and welding engineering in 1944. *Osaka daigaku 25-nen shi*, 4–5. As for private institutions, Waseda University established the Faculty of Engineering in 1920, and in 1928, Nihon University began offering civil engineering, architecture, mechanical engineering, and electrical engineering.

39. Christopher Madeley, "Britain and the World Engineering Congress: Tokyo 1929," in *Britain and Japan in the Twentieth Century: One Hundred Years of Trade and Prejudice* (London: I. B. Tauris, 2007), 46–61; and *Kōbe Shinbun*, 8 December 1926.

40. Tokyo daigaku 100-nenshi henshū iinkai, *Tokyo daigaku 100-nenshi bukyokushi*, 3:38.

41. Kagaku gijutsu seisakushi kenkyūkai, *Nihon no kagaku gijutsu seisakushi*, 32.

42. Sawai Minoru, "Daigaku (senzenki)," in *Nihon sangyō gijutsushi jiten* (Kyoto: Nihon shibunkaku shuppan, 2007), 473.

43. Tokyo daigaku 100-nenshi henshū iinkai, *Tokyo daigaku 100-nenshi bukyokushi*, 3: 39.

44. Uchimaru Saiichirō, "Jiron: Tokyo teikoku daigaku kikai kōgakuka ni okeru kyōiku no genjō," *Kikai gakkaishi* 40, no. 237 (January 1937): 2.

45. Kagaku gijutsu seisakushi kenkyūkai, *Nihon no kagaku gijutsu seisakushi*, 18–19; and Bartholomew, "Japanese Modernization and the Imperial Universities," 254–255.

46. Kagaku gijutsu seisakushi kenkyūkai, *Nihon no kagaku gijutsu seisakushi*, 37.

47. Ibid.; and Tokyo daigaku 100-nenshi henshū iinkai, *Tokyo daigaku 100-nenshi bukyokushi*, 3: 39.

48. Kagaku gijutsu seisakushi kenkyūkai, *Nihon no kagaku gijutsu seisakushi*, 37; Kōseisho jinkō mondai kenkyūjo, *Jinkō mondai kenkyū* 2, no. 12 (February 1941): 84; and Tokyo daigaku 100-nenshi henshū iinkai, *Tokyo daigaku 100-nenshi bukyokushi*, 3: 39.

49. Maema Takanori, *Man machine no shōwa densetsu: Kōkūki kara jidōsha e* (Tokyo: Kōdansha, 1996), 1: 91–92; and Maema Takanori, *Fugaku: Bei hondo o bakugeki seyo* (Tokyo: Kōdansha, 1995), 2: 273.

50. Bōeichō bōei kenshūjo seishishitsu, *Senshi sōsho: Rikugun kōkū heiki no kaihatsu · seisan · hokyū* (Tokyo: Asagumo shinbusha, 1975), 244.

51. *Chūgai shōgyō shinpō*, 4 October 1941.

52. Tokyo daigaku 100-nenshi henshū iinkai, *Tokyo daigaku 100-nenshi bukyokushi*, 3: 298–299.

53. *Chūgai shōgyō shinpō*, 24 July 1937.

54. *Osaka Asahi shinbun*, 1 October 1937.

55. *Tokyo Kogyōdaigaku 90-nen shi*, 520.

56. Kyūshū daigaku sōritsu 50-shūnen kinenkai, *Kyūshū daigaku 50-nen shi tsūshi*, 378; and Sawai Minoru, "Senjiki nihon teikoku ni okeru gijutsusha kyōkyū," in *Kindai higashi Asia keizei no shiteki kōzō: Higashi Asia shihon shugi keiseishi* (Tokyo: Nihon hyōronsha, 2007), 3: 326–333.

57. Tokyo daigakushi shiryōshitsu hen, *Tokyo daigaku no gakuto dōin gakuto shutsujin* (Tokyo: Tokyo daigaku, 1997), 181–198; Hiroshige Tetsu, *Kagaku no shakaishi: Sensō to kagaku* (Tokyo: Iwanami shoten, 2002), 1:202; and Byron Marshall, *Academic Freedom and the Japanese Imperial University, 1868–1939* (Berkeley: Univ. of California Press, 1992), 167–175.

58. Hiroshige, *Kagaku no shakaishi*, 1: 202–203; and Kagaku gijutsu seisakushi kenkyūkai, *Nihon no kagaku gijutsu seisakushi*, 36, 38.

59. Tokyo daigaku 100-nenshi henshū iinkai, *Tokyo daigaku 100-nenshi bukyokushi*, 3: 41.

60. Ibid., 24.

61. Ibid., 571, 573, 605.

62. Nagoya daigakushi henshū iinkai, *Nagoya daigaku 50-nenshi, bukyokushi* (Nagoya: Nagoya daigaku shuppankai, 1989), 2: 5, 8.

63. Waseda daigaku daigakushi henshūjo, *Waseda daigaku 100-nen shi* (Tokyo: Waseda daigaku, 1987,), 3: 990; and Sawai, "Senjiki nihon teikoku ni okeru gijutsusha kyōkyū," 326–333; and Kagaku gijutsu seisakushi kenkyūkai hen, *Nihon no kagaku gijutsu seisakushi*, 37.

64. Sawai, "Senjiki nihon teikoku ni okeru gijutsusha kyōkyū," 325, 333, 341; and *Jinkō mondai kenkyū* 2, no. 12 (February 1941): 84–86.

65. Tokyo daigakushi shiryōshitsu hen, *Tokyo daigaku no gakuto dōin gakuto shutsujin*, 27.

66. Maema, *Fugaku*, 1: 268, 277.

67. Yanagida Kunio, *Reishiki sentōki* (Tokyo: Bungei shunjū, 1977), 282–283.

68. Horikoshi Jirō and Okumiya Masatake, *Zerosen* (Tokyo: Asahi sonorama, 1992), 441–443, 461; Maema, *Man machine no shōwa densetsu*, 1: 126; and Horikoshi Jirō, *Zerosen: Sono tanjō to eikō no kiroku* (Tokyo: Kōbunsha, 1970), 184–186, 194–196.

69. Tessa-Morris Suzuki, *The Technological Transformation of Japan from the Seventeenth Century to the Twenty-First Century* (Cambridge: Cambridge Univ. Press, 1994), 71–104; and Yamamura Kōzō, "Success Illgotten? The Role of Meiji Militarism in Japan's Technological Progress," *Journal of Economic History* 37, no. 1 (1977): 113–135.

Chapter 2 · Navy Engineers and the Air War, 1919–1942

1. "Kūkan dai 28-gō, Shōwa 2-nen 6-gatsu 13-nichi, Hikō jigyō kakuchō ni kansuru kengian no ken," *Kōbunbikō: kōkū 1,* 57, 1927, NIDS.

2. Bartholomew, *Formation of Science in Japan,* 199–237.

3. Bōeichō bōei kenshūjo senshi shiryōshitu, *Rikugun kōkū no gunbi to un'yō 1* (Tokyo: Asagumo shinbunsha, 1971), 106.

4. Bōeichō bōei kenshūjo senshi shiryōshitu, *Rikugun kōkū heiki no kaihatsu · seisan · hokyū* (Tokyo: Asagumo shinbunsha, 1975), 43; Hara Takeshi and Yasuoka Akio, *Nihon rikukaigun jiten* (Tokyo: Shinjinbutsu ōraisha, 1997), 159.

5. Hara and Yasuoka, *Nihon rikukaigun jiten,* 158; and Bōeichō bōei kenshūjo senshi shiryōshitu, *Rikugun kōkū heiki no kaihatsu · seisan · hokyū,* 43.

6. Nihon kaigun kōkūshi hensan iinkai, *Nihon kaigun kōkushi* (Tokyo: Jiji tsūshisha, 1969) 3:9, 30–31; and Hara and Yasuoka, *Nihon rikukaigun jiten,* 202.

7. Kaigun gijutsu kenkyūjo genjō ippan furoku 1, *Kōbunbikō, kanshoku 7-kan 7,* NAJ.

8. Nihon kaigun kōkūshi hensan iinkai *Nihon kaigun kōkushi,* 3:31–32; and Kawamura Yutaka, "Kyūkaigun gijutsu kenkyūjo ni miru kenkyū kaihatsu no tokuchō," *Gijutsushi 2* (2001): 18–26.

9. Tomizuka Kiyoshi, *80-nen no shōgai no kiroku* (Tokyo: Tomizuka Kiyoshi, 1975), 106–108; and Tomizuka Kiyoshi, *Meiji umare no waga oitachi* (Tokyo: Tomizuka Kiyoshi, 1977), 350.

10. Tomizuka, *Meiji umare no waga oitachi,* 353.

11. Bartholomew, *Formation of Science in Japan,* 199–200, 217–223, 238–239.

12. National Physical Laboratory, NPL's History Highlights, accessed 18 April 2011, http://www.npl.co.uk/content/ConMediaFile/4279.

13. Tokyo daigaku 100-nenshi henshū iinkai, *Tokyo daigaku 100-nenshi: Shiryō 1* (Tokyo: Tokyo daigaku, 1984), 170–171.

14. Shiba Chūsaburō, "Kōkūkenkyūjo no jigyō ni tsuite," *Kikai gakkaishi* 27, no. 89 (1924): 818–819.

15. *Tokyo teikoku daigaku kōkū kenkyūjo yōran (1936),* 8, UTEO.

16. Tokyo teikoku daigaku kōkū kenkyūjo, *Tokyo teikoku daigaku kōkū kenkyūjo jigyō ichiran, taishō 15-nen* (1926), 2, 7, UTEO.

17. Sentanken tankendan, *Sentanken tankendan dai 3-kai hōkoku* (Tokyo: Sentanken tankendan, 1997), 11.

18. Tokyo daigaku 100-nenshi henshū iinkai, *Tokyo daigaku 100-nenshi: Tsūshi 2* (Tokyo: Tokyo daigaku, 1985), 320–321.

19. "Kaigun kyōju haken ni kansuru ken," *Monbushō oyobi shokō oufuku Showa18-nen,* 3, 1943, UT archive.

20. Tokyo teikoku daigaku gakujutsu taikan kankōkai, *Tokyo teikoku daigaku gakujutsu taikan kōgakubu kōkū kenkyūjo hen* (Tokyo: Tokyo teikoku daigaku, 1944), 397.

21. Nihon kaigun kōkūshi hensan iinkai, *Nihon kaigun kōkushi,* 3: 9.

22. Bōeichō bōei kenshūjo senshi shiryōshitu, *Rikugun kōkū heiki no kaihatsu · seisan · hokyū,* 7–12; Bōeichō bōei kenshūjo senshi shiryōshitu, *Rikugun kōkū no gunbi to un'yō 1,* 72; and Nihon kaigun kōkūshi hensan iinkai, *Nihon kaigun kōkushi,* 3: 5; Tomizuka, *80-nen no shōgai no kiroku,* 109; and Tomizuka Kiyoshi, *Kōkenki* (Tokyo: Mikishobō, 1998), 55.

23. Bōeichō bōei kenshūjo senshi shiryōshitu, *Rikugun kōkū no gunbi to un'yō 1,* 107.

24. Hashimoto Takehiko, "Theory, Experiment, and Design Practice: The Formation of

Aeronautical Research, 1909–1930" (PhD diss., Johns Hopkins Univ., 1991), 5; Paul Hanle, *Bringing Aerodynamics to America* (Cambridge, MA: MIT Press, 1982); and Theodore von Kármán, *The Wind and Beyond: Pioneer in Aviation and Pathfinder in Space* (Boston: Little, Brown, 1967).

25. "Gaikokujintaru gijutsusha toyō ni kansuruken," *1922 ōjūdainikki* 08-09, NAJ.

26. Tokyo daigaku 100-nenshi henshū iinkai, *Tokyo daigaku 100-nenshi: Bukyokushi 4* (Tokyo: Tokyo daigaku, 1987), 891; Hashimoto Takehiko, "Kōkūkenkyūjo," in *Nihon sangyōshi jiten* (Kyoto: Shibunkaku shuppan, 2007), 493.

27. "Kōkū kenkyūjo ni okeru Wieselsberger kōseki chōsho," *Monbushō oyobi shokō oufuku Showa 18-nen*, 3, 1943, UT archive; Hashimoto Takehiko, *Hikōki no tanjō to kūki rikigaku no keisei* (Tokyo: Tokyo Univ. Press, 2012), 253; and Nihon kaigun kōkūshi hensan iinkai, *Nihon kaigun kōkushi*, 3: 32.

28. Sakaue Shigeki, "Riku kaigun gunyōki," *Nihon sangyōshi jiten* (Tokyo: Shibunkaku, 2007), 247.

29. Bōeichō bōei kenshūjo senshi shiryōshitu, *Rikugun kōkū heiki no kaihatsu · seisan · hokyū* , 45–46, 67; and Nihon kōkū gakujutsushi henshū iinkai, *Nihon kōkū gakujutsushi, 1910–1945* (Tokyo: Maruzen, 1990), 203.

30. Hara and Yasuoka, 157; and Bōeichō bōei kenshūjo senshi shiryōshitu, *Rikugun kōkū heiki no kaihatsu · seisan · hokyū*, 61–63.

31. Nihon kōkū kyōkai, *Nihon kōkūshi: Showa zenki hen* (Tokyo: Nihon kōkū kyōkai, 1975), 6, 38; and Nihon kōkū gakujutsushi henshū iinkai, *Nihon kōkū gakujutsushi*, 203.

32. Bōeichō bōei kenshūjo senshi shiryōshitu, *Rikugun kōkū heiki no kaihatsu · seisan · hokyū* , 41–42.

33. Ibid., 77.

34. Kaigun henshū iinkai hen, *Kaigun* (Tokyo: Seibun tosho, 1981), 14: 170; Bōeichō bōei kenshūjo senshi shiryōshitu, *Kaigun kōkū gaishi*, 7–8; Nihon kaigun kōkūshi hensan iinkai, 3: 43, 76; and Mark Peattie, *Sunburst: The Rise of Japanese Naval Air Power, 1903–1942* (Annapolis, MD: Naval Institute Press, 2001), 26.

35. Richard Vogt, *Weltumspannende Memoiren eines Flugzeug-Konstrukterurs* (Steineback/Woerthsee: Flieger-Verlag, 1976), 65; Kawasaki kōkūki kogyō kabushiki kaisha, *Kōkūki seizō enkaku* (1946), 5; and Doi Takeo, *Hikōki sekkei 50-nen no kaisō* (Tokyo: Suitōsha, 1989), 45.

36. Maema, *Fugaku*, 1: 91.

37. Nihon kōkū gakujutsushi henshū iinkai, *Nihon kōkū gakujutsushi*, 203.

38. *Mitsubishi kōkūki kabushikigaisha Nagoya seisakujo*, Shōwa 5-nen hensan: shoshian, 121, MHI.

39. *Mitsubishi jukōgyō kabushikigaisha seisaku hikōki rekishi: kaigun kankei hikōki*, 121, MHI.

40. Ōjibōbō: Mitsubishi jūkō Nagoya 50-nen no kaiko (Nagoya: Ryōkōkai, 1970), 3: 67.

41. Doi Takeo, *Hikōki sekkei 50-nen no kaisō* (Tokyo: Suitōsha, 1989), 46.

42. *Mitsubishi jūkōgyō kabushikigaisha seisaku hikōki rekishi*, 3, MHI; and Kōkū kōgyōshi henshū iinkai, *Minkan kōkūki kōgyōshi* (1948), 152–156.

43. General Headquarters United States Army Forces, Pacific Scientific and Technical Advisory Section, "Reports on Scientific Intelligence Survey in Japan, September and October 1945," Vol. 1, 1 November 1945, 52, NA.

44. Kumagai Tadasu, "Kaigun gijutsu · rikugun gijutsu sono hito to soshiki," in *Nihon no gunji technology* (Tokyo: Kōjinsha, 2001), 233.

45. Tokyo daigaku 100-nenshi henshū iinkai, *Tokyo daigaku 100-nenshi: Bukyokushi 3* (Tokyo: Tokyo daigaku, 1987), 38–39.

46. Tokyo teikoku daigaku gakujutsu taikan kankōkai, *Tokyo teikoku daigaku gakujutsu taikan kōgakubu* kōkū kenkyūjo hen, 56–57; and Hata Ikuhiko, ed, *Nihon rikukaigun sogō jiten* (Tokyo: Tokyo daigaku shuppankai, 1991), 582–585, 635. Other institutions hosting future army technical officers from 1921 to 1945 included Tokyo Institute of Technology, Tōhoku University, and Kyoto University. By comparison, the navy system differed slightly. Graduates from Naval Academy later studied at imperial universities, and graduates from Naval Technical Academy were called *kikanka shikan* and assigned technical jobs without any additional education. There were 4,088 *kikanka shikan* by the war's end.

47. Bōeichō bōei kenshūjo senshi shiryōshitu, *Rikugun kōkū heiki no kaihatsu · seisan · hokyū*, 44–45.

48. Nihon kaigun kōkūshi hensan iinkai, *Nihon kaigun kōkushi*, 3: 372–373; Ujike Yasuhiro, "Kyū nihongun ni okeru bunkan nado no nin'yō ni tsuite: hannin bunkan o chūshin ni," *Bōei kenkyūjo kiyō* 8, no. 2 (2006): 74–75; and Ishikawa Junkichi, *Kokka sōdōinshi Shiryōhen 9* (Tokyo: Kokka sōdōinshi kankōkai, 1980), 973.

49. Kaigun henshū iinkai, *Kaigun: Kaigun gunsei, kyōiku, gijutsu, kaikei keiri, jinji* (Tokyo: Seibun tosho, 1981), 14: 251.

50. Nakagawa Ryōichi, "Watashi no senzen sengo (Hikōki kara jidōsha e)," in *Gunji gijutsu kara minsei gijutsu eno tenkan* (Tokyo: Gakujutsu shinkōkai, 1996), 1: 12; and Maema, *Man machine no Shōwa densetsu,* 1: 193–195.

51. Kaigun henshū iinkai hen, *Kaigun,* 14: 251; and Ishikawa Junkichi, *Kokka sōdōinshi (jō),* 1468–1476.

52. Tokyo daigakushi shiryōshitu, *Tokyo daigaku no gakuto dōin · gakuto shutsujin* (Tokyo: Tokyo daigaku shuppankai, 1997), 165–167.

53. Ikari Yoshirō, *Kaigun gijutsushatachi no taiheiyō sensō* (Tokyo: Kōjinsha, 1989), 275.

54. Naitō Hatsuho, *Kaigun gijutsu senki* (Tokyo: Tosho shuppan, 1976), 62.

55. Hatano Isamu, *Kindai nihon no gunsangaku fukugōtai* (Tokyo: Sōbunsha, 2005), 118–119.

56. Yoshiki Masao, "Omoidasu mamani 10," in *Senpaku kōgakka no 100-nen* (Tokyo: Tokyo daigaku, 1983), 105–106; and Naitō Hatsuho, *Gunkan sōchō Hiraga Yuzuru* (Tokyo: Bungei shunjū, 1987), 193–194.

57. Hatano, *Kindai nihon no gunsangaku fukugōtai,* 9.

58. Tokyo daigaku kōgakka senpaku kōgakka, *Senpaku kōgakka no 100-nen,* 12–14; and *Tokyo daigaku senpaku kōgakka sotsugyō meibo, Shōwa 29-nen 4-gatsu genzai,* UTN.

59. Nihon kaigun kōkūshi hensan iinkai, 378; and *Kaigun kōkūgijutsushō denkibu* (Tokyo: Kūgishō denkibu no kai, 1987), 9–10.

60. "1936 Shōjō menjo, Yūyo 2 rikugun, Kaigunbunai kinmu no rikugun gunjin (1)," *Kōbun bikō, Shōwa 11-nen jinji B-kan,* 35, NAJ; and "1936 Shōjō menjo, Yūyo 2 rikugun, kaigunbunai kinmu no rikugun gunjin (2)," *Kōbun bikō, Shōwa 11-nen jinji B-kan,* 35, NAJ.

61. "Shōwa 13-nendo chokurei dai566-gō kaigun shozoku no gishi mataha gite no shoku ni aritaru monoyori kaigun shikan ni ninyō nado ni kansuru ken o kaisei su," *Kōbun*

ruishū, dai 68-hen, Showa 19-nen, dai 39-kan, kanshoku 39, ninmen (naigaku · ōkurashō · rikukaigunshō- kantōkyoku), NDL.

62. "Gijutsukan seido kaisei · jisshi ni kansuru hōshin," *Kōkū kankei gijutsu gyōsei, 1937,* NIDS.

63. Naitō, *Kaigun gijutsu senki,* 55.

64. Nihon kaigun kōkūshi hensan iinkai, *Nihon kaigun kōkushi* 3: 375.

65. Bōeichō bōei kenshūjo senshi shiryōshitu, *Rikugun kōkū no gunbi to un'yō 1,* 260–261.

66. Bōeichō bōei kenshūjo senshi shiryōshitu, *Rikugun kōkū no gunbi to un'yō 2* (Tokyo: Asagumo shinbunsha, 1974), 212.

67. Bōeichō bōei kenshūjo senshi shiryōshitu, *Rikugun gunsenbi* (Tokyo: Asagumo shinbunsha, 1979), 405.

68. "Bunkan yori bukan ni tenkahsha kōho meibo sakusei kakusho," in *Showa 19-nen 3-gatsu kōkū kenkyū taisei tsuzuri sōmu buchō,* NIDS; and "Suehiro Takenobu rirekisho," in *Showa 19-nen 3-gatsu kōkū kenkyū taisei tsuzuri sōmu buchō,* NIDS.

69. Ishikawa, *Kokka sōdōinshi shiryōhen 9,* 900–901.

70. "Bunkan yori bukan ni tenkahsha kōho meibo sakusei kakusho," in *Showa 19-nen 3-gatsu kōkū kenkyū taisei tsuzuri sōmu buchō,* NIDS.

71. "Shōwa 13-nendo chokurei dai 566-go kaigun shozoku no gishi mataha gite no shoku ni aritaru monoyori kaigun shikan ni ninyō nado ni kansuru ken o kaiseisu," in Kōbun ruishū, dai 68-hen, Showa 19-nen, dai 39-kan, kanshoku 39, ninmen (naikaku · ōkurashō · rikukaigunshō kantōkyoku), NDL.

72. *Kaigun kōkū gijutsushō denkibu,*104.

73. Doi Zenjirō, *Kessen heiki maruyu: Rikugun sensuikan* (Tokyo: Kōjinsha, 2003).

74. Walter Grunden, *Secret Weapons and World War II: Japan in the Shadow of Big Science* (Lawrence: Univ. of Kansas Press, 2005), 41–47; and quote from General Headquarters United States Army Forces, Pacific Scientific and Technical Advisory Section, "Report on Scientific Intelligence Survey in Japan, September and October 1945," Vol. 1, 1 November 1945, 8, NA.

75. Nihon kōkū gakujutsushi henshū iinkai, *Nihon kōkū gakujutsushi,* 205.

76. Bōeichō bōei kenshūjo senshi shiryoshitu, *Rikugun kōkū no gunbi to un'yō 2,* 7.

77. Bōeichō bōei kenshūjo senshi shiryōshitu, *Rikugun kōkū heiki no kaihatsu · seisan · hokyū,* 340.

78. Nihon kōkū gakujutsushi henshū iinkai, *Nihon kōkū gakujutsushi,* 219.

79. Ibid., 208.

80. Bōeichō bōei kenshūjo senshi shiryōshitu, *Rikugun kōkū no gunbi to un'yō 2,* 211; Nihon kōkū gakujutsushi henshū iinkai, *Nihon kōkū gakujutsushi,* 205; and Bōeichō bōei kenshūjo senshi shiryōshitu, *Rikugun kōkū heiki no kaihatsu · seisan · hokyū,* 77–78.

81. Maema, *Man machine no shōwa densetsu,* 1: 348–350.

82. Mizusawa Hikari, "Asia taiheiyō sensōki ni okeru kyū rikugun no kōkū kenkyūkikan eno kitai" in *Kagakushi kenkyu* 43 (2004): 24.

83. British Intelligence Objective Sub-committee, *Structural Requirement and Techniques Used in Design of Japanese Aircraft,* 26 October 1945, 1, NASM.

84. Mizusawa, "Asia taiheiyō sensōki ni okeru kyū rikugun no kōkū kenkyūkikan eno kitai," 25.

85. Only the records of the years 1938 to 1940 were available at the time of my inquiry. The only comparison I can make sensibly is to observe the number of projects commissioned to Tokyo University. Neither the nature nor cost of many projects is known. "Itaku kenkyū jikōchō kōkū kenkyūjo," *Showa 15-nen kagaku kenkyū shōreikin kankei,* UT archive.

86. Tomizuka Kiyoshi, *Kōkenki* (Tokyo: Miki shobō, 1998), 110.

87. Ikari Yoshirō, *Sentōki hayabusa: Shōwa no meiki sono eikō to higeki* (Tokyo: kōjinsha, 2003), 130–131.

88. Bōeichō bōei kenshūjo senshi shiryōshitu, *Rikugun kōkū no gunbi to un'yō 2,* 212; and Bōeichō bōei kenshūjo senshi shiryōshitu, *Rikugun kōkū heiki no kaihatsu · seisan · hokyū,* 212.

89. Ikari, *Sentōki hayabusa,* 132–133.

90. Bōeichō bōei kenshūjo senshi shiryōshitu, *Rikugun kōkū no gunbi to un'yō 3* (Tokyo: Asagumo shinbunsha, 1976), 39; and Ikari, *Sentōki hayabusa,* 132–133.

91. Bōeichō bōei kenshūjo senshi shiryōshitu, *Rikugun kōkū heiki no kaihatsu · seisan · hokyū,* 77–78.

92. From 1932 to 1945, this research establishment earned different official names during organizational restructuring. It was originally named Kōkūshō (Air Arsenal), but it changed its name first to Kōkūgijutsushō, and later to Dai-ichi gijutsushō at the creation of its division Dai-ni gijutsushō. Writings in English conventionally refer to this research establishment as Navy Air Arsenal, but this translation is often confusing. By the war's end, across the country, the navy maintained several Kōkūshō (Air Arsenals) where aircraft production and maintenance were handled. Because I focus primarily on the navy's research and development in aeronautics, I refer to this establishment as the Institute for Navy Aeronautics. This translation is useful to contrast the establishment to its army counterpart, the Institute for Army Aeronautics (Rikugun kōkū gijutsu kenkyūjo), which differed from Army Air Arsenals (Rikugun kōkūshō). From 24 October to 24 November 1945, the Allies occupation authority examined aircraft research and development at the Army Air Arsenal in Tachikawa, Japan, and concluded that "the Japanese Air Arm was greatly inferior to that Naval Air Arm with regard to research and development policy." *Air Technical Intelligence Group Advanced Echelon FEAF, Report on Tachikawa Army Air Arsenal,* 24 November 1945, NASM.

93. *Chōsa kiroku kaigun kōkū kankei gijutsu chōsa kitai kankei sekkei shisaku no bu,* 797-5, NIDS.

94. *Nihon kōkū gakujutsushi, Nihon kōkū gakujutsushi,* 226.

95. *Shiryō boeichō kaijō bakuryō kanshi chōsabu, Nihon teikoku kaigun no kenkyū narabi ni kaihatsu (1925–45),* 34, Shōwakan.

96. Bōeichō bōei kenshūjo senshi shiryōshitu, *Kaigun kōkū gaishi* (Tokyo: Asagumo shinbunsha, 1976), 67.

97. *Kaigun kōkū gijutsushō* (Tokyo: Gakushū kenkyūsha, 2008), 105.

98. *Chōsa kiroku: Kaigun kōkū kankei gijutsu chōsa, kitai kankei sekkei shisaku no bu,* 797-5, NIDS.

99. *Kaigun kōkū gijutsushō,* 108.

100. *Shiryō boeichō kaijō bakuryō kanshi chōsabu, Nihon teikoku kaigun no kenkyū narabi ni kaihatsu (1925–45),* 36, Shōwakan.

101. *Kōkū kimitsu 6810-gō Shōwa 16-nen 7-gatsu 9-nichi Kōkū honbukei haiinhyō ni kansuru ken shōkai,* NIDS.

102. Nihon kōkū gakujutsushi henshū iinkai, *Nihon kōkū gakujutsushi*, 227; and *Shiryō boeichō kaijō bakuryō kanshi chōsabu, Nihon teikoku kaigun no kenkyū narabi ni kaihatsu (1925–45)*, 35, Shōwakan. According to the archival source, as of 15 August 1945, the INA employed 18,723 employees, including 1,009 engineers and technicians and 17,714 workmen. This variation, minor in nature, is understandable, given the administrative confusion that pervaded many surveys conducted by the GHQ or by the Japanese in the immediate postwar era. After expansion, the Department of Arms formed a separate institution in Yokohama, hosting 600 engineers and technicians as well as 12,000 workmen. *Shiryō boeichō kaijō bakuryō kanshi chōsabu, Nihon teikoku kaigun no kenkyū narabi ni kaihatsu (1925–45)*, 40, Shōwakan.

103. *Air Technical Intelligence Group Report No. 27 Aircraft Design and Development*, 1 November 1945, MUSAFB.

104. *Kaigun kōkūgijutsushō denkibu*, 11–12.

105. Ibid., 13.

106. Ikari Yoshiro, *Kaigun kūgishō* (Tokyo: Kōjinsha, 1985), 1: 30–33.

107. "Monbushō ōfuku・hōkoku Showa-16nen," R-196/A-218, *Monbu ōfuku* 5, UT archive.

108. "Monbushō ōfuku・hōkoku Showa-17-nen," R-201/A-225, *Monbu ōfuku* 6, UT archive.

109. Ikari, *Kaigun kugishō*, 1: 18–20.

110. "Dai 1729-go 7.5.10 Tochi kōnyū no ken: Kaigun kōkūshō," *Kōbun bikō Shōwa 7-nen K doboku・kenchiku kan-12*, NAJ.

111. *Kaigun kōkūgijutsushō denkibu*, 52.

112. *Chōsa kiroku: Kaigun kōkū kankei gijutsu chōsa, kitai kankei sekkei shisaku no bu*, 797-5, NIDS.

113. Walter Vincenti, *What Engineers Know and How They Know It* (Baltimore: Johns Hopkins Univ. Press, 1990), 51–111.

114. Nihon kaigun kōkūshi hensan iinkai, *Nihon kaigun kōkushi*, 3: 380.

115. Tsukada Hideo, "Yōhei to gijutsu no setten," in *Umiwashi no kōseki* (Tokyo: Hara shobō, 1982), 34.

116. *Chōsa kiroku: Kaigun kōkū kankei gijutsu chōsa, kitai kankei sekkei shisaku no bu*, 742-1, NIDS.

117. Kawasaki Motoo, "Kaigun de mananda kotodomo," in *Kaigun kōkūgijutsushō zairyōbu shūsen 50-shūnen kinenshi* (Yokohama: Kaigun kōkūgijutsushō zairyōbu no kai, 1996), 138.

118. *Kaigun kōkū gijutsushō kenkyū jikken seiseki hōkoku Kōgihō 04190 Sōryūyokugata chūshinsen hoi*, 15 May 1944, UTA.

119. Tani Ichirō, "Kenkyū kaihatsu to gakkai: Kaigun kōkū tono sōgū," in *Umiwashi no kōseki*, 39.

120. Bōeichō bōei kenshūjo senshi shiryōshitu, *Kaigun kōkū gaishi*, 147; and Nihon kaigun kōkūshi hensan iinkai, *Nihon kaigun kōkushi* 3: 377.

121. Okamura Jun, "Sōsetsuron," in *Kōkūgijutsu no zenbō* (Tokyo: Kōyōsha, 1953), 1: 43.

122. Igaki Kenzō, "Kūgishō zairyōbu kenkyū jikken seiseki hōkoku hakkutsu no ki," in *Kaigun kōkūgijutsushō zairyōbu shūsen 50-shūnen kinenshi*, 75.

123. Bōeichō bōei kenshūjo senshi shiryōshitu, *Kaigun kōkū gaishi*, 15.

124. Yamana Masao, "Suisei ni toritukareta saigetsu," in *Gunyōki kaihatsu monogatari* (Tokyo: Kōinsha, 2002), 1: 154.

125. Yamana Masao, "Suisei ga dekirumade," *Kōkūfan* 13, no. 15 (1964): 74.

126. *Chōsa kiroku: Kaigun kōkū kankei gijutsu chōsa, kitai kankei sekkei shisaku no bu*, 742–24, NIDS.

127. Yamana Masao, "Kanjō bakugekiki suisei," in *Sekkeisha no shōgen* (Tokyo: Suitōsha, 1994), 1: 324.

128. Yamana Masao, "Über die Elastische Stabilität der Metallflugzeugbauteile" (senior thesis, Tokyo University, 1929), UTA.

129. *Chōsa kiroku: Kaigun kōkū kankei gijutsu chōsa, kitai kankei sekkei shisaku no bu*, 742–24, NIDS.

130. Naitō Ichirō, "Hiniku na unmei o tadotta kanbaku suisei," in *Suisei/99-kanbaku* (Tokyo: Kōjinsha, 2000), 101.

131. Ibid., 96–97.

132. Yamana, "Kanjō bakugekiki suisei," 328.

133. Ibid., 332.

134. Zasshi maru henshūbu, ed., *Suisei/99-kanbaku*, 23.

135. Yamana, "Kanjō bakugekiki suisei," 333.

136. Ueyama Tadao, "Suisei kanjō gakugekiki suisei no sekkei," in *Suisei/99-kanbaku*, 158–159.

137. Yamana, "Suisei ga dekirumade," 77.

138. Yamana, "Kanjō bakugekiki suisei," 334; and Ueyama, "Suisei kanjō bakugekiki suisei no sekkei," 23.

139. Naitō Ichirō, "Hiniku na unmei o tadotta kanbaku suisei," 101. Without conclusive empirical evidence, Naitō writes that the D4Y Judy's figure was smaller than that of U.S. fighter Mustang P-51, suggesting that both aerodynamic designs, one developed in 1938 and the other in 1943, were equally sophisticated. This remains a moot point at best.

140. "Technical Air Intelligence Center Summary No. 18, December 1944: Judy," 1, NASM.

141. One of the two is Akatsuka Takeo, a technician who introduced the algebraic design method into the Shinkansen vehicle design. Another is Hara Tomoshige, who examined the drag on the aircraft's radiator in a wind tunnel before 1945 and later, conducted research on the aerodynamics of the high-speed train upon entering a tunnel. Ueyama, "Suisei no kūkirikigakuteki sekkei," 23.

142. Miki Tadanao, "Kōsoku densha no kaihatsu," in *Heisei 11-nendo Sangyō gijutsu no rekishi ni kansuru chōsa kenkyū hōkokusho* (Tokyo: Nihon kikai kōgyō rengōkai, 2000), 351–352; Miki Tadanao, interview by author, Zushi, 8 August 2003; and Nihon kōkū gakujutsushi henshū iinkai, *Nihon kōkū gakujutsushi*, 225; and Miki Tadanao, "Sensō no kanki" (senior thesis, Tokyo University, 1933), UTN.

143. "Dōtai to gisō," in *Ginga/Ichishiki rikkō* (Tokyo: Kōjinsha, 1994), 20.

144. "Technical Air Intelligence Center Summary No. 10, October 1944: New Japanese Navy Bomber Frances 11," 2, NASM.

145. Honjō Kirō, "Sotsugyō keikaku keisansho" (senior thesis, Tokyo University, 1926), UTA.

146. Miki Tadanao, "Kūgishō 'Ginga' rikujō bakugekiki no sekkei nado ni tsuite," in *Heisei 5-nendo Bunyabetsu kagakugijututaikei no genjō to shōrai ni kansuru chōsa kenkyū hōkokusho* (Tokyo: Nihon kikai kogyō rengōkai, 1994), 35; and Ikari, *Kaigun kūgishō*, 1: 142–143.

147. Miki Tadanao, "Kōsoku rikubaku ginga sekkei no tsuioku," in *Ginga/Ichishiki rikkō*, 108.

148. *Y-20 seisekihyō*, MT Papers.

149. Miki, "Kōsoku rikubaku ginga sekkei no tsuioku," 106, 108.

150. "Technical Air Intelligence Center Summary No. 10, October 1944: New Japanese Navy Bomber Frances 11," 8, NASM.

151. "Gisō," in *Ginga/Ichishiki rikkō*, 24, 45; and Kojima Masao, "Dōryoku · yuatsu sōchi," in *Ginga/Ichishiki rikkō*, 37.

152. "Kōchaku sōchi," in *Ginga/Ichishiki rikkō*, 42.

153. Horikoshi Jirō, *Eagles of Mitsubishi: The Story of the Zero Fighter* (Seattle: Univ. of Washington Press, 1981), 17–18; and Kofukuda Terufumi, *Zerosen kaihatsu monogatari: Nihon kaigun sentōki zenkishu no shōgai* (Tokyo: Kōjinsha, 1985), 87.

154. Horikoshi, *Eagles of Mitsubishi*, 15–16; Horikoshi Jirō, *Zerosen: sono eikō to kiroku* (Tokyo: Kōbunsha, 1995), 58–61; and Ikari Yoshirō, *Ikiteiru Zerosen* (Tokyo: Yomiuri shinbunsha, 1970), 57.

155. Richard Smith, "The Intercontinental Airliner and the Essence of Airplane Performance, 1929–1939," *Technology and Culture* 24, no. 33 (1983): 428–429, 434.

156. William Rodden, "Flutter (aeronautics)," in *McGraw-Hill Encyclopedia of Science and Engineering* (New York: McGraw Hill, 1997), 246–247.

157. Nihon kōkū gakujutsushi henshū iinkai, *Nihon kōkū gakujutsushi*, 390.

158. Horikoshi, *Eagles of Mitsubishi* , 117–118.

159. "Kaigun kōkū gijutsushō kenkyū jikken hōkoku, Kūgihō 029, Kitai no koyūshindōsū keisanho no kenkyū," 7 January 1942, UTA; "Kaigun kōkū gijutsushō kenkyū jikken hōkoku, Kūgihō 04347, Yokufure gendo sokudo kensanhō no kenkyū sono-3," 4 September 1944, UTA; Kaigun kōkū gijutsushō kenkyū jikken hōkoku, Kūgihō 0546, Yokufure gendo sokudo kensanhō no kenkyū sono-4," 20 February 1945, UTA; and Nihon kōkū gakujutsushi henshū iinkai, *Nihon kōkū gakujutsushi*, 60.

160. Shioda Toyoji, "Keikinzoku no kenkyū," in *Kaigun kōkūgijutsushō zairyōbu shūsen 50-shūnen kinenshi*, 102.

161. Ueda Kōzō, "Kaigun jidai no omoide," in *Kaigun kōkūgijutsushō zairyōbu shūsen 50-shūnen kinenshi*, 117.

Chapter 3 · *Engineers for the Kamikaze Air War, 1943–1945*

1. *Asahin shinbun*, 1 June 1945.

2. *Asahi shinbun*, 30 June 1945.

3. Nihon kōkū gakujutsushi henshū iinkai, *Nihon kōkū gakujutsushi*, 292.

4. The genesis of this research establishment before 1939 epitomized political bargaining between military and civilian officials. Mizusawa Hikaru, "Rikugun ni okeru kōkūkenkyūjo no setsuritsu kōsō to gijutsuin no kōkū jūtenka," *Kagakushi kenkyū* 42 (2003): 32–34.

5. Nihon kōkū gakujutsushi henshū iinkai, *Nihon kōkū gakujutsushi*, 292–293.

6. Bōeichō bōei kenshūjo senshi shiryōshitu, *Rikugun kōkū heiki no kaihatsu · seisan · hokyū*, 185–186; and Nihon kōkū kyōkai, *Nihon kōkūshi*, 543.

7. Nihon kōkū gakujutsushi henshū iinkai, *Nihon kōkū gakujutsushi*, 293–294.

8. Ueda Toshio, interview by author, Tokyo, 29 April 2004.

9. British Intelligence Objectives Sub-committee, *Japanese Aerodynamic Research and Research Equipment*, Report No. BIOS/JAP/PR/112, 2, BWM.

10. *Report on Organization and Equipment of Central Aeronautical Research Institute*, Air Technical Intelligence Group Report No. 82, 14 November 1945, 3, MUSAFB.

11. "Chūō kōkū kenkyōjo Kajima Kōichi hoka 42-mei," *Kōbunzassan Shōwa 17-nen 17-kan naikaku・kakuchō kōtōkan shōyo 1 (naikaku)*, NAJ.

12. "Chūō kōkū kenkyūjo shokutaku Miyoshi Kan'ichi hoka 26-mei," *Kōbunzassan Shōwa 17-nen 17-kan naikaku・kakuchō kōtōkan shōyo 1 (naikaku)*, NAJ.

13. *Chūō kōkū kenkyūjo hōkoku* 1, no. 1 (1942), NMRI; and *Chūō kōkū kenkyūjo ihō* 1 (1943), NMRI.

14. *Report on Organization and Equipment of Central Aeronautical Research Institute*, Air Technical Intelligence Group Report No. 82, 14 November 1945, 3, MUSAFB.

15. "Shōwa 17-nen 10-gatsu matsujitu genzai shokuin," *Chūō kōkū kenkyūjo shisetsu iinkai dai-3kai sōkai kankei shorui*, NIDS.

16. "Kōkū kenkyū taisei no seibi ni kansuru ken (kakugi kettei 17.10.22)," *Shōwa 19-nen 3-gatsu kōkū kenkyū taisei tsuzuri sōmu buchō*, NIDS.

17. Nihon kōkū gakujutsushi henshū iinkai, *Nihon kōkū gakujutsushi*, 292.

18. *Report on Organization and Equipment of Central Aeronautical Research Institute*, Air Technical Intelligence Group Report No. 82, 14 November 1945, 3, MUSAFB.

19. Tokyo daigaku 100-nenshi henshū iinkai, *Tokyo daigaku 100-nenshi shiryō 3*, 476.

20. Ibid., 466–467.

21. Kagaku gijutsu seisakushi kenkyūkai, *Nihon no kagaku gijutsu seisakushi*, 38.

22. *Tokyo Kogyōdaigaku 90-nen shi*, 567; and Tokyo daigaku 100-nenshi henshū iinkai, *Tokyo daigaku 100-nenshi bukyokushi 3*, 41; and Monbu kagakushō, *Gakusei 100-nenshi*, accessed 4 October 2011, http://www.mext.go.jp/b_menu/hakusho/html/hpbz198101/hpbz198101_2_121.html.

23. Kyoto daigaku 70-nen shi henshū iinkai, *Kyoto daigaku 70-nen shi*, 127.

24. Ishiwari Kōtarō, "Omoide," in *Kaigun kōkūgijutsushō zairyōbu shūsen 50-shūnen ki-nenshi*, 115.

25. Tokyo daigakushi shiryōshitsu hen, *Tokyo daigaku no gakuto dōin, gakuto shutsujin*, 153–154.

26. Tokyo daigaku seisan gijutsu kenkyūjo-hen, *Tokyo daigaku dai-2 kōgakubushi* (Tokyo: Tokyo daigaku seisan gijutsu kenkyūjo, 1968), 72; and Tokyo daigaku 100-nenshi henshū iinkai, *Tokyo daigaku 100-nenshi shiryō 3*, 513.

27. Tokyo daigaku 100-nenshi henshū iinkai, *Tokyo daigaku 100-nenshi shiryō 3*, 511.

28. General Headquarters United States Army Forces, Pacific Scientific and Technical Advisory Section, *Report on Scientific Intelligence Survey in Japan, September and October 1945*, 1 November 1945, 1: 52, NA.

29. Nihon kōkū kyōkai, *Nihon kōkūshi: Showa zenki-hen* (Tokyo: Nihon kōkū kyōkai, 1975), 541; *Shōwa 16・17-nen Kōkūnenkan* (Tokyo: Dainihon hikō kyōkai, 1943), 325; Google map was used for the distance calculation.

30. Tokyo daigaku 100-nenshi henshū iinkai, *Tokyo daigaku 100-nenshi bukyokushi 3*, 595–596; Kobashi Yasujirō, *Tani Ichirō sensei koki kinen kōenshū*, 1977; and Kōkū uchū kōgakuka kōkū uchū kōgaku senkō gakui ronbun, UTA, accessed 29 September 2011, http://133.11.88.138/FMRes/FMPro. Google map was used for the distance calculation.

31. General Headquarters, United States Army Forces, Pacific Scientific and Technical Advisory Section, *Report on Scientific Intelligence Survey in Japan, September and October 1945*, November 1945, 1: 16, NA.

32. Tokyo daigaku seisan gijutsu kenkyūjo hen, *Tokyo daigaku dai-2 kōgakubushi*, 35, 69.

33. Tokyo daigaku 100-nenshi henshū iinkai, *Tokyo daigaku 100-nenshi bukyokushi 3*, 606.

34. Tokyo daigaku seisan gijutsu kenkyūjo hen, *Tokyo daigaku dai-2 kōgakubushi*, 74–75.

35. Ibid., 72.

36. *Osaka daigaku 25-nen shi*, 6, 355.

37. Sōmushō, Nagoyashi ni okeru sensai no jyōkyō: Ippan sensai homepage, last accessed 14 November 2011, http://www.soumu.go.jp/main_sosiki/daijinkanbou/sensai/situ ation/state/tokai_06.html.

38. Nagoya daigakushi henshū iinkai, *Nagoya daigaku 50-nenshi, bukyokushi 2* (Nagoya: Nagoya daigaku shuppankai, 1989), 10.

39. Among the 10 faculties, that of medicine ranked the highest with 4.6 percent. All the percentages were relative to the number of freshmen during the years under study. Tokyo daigakushi shiryōshitsu hen, 129.

40. Sōmushō, "Tachikawa-shi ni okeru sensai no jōkyō (Tokyo-to)," Ippan sensai homepage, accessed 14 September 2011, http://www.soumu.go.jp/main_sosiki/daijinkanbou /sensai/situation/state/kanto_21.html.

41. *Rikugun kōkū heiki shinsa kenkyū kikan · sokai jyōkyō ichiranhyō*, 15 August 1945, NIDS. One exception to the pattern of evacuation was Tama Research Division, which became independent from the fourth division in June 1943. While two sections were relocated from the city of Kunitachi, Tokyo, to the distant cities of Fujioka (in Gunma prefecture) and Kamisuwa (in Nagano prefecture), two other sections remained in Tokyo: one in the city of Ōme and another in the city of Kugayama.

42. General Headquarters, United States Army Forces, Pacific Scientific and Technical Advisory Section, *Report on Scientific Intelligence Survey in Japan, September and October 1945*, Vol. 2, 1 November 1945, Appendix 1-F-1, NA.

43. Shiryō bōeichō kaijō bakuryōkanbu chōsabu, *Nihon teikoku kaigun no kenkyū narubi ni kaihatsu (1925–45)*, 1957, Shōwakan.

44. United States Pacific Fleet and Pacific Ocean Areas, *Air Information Summary: First Supplement to Tokyo Bay Area Air Information Summary*, 30 December 1944, MUSAFB.

45. Hirayama Shin'ichi, "Kakō mokuzai shūsen zengo no koto," in *Kaigun kōkūgijutsushō zairyōbu shūsen 50-shūnen kinenshi*, 111.

46. Ueno Kagehira, "Kūgishō zairyōbu deno 200-nichi to sengo," in *Kaigun kōkūgijutsushō zairyōbu shūsen 50-shūnen kinenshi*, 118.

47. Kawasaki Motoo, "Shidei buin o shinobu," in *Kaigun kōkūgijutsushō zairyōbu shūsen 50-shūnen kinenshi*, 140–141.

48. Ōta Shimizu, "Kaigun kiryū shucchōjo, in *Kaigun kōkūgijutsushō zairyōbu shūsen 50-shūnen kinenshi*, 33–34; and Kikuchi Teizō, "Omoide," in *Kaigun kōkūgijutsushō zairyōbu shūsen 50-shūnen kinenshi*, 143.

49. *Chōsa kiroku kyū kaigun shiryō kōkū kankei sono 1*, 742-2, 742-4, 742-5, NIDS.

50. Nihon kōkū gakujutsushi henshū iinkai, *Nihon kōkū gakujutsushi*, 419–428.

51. *Chōsa kiroku: Kaigun kōkū kankei gijutsu chōsa, kitai kankei sekei shisaku no bu,* 797-5, NIDS.

52. *Air Technical Intelligence Group No. 27, Aircraft Design and Development,* 1, 1 November 1945, MUSAFB.

53. *Kaigun kōkū gijutsushō kenkyū jikken seiseki hōkoku Kōgyōhō 029 kitai no koyū shindō keisanhō no kenkyū,* 7 January 1942, UTA; *Kaigun kōkū gijutsushō kenkyū jikken seiseki hōkoku Kōgyōhō 04347 Tstubasa fure genkai sokudo kensanhō no kenkyū sono 3,* 4 September 1944, UTA; *Kaigun kōkū gijutsushō kenkyū jikken seiseki hōkoku Kōgyōhō 0546 tsubasa fure genkai sokudo keisanhō kenkyū sono 4,* 20 February 1945, UTA; and Nihon kōkū gakujutsushi henshū iinkai, *Nihon kōkū gakujutsushi,* 60.

54. Shioda Toyoharu, "Keikinzoku no kenkyū," in *Kaigun kōkūgijutsushō zairyōbu shūsen 50-shūnen kinenshi,* 102; and *Chōsa kiroku kyū kaigun shiryō kōkū kankei sono 1,* 742-7, NIDS

55. *Chōsa kiroku kyū kaigun shiryō kōkū kankei sono 1,* 7, NIDS.

56. Naitō Ichirō, "Hiniku na unmei o tadotta kanbaku suisei," in *Suisei/99-kanbaku,*103.

57. Miki Tadanao, "Kōsoku rikubaku ginga sekkei no tsuioku," in *Ginga/Ichishiki rikkō,* 108.

58. Ibid., 107.

59. Kojima Masao, "Dōryoku · Yuatsu sōbi," in *Ginga/Ichishiki rikkō,* 39.

60. *Chōsa kiroku kyū kaigun shiryō kōkū kankei sono 1,* 742, NIDS.

61. Yamana Masao, "Suisei ga dekirumade," *Kōkūfan*13, no. 15 (1964): 78; and *Chōsa kiroku kyū kaigun shiryō kōkū kankei sono 1,* 742–82, NIDS.

62. "Kaigun kōkū gijutsushō shōhō, Showa 18-nen 10-gatsu," NIDS.

63. Shioda Toyoji, "Keikinzoku no kenkyū," in *Kaigun kōkūgijutsushō zairyōbu shūsen 50-shūnen kinenshi,* 102.

64. Ueda Kōzo," Kaigun jidai no omoide," in *Kaigun kōkūgijutsushō zairyōbu shūsen 50-shūnen kinenshi,* 117.

65. United States Pacific Fleet and Pacific Ocean Areas, *Quarterly Report on Research Experiments,* 1, Special Translation no. 52 (1945): 10, MUSAFB.

66. Kawasaki Motoo, "3-ka no katsudō," in *Kaigun kōkūgijutsushō zairyōbu shūsen 50-shūnen kinenshi,* 15, 17; and Bōeichō bōei kenshūjo senshi shiryōshitu, *Kaigun kōkū gaishi,* 433.

67. Iwaya Eiichi, "Arishi hino waga kaigun kōkūki no zenbō to sono jittai," in *Kōkūgijutsu no zenbō* (Tokyo: Kōyōsha, 1953), 1: 256–257.

68. Peattie, *Sunburst,* 189.

69. Vincenti, *What Engineers Know and How They Know It,* 9, 12.

70. Michael S. Sherry, *The Rise of American Air Power: The Creation of Armageddon* (New Haven, CT: Yale Univ. Press, 1987). The author examines technological, strategic, and bureaucratic imperatives as chief sources of wartime fanaticism in support of American strategic bombing. Hugh Gusterson, *Nuclear Rites: A Weapons Laboratory at the End of the Cold War* (Berkeley: Univ. of California Press, 1996) is an ethnographic study of a nuclear weapons laboratory. Gusterson shows how scientists and engineers internally solved nuclear deterrence, which was by nature suicidal. The quote is from page 3.

71. Naitō, Hatsuho, *Ōka: kyokugen no tokkōki* (Tokyo: Chūōbunko, 1999), iii–iv; Ozaki Toshio, "Ōka 43-otsu gata sekkei no omoide, *Kōkūfan* 13, no. 14 (1964): 68; Rene J. Francillon, *Japanese Aircraft of the Pacific War* (Annapolis, MD: Naval Institute Press, 1970),

476–477; and "Structural Features of the Oka or Baka Japanese Suicide Glider Bomb," 6, WRAFB.

72. Assistant Chief of Air Staff, Intelligence, *The Japanese Piloted Rocket-Propelled Suicide Aircraft (BAKA)*, 1 May 1945, 1, MUSAFB.

73. Miki Tadanao, *Ōka 11-gata (genkei) shisaku keika gaiyō 1: Showa 20-nen 10-gatsu*, NIDS.

74. Onda Shigetaka, *Tokkō* (Tokyo: Kōdansha, 1991), 404–408, 410; Hata Ikuhiko, *Shōwa no nazo o ou* (Tokyo: Bungei shunjū, 1999), 1: 510–515; Naitō Hatsuho, *Thunder Gods: The Kamikaze Pilots Tell Their Story* (New York: Dell Book, 1982), 14–17. The extent to which Ōta was involved in—and culpable of—the sin of the MXY7 project has been debated. A few detailed studies chronicle the navy's project. Many agree that the start of the project required far more than Ōta's individual initiative. Deeply suspicious of the lone-man theory, Naitō Hatsuho describes Ōta as the "puppet" of the navy's leadership. Probably an escape goat, Ōta remains portrayed as the villain because of a lack of extant, empirical evidence refuting this loner theory. A more sensible way to reexamine the process may be to focus on the merger of interests between the navy's high command at the top and Ōta at the bottom. By one account, civilian firms were involved at this early stage of the MXY7 blueprint design. Mitsubishi purportedly contributed to the development of the aero-engine through some collaboration with Tokyo University.

75. Kawasaki Motoo, "Kaigun de mananda kotodomo," in *Kaigun kōkūgijutsushō zairyōbu shūsen 50-shūnen kinenshi*, 138.

76. Miki Tadanao, "Kōsoku densha no kaihatsu," in *Heisei 11-nendo Sangyō gijutsu no rekishi ni kansuru chōsa kenkyū hōkokusho* (Tokyo: Nihon kikai kōgyō rengōkai, 2000), 354.

77. Nazuka Iwao, "Ōka no seizō genba de," in *Ningen bakudan to yobarete: Shōgen ōka tokkō* (Tokyo: Bungei shunjū, 2005), 331.

78. Naitō, *Thunder Gods*, 30.

79. *Technical Air Intelligence Center Summary #31 BAKA, June 1945*, 1, MUSAFB.

80. Naitō, *Thunder Gods*, 18.

81. *MXY ridatsu shiken (k1 oyobi k2) 19.10.25 dai-3 fūdō*, MT Papers; and Naitō Hatsuho, *Thunder Gods*, 34.

82. *Ōka no shisaku jikken ni kansuru meirei oyobi keikakusho*, NIDS.

83. Nazuka, "Ōka no seizō genba de," 333.

84. Miki Tadanao, "ōka 11-gata (genkei) shisaku keika gaiyo 1, Shōwa 20-nen 10-gatsu," NIDS.

85. Miki Tadanao, *Jinrai tokubetsu kōgekitai* (Tokyo: Sannō shobō, 1968), 35.

86. Naitō, *Thunder Gods*, 18.

87. Ibid., 24.

88. Nazuka, "Ōka no seizō genba de," 337.

89. Naitō, *Thunder Gods*, 18.

90. Yokosukashi, ed., *Senryōka no Yokosuka: Rengō kokugun no jōriku to sono jidai* (Yokosuka: Yokosukashi, 2005), 97.

91. *Kaigun kōkū gijutsushō denkibu*, 32.

92. *Time*, 30 April 1945, 7 May 1945, 14 May 1945, and 25 June 1945.

93. Itō Hiromitsu, "Haikyo no entotsu," *Suikō* 343 (1982): 24.

94. "Shisei ōka toriatsukai setsuieisho Showa 19-nen 11-gatsu, Kaigun kōkū gijutsushō," NASM.

95. "Kūgishō kimitsu dai 9269-gō," NIDS.

96. Kūgishō kimitsu dai 9509-gō, Showa 19-nen 11-gatsu 3-nichi, Marudai tanza renshūki kansei kenkyūkai oboegaki," NIDS.

97. "Kaigi teki dai-796gō, 20-10-1944 Marudai heiki yō kaizō jikō uchiawasekai shingi jikō tekiyō," MT Papers.

98. Royal Aircraft Establishment, Farnborough, *Foreign Aircraft: "Oka" or "Baka" Japanese Suicide Glider Bomb*, May 1946, 4, WPAFB.

99. "MXY7 kaizō yōkō," MT Papers.

100. "Technical Air Intelligence Center Summary # 31 BAKA, June 1945," 13, MUSAFB.

101. Matsuura Yoshinari, "Tokushu kōgekiki ōka fukugen ni tsuite" (unpublished manuscript, December 2001), Appendix 1.

102. "Ōka 22-gata keikaku yōkō 20-2-28," MT Papers.

103. "Kaigun kōkū gijutsushō hikoukibu sekkeigakari, jyūryō jyūshin keisan 20-1-20," MT Papers.

104. "Daiichi kaigun gijutsushō shisei ōka 22-gata toriatsukai setsumeisho (an) Showa 20-nen 5-gatsu," NIDS.

105. "43-gata jyūryō mitsumori 20-3-22," MT Papers.

106. Aichi kōkūki kabushiki kaisha gijutsubu, "Ōka 43 otsu-gata jyūshin keisansho Showa 20-nen 4-gatsu 5-nichi," MT Papers.

107. "Shisen ōka 43 otsu-gata keisan yōhyū (20.4.26 kettei)," MT Papers.

108. "Showa 20-nen 7-gatsu 23-nichi ōka 43 otsu-gata kōsaku kan'ika ni kansuru uchiawase oboe," MT Papers.

109. *Chōsa kiroku kyū kaigun shiryō kōkū kankei sono 1*, 742-10, NIDS.

110. Iwaya Eiichi, "Arishi hino waga kaigun kōkūki no zenbō to sono jittai," 258.

111. Nihon kōkū gakujutsushi henshū iinkai, 413–428; and *Chōsa kiroku kyū kaigun shiryō kōkū kankei sono 1*, 742-16, NIDS.

112. Kaigun jinrai butai senyūkai henshūiinkai, *Kaigun jinrai butai* (Tokyo: Kaigun jinrai butai senyūkai, 1996), 37; Nihon kōkū gakujutsushi henshū iinkai, *Nihon kōkū gakujutsushi*, 416; and Maema, *Fugaku*, 2: 124–126.

113. Kaigun jinrai butai senyūkai henshūiinkai, *Kaigun jinrai butai*, 3, 17–44.

114. Nihon kaigun kōkūshi hensan iinkai, *Kakgun kōkūshi* (Tokyo: Jiji tsūshinsha, 1969), 1: 513. There is no agreement on the actual number of deaths of Japanese airmen and physical damages inflicted on the Allied forces. According to one report, 3,913 Japanese kamikaze pilots died altogether in the "special attack" missions against the Allies during the war. Of these, 2,525 were navy men, most of whom were between the ages of 18 and 20. Some of them were 17 years old. The remaining 1,388 were army pilots, most of whom were ages between 18 and 24. Naitō Hatsuho, *Thunder Gods*, xxvii.

115. Peter Hill, "Kamikaze, 1943–5," in *Making Sense of Suicide Missions*, ed. Diego Gambetta (Oxford: Oxford Univ. Press, 2005), 8–11.

116. Itō, "Haikyo no entotsu," 23.

117. Yanagida Kunio, *Zerosen moyu* (Tokyo: Bungei shunjū, 1990), 5: 69.

118. Iwaya, "Arishi hino waga kaigun kōkūki no zenbō to sono jittai," 236.

119. Advanced Echelon Far East Air Forces Air Technical Intelligence Group, *Supplement to Report No. 70 High Speed Wind Tunnel Research*, 23 November 1945, 2, MUSAFB.

120. F. W. Williams, "Japan's Aeronautical Research Program and Achievements," in *Technical Intelligence Supplement: A Report Prepared for the AAF Scientific Advisory Group,* May 1946, 151, MUSAFB165-166.

121. Headquarters, Arnold Engineering Development Center Air Research and Development Command, United States, Air Force, *History of the Arnold Engineering Development Center, 1 January 1952–30 June 1952,* 68–69, MUSAFB.

Chapter 4 · Integrating Wartime Experience in Postwar Japan, 1945–1952

1. *Tokyo shinbun,* 15 August 1945.

2. Kimura Hidemasa, *Waga hikōki jinsei* (Tokyo: Nihon tosho center, 1997), 155–157; and Maema Takanori, *YS-11 Kokusan ryokakki o tsukutta otokotachi* (Tokyo: Kōdansha, 1994), 82.

3. Kagaku gijutsu seisakushi kenkyūkai, *Nihon no kagaku gijutsu seisakushi,* 50; and Nakayama Shigeru, "Introduction: Occupation Period 1945–<AP>52," in *A Social History of Science and Technology in Contemporary Japan* (Melbourne: Trans Pacific Press, 2001), 1: 23.

4. *Kagaku bunka shinbun,* 7 September 1946.

5. Nakayama Shigeru, "The Scientific Community Post-Defeat," in *A Social History of Science and Technology in Contemporary Japan,* 1: 270.

6. *Kagaku bunka shinbun,* 29 April 1946.

7. Ibid., 17 December 1946.

8. Ibid., 25 June 1947.

9. Ibid., 7 September 1946.

10. Torigata Hirotoshi, "Kaigun to nihon no sangyō," *Gunji gijutsu kara minsei gijutsu eno tenkan: Dainiji sekai taisen kara sengo eno wagakuni no keiken* (Tokyo: Nihon gakujutsu shinkōkai, 1994), 1: 161–178.

11. Yamashita Sachio, "Nihon zōsengyō ni miru gijutsu no keishō : Senzen kara sengo e," in *Kigyō keiei no rekishiteki kenkyū* (Tokyo: Iwanami shoten, 1990), 364–389; Ikeda Tomohira, "Nihon zōsengyō no sengo 10-nen: Yushutsu sangyō eno doutei," in *Kigyō keiei no rekishiteki kenkyū* , 412–432; Maema Takanori, *Senkan Yamato no iseki,* vol. 1 (Tokyo: Kōdansha, 2005); and Kohagura Yasuyoshi, *Kōseki: Zōsen shikan Fukuda Tadashi no tatakai* (Tokyo: Kōjinsha, 1996).

12. Nakagawa Yasuzō , *Kaigun gijutsu kenkyūjo: Electronics ōkoku no senkusha tachi* (Tokyo: Nihon keizai shinbunsha, 1987), 13 14; Akio Morita Library, "2010 Morita Asset Management," accessed 23 September 2010, http://www.akiomorita.net/profile/life.html; Morita Akio and Ibuka Masaru, "Ibuka taidan, Innen ni michibikareruyō ni (2) Guest: Morita Akio," accessed 23 September 2010, http://www.sony-ef.or.jp/library/ibuka/pdf /taidan_no62_2.pdf; and John Nathan, *Sony: The Private Life* (New York: Houghton Mifflin, 1996), 4–5.

13. Suzukawa Hiroshi, "Gunji gijutsu no heiwa sangyō eno kakawari: Seimitsu kikai sangyō ni okeru jirei," in *Gunji gijutsu kara minsei gijutsu eno tenkan: Dainiji sekai taisen kara sengo eno waga kuni no taiken,* 1: 78–93.

14. Kōgaku kōgyōshi henshū kai, ed., *Heiki o chūshin to shita nihon no kōgaku kogyōshi* (Tokyo: Kōgaku kogyōshi henshū kai, 1955), 610–611.

15. Tomita Tetsuo, "Gijutsu no juyō ni oyobosu shijōkōzō oyobi fūdo kankyō ni kansuru jisshōteki bunseki" (PhD diss., Tokyo Institute of Technology, 1999).

16. Nakajima Shigeru, "Hisenryō ka ni okeru seisan hinmoku settei to sonogo no seisan jōkyō (nihon musen no baai)," in *Gunji gijutsu kara minsei gijutsu eno tenkan: Dainiji*

sekai taisen kara sengo eno waga kuni no taiken, 1: 94–113; and Tsurugaya Takeo, "SONAR Gunjugijutsu kara heiwa sangyō eno tenkan," in *Gunji gijutsu kara minsei gijutsu eno tenkan: Dainiji sekai taisen kara sengo eno waga kuni no taiken*, 2: 64–76.

17. Matsuyama Kihachirō, "Televison no gunji gijutsu kara minju sangyō eno tenkan," in *Gunji gijutsu kara minsei gijutsu eno tenkan: Dainiji sekai taisen kara sengo eno waga kuni no taiken*, 1: 114–137.

18. *Kagaku bunka shinbun*, 20 December 1948.

19. Shigeru Nakayama, "The Scientific Community Post-Defeat," 1: 267.

20. Doi Takeo, *Hikōki sekkei gojūnen no kaisō* (Tokyo: Kantōsha, 1989), 250–253; and Maema, *YS-11 Kokusan ryokakki o tsukutta otokotachi*, 108.

21. Watanabe Saburō, conversation with author, Tokyo, 16 July 2003.

22. Maema, *Gijutsusha tachino haisen*, 39.

23. Nakagawa Ryōichi, "Watashi no senzen, sengo (hikōki kara jidōsha e) Dual Use Technology," in *Gunji gijutsu kara minsei gijutsu eno tenkan: Dainiji sekai taisen kara sengo eno waga kuni no taiken* (Tokyo: Nihon gakujutsu shinkō kai, 1994), 1: 16.

24. Maema, *Man Machine no Shōwa densetsu*, 1: 569.

25. Itokawa Hideo, *Kyōi no jikan katsuyōjutsu: Naze koredake saga tsukunoka?* (Tokyo: PHP kenkyūjo, 1985), 162–163.

26. The wartime aircraft industry and engineering formed the basis for the postwar success in the civilian automobile industry. Matthias Koch, *Rüstungskonversion in Japan nach dem Zweiten Weltkrieg: Von der Kriegswirtschaft zu einer Weltwirtschaftsmacht* (Tokyo: Deutsches Institut für Japanstudien, 1998), 152–175; and Yamaoka Shigeki, "Mitsubishi ZC707 Chijō ni orita engine," *Tetsudōshigaku* 11 (December 1992): 7–13.

27. Teruo Ikehara, "Corolla kaihatsu no Hasegawa Tatsuo-shi ni okeru 'Shusa 10-kajō," *Nikkei Business Online*, accessed 22 September 2010, http://business.nikkeibp.co.jp/article/tech/20060825/108606/; and Maema, *Man machine no Shōwa densetsu*, 1: 603–615.

28. Hasegawa Akio, *My Father Tatsuo Hasegawa (1916–2008)*, accessed 22 September 2010, http://www.geocities.jp/pinealguy/tatsuo.tatsuo.htm; and Toyota Motors, *Corolla no tetsugaku*, accessed 22 September 2010, http://toyota.jp/information/philosophy/corolla/history/index.html.

29. Subaru Museum, "Saishō gen no tuning ni yoru chōsen: Subaru hakubutsukan," accessed 18 February 2011, http://members.subaru.jp/know/museum/subaru360/; and Maemai, *Gijutsusha tachi no haisen*, 250; and Maema, *Man machine no Showa densetsu*, 2: 91–94.

30. Maema, *Gijutsusha tachi no haisen*, 210–255; and Maema, *Man machine no Showa densetsu*, 1: 24–70, 751–752.

31. Ryōichi Nakagawa, "Watashi no senzen, sengo (hikōki kara jidōsha e) Dual Use Technology," 1: 9–34 (quotation from p. 32).

32. Fukushima Mutsuo, "Zero Inspired Today's Innovations," *Japan Times*, 14 January 2004.

33. Maema, *YS-11 Kokusan ryokakki o tsukutta otokotachi*, 88.

34. *Meikū kōsakubu no senzen sengoshi: Moriya sōdanyaku, watashi to kōkūki seisan* (Nagoya: Mitsubishi jūkōgyō kabushiki kaisha Nagoya kōkūki seisakujo, 1988), 49–50; *Kōsoku daisha shindō kenkyūkai kiroku, dai ikkai-dairokkai*, RTRI archive; and Japan Automobile Hall of Fame, 2009, "Kūriki no tokusei to kihon jūshi no kōseinō o kaihatsu: Kubo Tomio," accessed 22 September 2010, http://www.jahfa.jp/JAHFA_PR2_2009.pdf.

35. *Kōkūkai kaiin meibo 1973*; and *Kōkūkai kaiin meibo 1976*.

36. Gary Coombs, "Opportunities, Information Networks and the Migration-Distance Relationship," *Social Networks* 1 (1979): 257–276.

37. Louise Young, *Japan's Total Empire: Manchuria and the Cultural Wartime Imperialism* (Berkeley: Univ. of California Press, 1998).

38. Hirota Kōzō , *Mantetsu no shūen to sonogo: Aru chūō shikenjoin no hōkoku* (Tokyo: Sōgensha, 1990).

39. Daqing Yang, "Chūgoku ni todomaru nihon gijutsusha: Seiji to gijutsu no aida," in *1945-nen no rekishi ninshiki: Shūsen o meguru nicchū taiwa no kokoromi*, ed. Kawashima Makoto (Tokyo: Tokyo Univ. Press, 2009), 113–139; Zhongguo zhongri guanxi shixuehui, ed., *Youyizhuchunqiu: Weixin zhongguo zuochu gongxian de ribenren* (Beijing: Xinhua chuban, 2002); and Nishikawa Akiji no omoide henshū iinkai, *Nishikawa Akiji no omoide* (Nagoya, 1964).

40. Robert Kane, *All Transportation*, 14th ed. (Dubuque, IA: Kendall Hunt Publishing, 2003), 82–83; Jean-Denis G. G. Lepage, *Aircraft of the Luftwaffe, 1935–1945: An Illustrated Guide* (Jefferson, NC: McFarland, 2009), 23; and Young, *Japan's Total Empire*, 312–313.

41. Saitō Michinori, *Chōhōsen: Rikugun noborito kenkyūjo* (Tokyo: Gakushū kenkyūsha, 2001), 7–29.

42. *Pacific Air Command Occupation Directive Number 3*, 28 June 1946, MUSAFB.

43. John Anderson, Jr., *A History of Aerodynamics and Its Impact on Flying Machines* (Cambridge: Cambridge Univ. Press, 2000), 295–296; and Michael Eckert, "Strategic Internationalism and the Transfer of Technical Knowledge: The United States, Germany, and Aerodynamics after World War I," *Technology and Culture* 46, no. 1 (2005): 104–131.

44. Timothy Hatton and Jeffrey Williamson, *Global Migration and the World Economy: Two Centuries of Policy and Performance* (Cambridge, MA: MIT Press, 2005), 60–67; Coombs, "Opportunities," 259–261; and Davor Jedlicka, "Opportunities, Information Networks and International Migration Streams," *Social Networks*, 1 (1979): 277–284.

45. Readily available examples of such exceptions include, for instance, biochemist Takamine Jōkichi (1854–1922). In the late 1880s, he emigrated from Japan to the United States where he acquired an American wife. In New York City, he established his own research laboratory and conducted research on the hormone adrenaline. Also relevant is the case of Yuasa Toshiko (1909–80), Japan's female physicist who stayed in France during 1940–44 and after five years in Tokyo, returned to France in January 1949. Itō Kenji, "Gender and Physics in Early 20th Century Japan: Yuasa Toshiko's Case," *Historia Scientiarum* 14 (2004): 118–135.

46. John Gimbel, *Science, Technology, and Reparations: Exploitation and Plunder in Postwar Germany* (Stanford, CA: Stanford Univ. Press, 1990); and Roger E. Bilstein, *Orders of Magnitude: A History of the NACA and NASA, 1915–1990* (Washington, DC: National Aeronautics and Space Administration, 1989).

47. Burghard Ciesla, "Das 'Project Paperclip'—deutsche Naturwissenshaftler und Techniker in den USA (1946 bis 1952), in *Historische DDR-Forschung : Aufsätze und Studien* (Berlin: Academie-Verlag, 1994), 1: 287–301.

48. Bowen C. Dees, *The Allied Occupation and Japan's Economic Miracle: Building the Foundations of Japanese Science and Technology* (Surrey, England: Japan Library, 1997), 49–56.

49. Ibid.

50. Recent studies about infamous Japanese Army Unit 731 are revealing. Military scientists conducted hideous experiments on human subjects, helped develop biochemi-

cal weapons, and provided the United States with their medical expertise after the war. In exchange for their wartime research reports, they were exempt from indictment from the Occupation Authority during the Tokyo Military Tribunal. Apparently, none of them emigrated outside of Japan. Sheldon Harris, *Factories of Death: Japanese Biological Warfare, 1932–1945, and the American Cover-Up* (New York: Routledge, 2002).

51. Saitō, *Chōhōsen*, 131–152, 186–222.

52. John Dower, *War without Mercy: Race and Power in the Pacific War* (New York: Pantheon Books, 1986).

53. Mae M. Ngai, "The Architecture of Race in American Immigration Law: A Reexamination of the Immigration Act of 1924," *Journal of American History* 86, no. 1 (1999): 67–92.

54. Marion Bennett, "The Immigration and Nationality (McCarran-Walter) Act of 1952, as Amended to 1965, *Annals of the American Academy of Political and Social Science* 367 (1966): 131; and Helen Eckerson, "Immigration and National Origins," *Annals of the American Academy of Political and Social Science* 367 (1966): 8.

55. Suga Miya, "Beikoku 1952-nen imin kikahō to nihon ni okeru 'imin mondai' kan no henyō," *Tokyo gakugei daigaku kiyō jinbun shakai kagaku-kei II*, 61 (2010): 129–130.

56. Coombs, "Opportunities," 258–259; and Hatton and Williamson, *Global Migration and the World Economy*, 225–229.

57. Shigeru Nakayama, "The Sending of Scientists Overseas," in *A Social History of Science and Technology in Contemporary Japan*, 1: 249.

58. Ibid., 250, 256.

59. Ernest Rubin, "The Demography of Immigration to the United States," *Annals of the American Academy of Political and Social Science* 367 (1966): 20–21; Adam McKeown, *Melancholy Order: Asian Migration and the Globalization of Borders* (New York: Columbia Univ. Press, 2008), 91, 361; Marion Houstoun, Roger Kramer, and Joan Barrett, "Female Predominance in Immigration to the United States since 1930: A First Look," *International Migration Review* 18, no. 4 (1984): 913–920; and Yasutomi Shigeyoshi, "'Sensō hanayome' to Nikkei community (III) stereotypes ni motozuku haiseki kara juyō e, *Kaetsu daigaku kenkyū ronshū* 44, no. 2 (2002): 57–64.

60. Nakayama Shigeru, "Sending Scientists Overseas," 1: 249–252.

61. Itō, "Gender and Physics," 118–135.

62. The birth-order theory has been a controversial subject since the turn of the twentieth century. While nature-versus-nurture debates have elicited responses from well-known psychiatrists, including Alfred Adler (1870–1937), advocates suggest that birth order tends to leave a profound, lasting influence on the person's disposition. An individual's personality could be malleable to multiple intervening factors, including the changing economic circumstances of the parent(s) as well as gender dynamics and living conditions of the family. Accompanied by statistical data, this theory has been a useful tool for academic scholars to examine the social, political, and demographic changes associated with birth order; for instance, firstborns had historically tended to be more reserved, conservative, adamant, conscientious, risk averse, and domineering in society than younger siblings. Birth order could leave a social impact on patterns of migration in Europe. In nineteenth-century Norway, where older sons tended to legally inherit the entire family estate, the younger siblings with neither property nor obligations seemed more likely to immigrate

to the United States. Similarly, in rural villages of Alsace from 1750 to 1885, firstborn children were overall more likely to marry and remained in their villages of residence longer than later-born children. More complicated dynamics could operate in Japan. For instance, in the rural, agriculture-based Echizen region from 1823 to 1871, among various personal and social characteristics of migrants, such as gender, birth order, age, marital status, and class (landed households versus landless households), socioeconomic class was a determining factor in shaping the demographical composition of migrants and their destinations. Frank Sulloway, *Born to Rebel: Birth Order, Family Dynamics, and Creative Lives* (New York: Pantheon Books, 1996); Ran Abramitzky, Leah Boustan, and Katherine Eriksson, "Productivity and Migration: New Insights from the 19th Century," *Stanford News*, 1 May 2010, accessed 7 March 2011, http://news.stanford.edu/news/2010/may/siepr-productivity -migration-050310.html; Kevin McQuillan, "Family Composition, Birth Order and Marriage Patterns: Evidence from Rural Alsace, 1750–1885," in *Annales de démographie historique* 1 (2008): 57–71; and Mark Fruin, "Peasant Migrants in the Economic Development of Nineteenth-Century Japan," *Agricultural History* 54, no. 2 (1980): 261–277.

63. Rubin, "Demography of Immigration," 21.

64. This life transition was not unique to Japan during the twentieth century; it could be well observed among migrants in Africa, Latin America, and Europe. James White, "Internal Migration in Prewar Japan, *Journal of Japanese Studies* 4, no. 1 (1978): 89–91.

65. Saitō, *Chōhōsen*, 193–208.

Chapter 5 • Former Military Engineers in the Postwar Japanese National Railways, 1945–1955

1. Fritz Zwicky, "Remarks on the Japanese War Effort," in *Technical Intelligence Supplement: A Report Prepared for the AAF Scientific Advisory Group*, May 1946, 165–166, MUSAFB.

2. Nihon kokuyū tetsudō tetsudō gijutsu kenkyūjo 50nen-shi kankō iinkai, *Nihon kokuyū tetsudō tetsudō gijutsu kenkyūjo 50nen-shi* (Tokyo: Kenyūsha, 1957), 274; *Nihon kokuyū tetsudō, 100-nen shi* (Tokyo: Nihon kokuyū tetsudō, 1973), 10: 127; and Aoki et al., *A History of Japanese Railways*, 118.

3. Nihon kokuyū tetsudō tetsudō gijutsu kenkyūjo 50nen-shi kankō iinkai, *Nihon kokuyū tetsudō tetsudō gijutsu kenkyūjo 50nen-shi*, 666.

4. *Tetsudō 80-nen no Ayumi, 1872–1952* (Tokyo: Nihon kokuyū tetsudō, 1952), 100.

5. Nihon kokuyū tetsudō, *Tetsudō gijutsu hattatsushi dai 6-hen (senpaku), dai 7-hen (kenkyū) dai 8-hen (nenpyō)* (Tokyo: Nihon kokuyū tetsudō, 1958), 362.

6. Noma Sawako, ed., *Shōwa 20-21 nen, Shōwa 20,000 nichi no zenkiroku* (Tokyo: Kōdahsha, 1989), 7:185.

7. Un'yushō, *Kokuyū tetsudōno genjō: Kokuyū tetsudō jissō hōkokusho* (Tokyo, 1947), 27.

8. Nihon kokuyū tetsudō, *100-nen shi*, 10:127.

9. Kubota Hiroshi, *Tetsudō jūdai jiko no rekishi* (Tokyo: Grandpri shuppan, 2000), 54.

10. Nishii Kazuo, ed., *Shōwa-shi zenkiroku* (Tokyo: Mainichi shinbunsha, 1989), 8: 373.

11. Kubota, *Tetsudō jūdai jiko no rekishi*, 83.

12. Uno Hiroshi, ed., *Asahi shinbun ni miru nihon no ayumi: Shōdo ni kizuku minshu shugi* (Tokyo: Asahi shinbunsha, 1973), 2: 40.

13. *Asahi shinbun*, 26 February 1947.

14. Editorial, *Asahi shinbun*, 27 February 1947.

15. *Nihon kokuyū tetsudō, 100-nen shi* 11: 707–8; and Izawa Katsumi, "Kyakusha Kōtaika," *Kōtsū gijutsu* 40 (1949): 15.

16. Uno Hiroshi, ed., *Asahi shinbun ni miru Nihon no ayumi: shōdo ni kizuku minshu shugi* (Tokyo: Asahi shinbunsha, 1973), 3: 85.

17. Noma Sawako, ed., *Shōwa ni man nichi no zenkiroku* (Tokyo: Kōdahsha, 1989), 8: 302–305. After a series of trials that attracted media attention, in 1963 the Supreme Court exonerated those arraigned for the derailment.

18. Nihon kokuyū tetsudō, *Tetsudō sengo shorishi* (Tokyo: Taishō shuppan, 1981), 799.

19. Un'yushō, *Kokuyū tetsudō no genjō: Kokuyū tetsudō jissō hōkokusho* (1947), 42–45.

20. Ibid., 46–47; and Aoki et al., *A History of Japanese Railways*, 122.

21. *Kokutetsu shokuin meibo (gijutsu gakushi) shōwa 25-nen 8-gatsu 10-ka genzai* (1950).

22. Kanematsu Manabu, *Shūsen zengo no ichi shōgen: Aru tetsudōjin no kaisō* (Tokyo: Kōtsū kyōkai, 1986), 46–47.

23. Aoki et al., *A History of Japanese Railways*, 118.

24. *Kokutetsu shokuin meibo (gijutsu gakushi) shōwa 25-nen 8-gatsu 10-ka genzai* (1950).

25. Un'yushō, *Kokuyū tetsudō no genjō*, 47–48.

26. Nihon kokuyū tetsudō tetsudō gijutsu kenkyūjo 50nen-shi kankō iinkai, *Nihon kokuyū tetsudō tetsudō gijutsu kenkyūjo 50nen-shi*, 1–35.

27. Ibid., 3, 40–43, 51–52, 204–205, 828.

28. *Kokutetsu shokuin meibo (gijutsu gakushi) shōwa 25-nen 8-gatsu 10-ku genzai* (1950).

29. Nihon kokuyū tetsudō tetsudō gijutsu kenkyūjo 50nen-shi kankō iinkai, *Nihon kokuyū tetsudō tetsudō gijutsu kenkyūjo 50nen-shi*, 42; and Nihon kōkū gakujutsushi henshū iinkai-hen, *Nihon kōkū gakujutsushi, 1910–1945* (Tokyo: Maruzen, 1990), 292.

30. Takabayashi Morihisa, personal communication, 14 December 2002.

31. Nihon kokuyū tetsudō tetsudō gijutsu kenkyūjo 50nen-shi kankō iinkai, *Nihon kokuyū tetsudō tetsudō gijutsu kenkyūjo 50nen-shi*, 84–85.

32. Ibid., 148–150.

33. Ibid., 150–151.

34. Ibid., 164–171.

35. Ikari, *Kaigun gijutsushatachi no taiheiyō sensō*, 262.

36. Nakamura Hiroshi, telephone conversation with author, 24 March 2004.

37. Tetsudō gijutsu kenkyūjo, *Tetsudō gijutsu kenkyūjo sōritsu 70-shūnen: 10-nen no ayumi* (Tokyo: Tetsudō gijutsu kenkyūjo, 1977), 290–291; Miki Tadanao, "Monorail 45-nen no tsuioku," *Monorail 82* (1994): 1; Hayashi Masami, "Watashi no sengo," *Kaigun kōkū gijutsushō denkibu* (Tokyo: Kūgishō denkibu no kai, 1987), 97; and interview with Nakamura Hiroshi, 24 March 2004.

38. Kubo Masaki, "Tetsudō gijutsu kenkyūjo no genzai to shōrai eno michi," *Kōtsū gijutsu* 30 (1949): 10–15.

39. Un'yushō, *Kokuyū tetsudō no fukkō: Tetsudō 75-nen kinen shuppan daiisshū* (1948), 203.

40. Aoki et al., *A History of Japanese Railways*, 121.

41. *Kokutetsu shokuin meibo (gijutsu gakushi) shōwa 25-nen 8-gatsu 10-ka genzai* (1950); and *Shōwa 25-nen 12-gatsu 15-nichi genzai (honchō) Nihon kokuyū tetsudō shokuinroku* (1951), 49, TM.

42. Ibid.

43. Uchihashi Katsuto, *Zoku zoku takumi no jidai: Kokutetsu gijutsujin zero hyōshiki kara no nagai tabi* (Tokyo: Sankei shuppan, 1979), 24, 35.

44. Satō Yasushi, "*Dainiji sekai taisen zengo no kokutetsu gijutsu bunka,*" *Kagakushi Kenkyū* 46 (2007): 211.

45. Nihon kokuyū tetsudō tetsudō gijutsu kenkyūjo 50nen-shi kankō iinkai, *Nihon kokuyū tetsudō tetsudō gijutsu kenkyūjo 50nen-shi*, 49, 416.

46. Matsudaira Tadashi, "Kōsoku tetsudō gijutsu no raimei II," *Railway Research Review* 50, no.4 (1993): 30, 32.

47. Un'yushō, *Tetsudō gijutsu kenkyūjo: Shōwa 22-nendo nenpō* (Tokyo, 1947), 6, RTRI archive.

48. *Kōsoku daisha shindō kenkyūkai kiroku dai 1-kai—dai 6-kai*, RTRI archive.

49. Ibid.

50. Matsudaira Tadashi, "Kōsoku tetsudō gijutsu no raimei I," *Railway Research Review* 50, no. 3 (1993): 28.

51. Shima Hideo, *D-51 kara Shinkansen made: Gijutsusha no mita kokutetsu* (Tokyo: Nihon keizai shinbunsha, 1977), 119.

52. Un'yushō Tetsudō gijutsu kenkyūjo, *Shōwa 22-nendo nenpō* (1947), 37.

53. Matsudaira, "Kōsoku tetsudō gijutsu no raimei I," 29.

54. Matsudaira, "Kōsoku tetsudō gijutsu no raimei II," 32.

55. *Kenkyū happyō kōenkai kōen gaiyō, Shōwa 23-nen 4-gatsu 19–20 nichi, Tetsudō gijutsu kenkyūjo*, RTRI archive.

56. Matsudaira Tadashi, "Kyakusha oyobi densha no koyū shindōsū," *Tetsudō gyōmu kenkyū shiryō* 6, no. 2 (1949): 3–14.

57. Matsudaira Tadashi, "Sharinjiku no dakōdō," *Tetsudō gyōmu kenkyū shiryō* 9, no. 1 (1952): 16–26; and Nihon kokuyū tetsudō tetsudō gijutsu kenkyūjo 50nen-shi kankō iinkai, *Nihon kokuyū tetsudō tetsudō gijutsu kenkyūjo 50nen-shi*, 43.

58. Matsudaira Tadashi et al., "Nijiku kasha no banetsuri sōchi kaizō ni yoru kōsokuka," *Tetsudō gyōmu kenkyū shiryō* 10, no. 18 (1953): 5–9.

59. Nihon kokuyū tetsudō tetsudō gijutsu kenkyūjo 50nen-shi kankō iinkai, *Nihon kokuyū tetsudō tetsudō gijutsu kenkyūjo 50nen-shi*, 201–206; Hashimoto Kōichi, "Kokutetsu ni okeru kyōryō kyōdo shindō shiken no genjō to shōrai," *Doboku gakkaishi* 33, 110. 5–6 (1948): 31–34; and Hashimoto Kōichi and Itō Fumihito, "Rosen dōro Miyagino-bashi no kyōdo sokutei," *Doboku gakkaishi* 37, no. 4 (1952): 13–17.

60. Enomoto Shinsuke, "Kinzoku zairyo no hirō to naibu masatsu ni kansuru kenkyū," *Chūō kōkū kenkyūjo ihō* 2, no.7 (1943): 177–189; and Enomoto Shinsuke, "Kinzoku zairyo no hirō to naibu masatsu ni kansuru kenkyū," *Chūō kōkū kenkyūjo ihō* 2, no. 10 (1943): 305–324; and Enomoto Shinsuke, letter to author, 5 April 2004.

61. Nihon kokuyū tetsudō tetsudō gijutsu kenkyūjo 50nen-shi kankō iinkai, *Nihon kokuyū tetsudō tetsudō gijutsu kenkyūjo 50nen-shi*, 526.

62. *Kokutetsu shokuin meibo (gijutsu gakushi) shōwa 25-nen 8-gatsu 10-ka genzai* (1950); Tetsudōgijutsu kenkyūjo, *Senpai genshokusha meibo, Shouwa 52-nen 1-gatsu 1-nichi genzai*; and Kaigun kōkū gijutsushō zairyōbu no kai, *Kaigun kōkū gijutsushō zairyōbu shūsen 50-shūnen kinenshi* (Tokyo: Kaigun kōkū gijutsushō zairyōbu no kai, 1996), 4.

63. Ikari Yoshirō, *Kōkū technology no tatakai* (Tokyo: Kōjinsha, 1996), 261.

64. Kaigun kōkū gijutsushō zairyōbu no kai, *Kaigun kōkū gijutsushō zairyōbu shūsen 50-shūnen kinenshi*, 152.

65. Ibid., 134; and Nihon kōkū gakujutsushi hensan iinkai-hen, *Nihon kōkū gakujutsushi*, 145–146.

66. Nihon kōkū gakujutsushi hensan iinkai-hen, *Nihon kōkū gakujutsushi*, 142–145.

67. Nihon kokuyū tetsudō tetsudō gijutsu kenkyūjo 50nen-shi kankō iinkai, *Nihon kokuyū tetsudō tetsudō gijutsu kenkyūjo 50nen-shi*, 536.

68. Nihon kōkū gakujutsushi hensan iinkai-hen, *Nihon kōkū gakujutsushi*, 161.

69. Kaigun kōkū gijutsushō zairyōbu no kai, *Kaigun kōkū gijutsushō zairyōbu shūsen 50-shūnen kinenshi*, 182.

70. Ibid., 37.

71. Nihon kōkū gakujutsushi hensan iinkai-hen, *Nihon kōkū gakujutsushi*, 162.

72. Nihon kokuyū tetsudō tetsudō gijutsu kenkyūjo 50nen-shi kankō iinkai, *Nihon kokuyū tetsudō tetsudō gijutsu kenkyūjo 50nen-shi*, 668.

73. Ibid.

74. Ibid., 669.

75. Ibid., 670. A few railway companies developed a similar method at the time. Tani Seiichirō, "Bōfu makuragi ni tsuite," *Kōtsū gijutsu* 64 (1951): 30–31.

76. Nihon kokuyū tetsudō tetsudō gijutsu kenkyūjo 50nen-shi kankō iinkai, *Nihon kokuyū tetsudō tetsudō gijutsu kenkyūjo 50nen-shi*, 12, 416–420; and *Kokutetsu shokuin meibo (gijutsu gakushi) shōwa 25-nen 8-gatsu 10-ka genzai*.

77. Sagawa Shun'ichi, "Ressha tono tsūshin I," *Japan Railway Engineer's Association* 5, no. 1 (1962): 50.

78. Shinohara Osamu, letter to author, 11 March 2004; *Kokutetsu shokuin meibo (gijutsu gakushi) shōwa 25-nen 8-gatsu 10-ka genzai* (1950); and Tetsudō gijutsu kenkyūjo, *Senpai genshokusha meibo, Shouwa 52-nen 1-gatsu 1-nichi genzai*.

79. Nihon kokuyū tetsudō tetsudō gijutsu kenkyūjo 50nen-shi kankō iinkai, *Nihon kokuyū tetsudō tetsudō gijutsu kenkyūjo 50nen-shi*, 423.

80. Sagawa, "Ressha tono tsūshin I," 51.

81. Nihon kokuyū tetsudō tetsudō gijutsu kenkyūjo 50nen-shi kankō iinkai, *Nihon kokuyū tetsudō tetsudō gijutsu kenkyūjo 50nen-shi*, 422–431; Amamiya Yoshifumi, "Kokutetsu densha yori hassei suru chūtanpa musen zatsuon ni kansuru kenkyū," *Tetsudō gijutsu kenkyū hōkoku* 58, no.9 (1959): 1–10; and Amamiya Yoshifumi and Maki Yoshikata, "Zatsuon denryoku ni chakumoku shita zatsuongen tanchiki," *Tetsudō gijutsu kenkyū hōkoku* 132, no. 23 (1960): 1; *Kaigun kōkū honbu Shōwa 20-nen 3-gatsu denki kankei gijutsushikan meibo*, Shōwakan archive; and Shinohara Osamu, letter to author, 11 March 2004.

82. Maruhama Tetsurō, "Kokutetsu shuyō kansenkei SHF kaisen wo kaerimite," *Kōtsū gijutsu* 176 (1960): 22.

83. Shinohara Yasushi and Hiroyuki Kimoto, "Osaka-Himeji kan S.H.F. no sekkei to shiken kekka," *Kōtsū gijutsu* 99 (1954): 32–35; and *Hattensuru tetsudō gijutsu: Saikin 10-nen no ayumi* (Tokyo: Nihon tetsudō gijutsu kyōkai, 1965), 70.

84. Un'yu gijutsu kenkyūjo, *10-nen shi* (Tokyo: Transportation Technology Research Center, 1960), 275–276.

85. Nakata Kin'ichi, "Tetsudō gijutsu kenkyūjonai ni okeru gas turgine no kenkyū," *Tetsudō gyōmu kenkyū shiryō* 7, no. 17 (1950): 4.

86. Nihon kōkū gakujutsushi hensan iinkai-hen, *Nihon kōkū gakujutsushi*, 152.

87. Nihon kokuyū tetsudō tetsudō gijutsu kenkyūjo 50nen-shi kankō iinkai, *Nihon kokuyū tetsudō tetsudō gijutsu kenkyūjo 50nen-shi*, 375–376.

88. Interview with Ueda Toshio, 29 April 2004.

Chapter 6 · Opposition Movements of Former Military Engineers in the Postwar Railway Industry, 1945–1957

1. *Asahi shinbun,* 21–28 September 1957.

2. Imamura Yōichi, "Yokosuka, Kure, Sasebo, Maizuru ni okeru kyūgunyōchi no tenyō ni tsuite: 1950–1976 nendo no kyūgunkōshi kokuyū zaisan shori shingikai ni okeru kettei jikō no kōsatsu o tōshite," *Nihon toshi keikaku gakkai toshikeikaku ronbunshū* 43, no. 3 (2008): 194.

3. Ibid., 194–195.

4. Yokosukashi, ed., *Yokosukashishi* (Yokosuka: Yokosukashi, 1988), 1: 566.

5. Yokosukashi, ed., *Senryōka no Yokosuka: Rengō kokugun no jōriku to sono jidai* (Yokosuka 2005), 74; and Yokosukashishi hensan iinkai, *Yokosukashishi* (Yokosuka: Yokosuka shiyakusho, 1957), 1286.

6. Yokosukashi, *Senryōka no Yokosuka,* 52; and Yokosukashishi hensan iinkai, *Yokosukashishi,* 1269.

7. Yokosukashi, *Yokosukashishi,* 1: 579.

8. Ibid., 571.

9. Makita Kōji and Fujita Yōetsu, "Kyū Yokosuka kaigun kōshō kōin kishukusha o tenyō shita shiei jūtaku ni tsuite: Yokosukashi ni okeru sengo shiei jūtaku ni kansuru kenkyū sono 5," accessed 7 September 2010, http://www.cit.nihonu.ac.jp/kenkyu/kouennkai/ref erence/No_38/4_kenchiku/4-050.pdf.

10. Imamura, "Yokosuka, Kure, Sasebo, Maizuru," 194.

11. Yokosukashi, *Yokosukashishi,* 1: 522; and Yokosukashi, *Senryōka no Yokosuka,* 6.

12. Matsuura Yoshinari, a navy technician and a local historian, interview by author, Yokosuka, Japan, 1 August 2006.

13. Yokosukashishi hensan iinkai, *Yokosukashishi* (1957), 1254.

14. Yokosukashi, *Yokosukashishi* (1988), 1: 587–589.

15. Ibid., 534.

16. Ibid., 568.

17. Yokosukashi, *Senryōka no Yokosuka,* 74.

18. Yokosukashi, *Yokosukashishi,* 1: 633.

19. Miki Tadanao, interview by author, Zushi, 19 June 2003.

20. Yokosukashi, ed., *Yokosukashishi,* 1: 625.

21. Ibid., 630.

22. British Intelligence Objectives Sub-committee, *Japanese Aerodynamic Research and Research Equipment,* 4 November 1945, BWM.

23. *Air Technical Intelligence Report No. 27, Aircraft Design and Development,* 1 November 1945, 1, MUSAFB.

24. Miki Tadanao, "ōka 11-gata (genkei) shisaku keika gaiyō (1) shōwa 20-nen 10-gatsu," NIDS.

25. Yokosukashi, *Yokosukashishi,* 1: 579.

26. Ibid., 580, 582.

27. Ibid., 580, 634.

28. Sangiin kaigiroku, "Sangiin kaigi gijiroku jōhō dai001kai kokkai, zaisei oyobi kinyū iinkai dai 50-gō," accessed 6 September 201, NDL, http://kokkai.ndl.go.jp/SENTAKU/san giin/001/1362/00112071362050c.html.

29. Sōmushō, "Kyū gunkōshi tenkanhō, Shōwa 25-nen 6-gatsu 28-nichi hōritsu dai 220-gō," accessed 18 April 201, http://law.egov.go.jp/htmldata/S25/S25HO220.html.

30. Sangiin kaigiroku, "Shōwa 25-nen 3-gatsu 24-ka Sangiin ōkura iinkai dai 29-gō," NDL, accessed 10 September 2010, http://kokkai.ndl.go.jp/cgi-bin/KENSAKU/ swk_dis pdoc.cgi?SESSION=18564&SAVED_RID=2&PAGE=0&POS=0&TOTAL=0&SRV _ID=1&DOC_ID=6815&DPAGE=2&DTOTAL=42&DPOS=40&SORT_DIR=1&SORT _TYPE=0&MODE=1&DMY=18949.

31. Schūgiin kaigiroku, "Showa 25-nen 4-gatsu 10-ka dai 7-kai kokkai ōkura iinkai dai 48-go," NDL, accessed 10 September 2010, http://kokkai.ndl.go.jp/SENTAKU/syugiin/007 /0284/00704100284048c.html.

32. Shūgiin kaigiroku, "Shōwa 25-nen 4-gatsu 11-nichi dai 7-kai kokkai honkaigi dai 36-gō," accessed 10 September 2010, NDL, http://kokkai.ndl.go.jp/SENTAKU/syugiin/007 /0512/00704110512036c.html.

33. Yokosukashi, *Yokosukashishi*, 1: 584.

34. Nihon kokuyū tetsudō tetsudō gijutsu kenkyūjo 50nen-shi kankō iinkai, *Nihon kokuyū tetsudō tetsudō gijutsu kenkyūjo 50nen-shi*, 63.

35. *Kokutetsu shokuin meibo (gijutsu gakushi) shōwa 25-nen 8-gatsu 10-ka genzai* (1950); *Tetsudōgijutsu kenkyūjo, Senpai genshokusha meibo, Showa 59-nen 9-gatsu 1-nichi genzai*, JRE; Akatsuka Takeo, letter to author, 10 December 2002; Takabayashi Morihisa, letter to author, 14 December 2002; Nakamura Kazuo, conversation with author, 16 January 2003; and Miki Tadanao, interview by author, Zushi, 6 February 2003.

36. *Kokutetsu shokuin meibo (gijutsu gakushi) shōwa 25-nen 8-gatsu 10-ka genzai* (1950); *Tetsudōgijutsu kenkyūjo, Senpai genshokusha meibo, Shouwa 52-nen 1-gatsu 1-nichi genzai* (1977); and Fukuhara Shun'ichi, *Business tokkyū o hashiraseta otokotachi* (Tokyo: JTB, 2003), 71.

37. Nihon kokuyū tetsudō, *Tetsudō gijutsu hattatsu shi, dai 4-hen (Sharyō to kikai) I* (Tokyo: Nihon kokuyū tetsudō, October 1958), 2–6.

38. Ibid., 37.

39. "Itaku shaken itaku kenkyū," in *Tetsudō gijutsu kenkyūjo gaiyō 1959*, 20, RTRI archive.

40. The section leader Shima Hideo has been widely known as the builder of the integrated Shinkansen rail system. He and his assistant, Hoshi Akira, graduated from the Department of Mechanical Engineering at Tokyo University.

41. Nihon kokuyū tetsudō, *Tetsudō gijutsu hattatsu shi, dai 4-hen (Sharyō to kikai) II* (Tokyo: Nihon kokuyū tetsudō, October 1958), 802.

42. Nihon kokuyū tetsudō, *Tetsudō 80-nen no Ayumi, 1872–1952* (Tokyo: Nihon kokuyū tetsudō, 1952), 42.

43. Kawamura Atsuo, *Kyaku kasha no kōzō oyobi riron* (Tokyo: Kōyūsha, 1952), 385.

44. Kubota Hiroshi, *Nihon no tetsudō sharyō shi* (Tokyo: Grandpri shuppan, 2001), 84.

45. Kawamura, *Kyaku kasha no kōzō oyobi riron*, 5.

46. Nihon kokuyū tetsudō, *Tetsudō gijutsu hattatsu shi, dai 4-hen (Sharyō to kikai) I*, 152.

47. Izawa Katsumi, "Kyakusha kōtai ka," *Kōtsū gijutsu*, 40 (1949): 15.

48. Kubota, *Nihon no tetsudō sharyō shi*, 153.

49. Jacob Meunier, *On the Fast Track French Railway Modernization and the Origins of the TGV, 1944–1983* (Westport, CT: Praeger, 2002), 82–83.

50. Takabayashi Morihisa, letter to author, 7 July 2003.

51. Hoshi Akira, *Sharyō no keiryōka* (Tokyo: Nihontosho kankōkai, 1956); *Kokutetsu shokuin meibo (gijutsu gakushi) shōwa 25-nen 8-gatsu 10-ka genzai* (1950); Tetsudōgijutsu kenkyūjo, *Senpai genshokusha meibo, Shouwa 52-nen 1-gatsu 1-nichi genzai* (1979); and Letter from Hoshi Akira to Miki Tadanao, 3 February 1954, MT Papers.

52. Nakamura Kazuo, "Hizumi gauge tanjō 50-nen," *Kyowa gihō* 370 (1988): 1–11.

53. Yokobori, "Kyaku densha no gijutsu teki mondai ten," 438–439; and Tetsudō gijutsu kenkyūjo, *Tōkaidō shinkansen ni kansuru kenkyū: sōron* (Tokyo: Japan National Railways, 1960), 42.

54. Kubota, *Nihon no tetsudō sharyō shi*, 201; Hayashi Shōzō, "Keiryō 3-tōsha naha 10-keishiki," *Kōtsū gijutsu* 108 (1955): 260–263; and Unoki Jūzō, "Zoku keiryō kyakusha sonogo," *Kōtsū gijutsu* 136 (1957): 306–310.

55. Kubota, *Nihon no tetsudō sharyō shi*, 141.

56. Miki Tadanao, "Kōzō kyōdo kara mi ta densha no dōkō," *Denkisha no kagaku* 10, no. 4 (April 1957): 7.

57. Nihon kokuyū tetsudō tetsudō gijutsu kenkyūjo 50nen-shi kankō iinkai, *Nihon kokuyū tetsudō tetsudō gijutsu kenkyūjo 50nen-shi*, 475–489, 518–520.

58. "Showa 29-nendo kenkyū seika gaiyō (July 1955)," RTRI archive; Miki Tadanao and others, "ōgata trailer bus shaken hōkoku," *Tetsudō gyōmu kenkyū shiryō* 6, no. 1(1949):5–12; Miki Tadanao et al., "Shōnan denshayō tsūfūki shaken," *Tetsudō gyōmu kenkyū shiryō* 7, no. 15 (1950): 4–10; Miki Tadanao et al., "Jiko kara mita kyakusha · densha no kōzō sekkei shiryō," *Tetsudō gyōmu kenkyū shiryō* 7, no. 4 (1950): 4–10; and Nihon kokuyū tetsudō tetsudō gijutsu kenkyūjo 50nen-shi kankō iinkai, *Nihon kokuyū tetsudō tetsudō gijutsu kenkyūjo 50nen-shi*, 475–489, 518–520.

59. *Tetsudō gijutsu kenkyūjo kotei shisan ichiranhyō Showa 46-nen 3-gatsu 31-nichi genzai*, 96, RTRI archive.

60. Miki, "Kōzō kyōdo kara mi ta densha no dōkō," 7.

61. Ibid., 9–10.

62. *Un'yu tsūshinshō tetsudō gijutsu kenkyūjo gaiyō Showa 18-nen 12-gatsu*, 4, RTRI archive.

63. Sumida Shunsuke, *Sekai no kōsoku tetsudō to speed up* (Tokyo: Nihon tetsudō tosho, 1994), 55.

64. "Rail Plane keikaku setsumeisho," MT Papers; and "Shōwa 25.1.31 Rail Plane 1/30 mokei kumitate," MT Papers.

65. *Mainichi shinbun*, 19 February 1950; and *Nihon keizai shinbun*, 19 February 1950.

66. "Kūchū densha shatai oyobi kensui hashiri sōchi shiyōsho, Shōwa 26-nen 2-gatsu 8-nichi, Tetsudō gijutsu kenkyūjo kyakkasha kenkyūshitsu," MT Papers; "Rail Plane Memo," MT Papers; and Miki Tadanao, "Monorail 45-nen no tsuioku," *Monorail* 82 (1994): 1–5.

67. Letter from Yamamoto Risaburō to Ōtsuka Seishi, 11 July 1953, MT Papers; "Mukō-gaokayūen yakyūjō fukinzu," MT Papers; "Keikaku setsumeisho, Shōwa 28-nen 9-gatsu," MT Papers; "Mukōgaokayūenchiyo TM-shiki teishōshiki tan'itsu kijōshiki dendōsha mitsumori shiyōsho, Shōwa 29-nen 3gatsu 17-nichi, Kabushiki geisha Hitachi seisakujo," MT Papers; and letter from Hitachi Department of Rail Car Enterprise to Miki Tadanao, 2 April 1956, MT Papers.

68. "Kōrakuen Roller-Coaster Sekkei Keikakusho, Showa 30-nen 2-gatsu 20-nichi" and "Kōrakuen Coaster ippan shatai kōzōzu," 17 February 1955, MT Papers.

69. Miki Tadanao, interview by author, Zushi, 1 August 2003.

70. *Yokuyūkai meibo, Showa 44-nen* (1969), UTA.

71. Miki Tadanao, "Chō tokkyū ressha (Tokyo-Osaka 4-jikan han) no ichi kōsō," *Kōtsū gijutsu* 89 (1954): 2–6.

72. "SE-sha (kyōki sekai kiroku) 20-shūnen kinen zadankai," *Denkisha no kagaku* 31, no. 1 (July 1978): 22; and *Asahi shinbun*, 17 October 1953.

73. "Tokyo-Osaka 4-jikan 45-fun Tōkaidōsen dangan ressha kakū shijōki," *Popular Science* 2, no. 1 (January 1954): 69–74.

74. "Odakyū SE-sha kōsoku shiken kara 20-nen," *Denkisha no kagaku* 30, no. 11 (October 1977): 16.

75. Yamamoto Risaburō, "Kōsoku kansetsu densha SE ni tsuite," *Sharyō gijutsu* 40 (1958): 312; and Ubukata Yoshio and Morokawa Hisashi, *Odakyū Romance Car monogatari* (Osaka: Hoikusha, 1994), 72, 140.

76. Thomas R. Havens, *Architects of Affluence: The Tsutsumi Family and the Seibu-Saison Enterprise in Twentieth-Century Japan* (Cambridge, MA: Harvard Univ. Press, 1994), 21; "Hakone Yumoto onsen, Hakone onsen no rekishi," accessed 28 October 2011, http://www.hakone-yado.jp/hakone-rekishi.html; and Hakone onsen ryokan kyōdō kumiai, "Hakone onsei kōshiki gaido: Hakopita," accessed 28 October 2011, http://www.hakoneryokan.or.jp/002_rekishi.html.

77. Odakyū dentetsu, "Kaisha gaiyō," accessed 24 November 2011, http://www.odakyu.jp/company/about/outline; Odakyū dentetsu, "Odakyū 80-nenshi," accessed 20 November 2011, http://www.odakyu.jp/company/history80/01.html; and Ubukata Yoshio, *Odakyū monogatari* (Kawasaki: Tamagawa shinbunsha, 2000), 40.

78. Hakone onsen ryokan kyōdō kumiai, *Hakone onsenshi* (Hakone: Hakone onsen ryokan kyōdō kumiai, 1986), 234–237; and Eighth Army Special Services Section, *Leave Hotels in Japan*, 1 January 1945, MUSAFB.

79. Tōkyū sharyō seizō kabushiki kaisha, *Tōkyū sharyō 30-nen no ayumi* (Yokohama: Tōkyū sharyō seizō kabushiki kaisha, 1978), 14.

80. Ubukata, *Odakyū monogatari* , 39.

81. Odakyū dentetsu, "Odakyū 80-nenshi," accessed 20 November 2011, http://www.odakyu.jp/company/history80/01.html.

82. Hakone onsen ryokan kyōdō kumiai, "Hakone onsei kōshiki gaido: Hakopita."

83. "5-ryō kotei hensei chōtokkyūsha (SE sha) shiyō taiyō (an) Shōwa 30-nen 1-gatsu 25-nichi, Tetsudō gijutsu kenkyūjo shidō, Odakyū dentetsu kabushiki kaisha," RTRI archive.

84. "Keiryō tokkyū denshaan 29-10-22 tetsudō giken kyakkasha kenkyūshitu," RTRI archive; and "Kyokusen ni yoru seigen sokudo," RTRI archive.

85. "Dai 2-kai SE-sha sōgō kaigi gijiroku, 4-2-1955," MT Papers; "Dai 3-kai SE-sha sōgō kaigi gijiroku, 5-21-1955," MT Papers; "Dai 7-kai SE-sha sōgō kaigi gijiroku, 4-14-1955," MT Papers; "Dai 8-kai SE-sha sōgō kaigi gijiroku, 4-22-1955," MT Papers; "Dai 9-kai SE-sha sōgō kaigi gijiroku, 5-6-1955," MT Papers; "Dai 12-kai SE-sha sōgō kaigi gijiroku, 5-30-1955," MT Papers; "Dai 16-kai SE-sha sōgō kaigi gijiroku, 6-20-1955," MT Papers; "Dai 18-kai SE-sha sōgō kaigi gijiroku, 7-4-1955," MT Papers; "Dai 20-kai SE-sha sōgō kaigi gijiroku, 7-29-1955," MT Papers; and "Dai 25-kai SE-sha sōgō kaigi gijiroku, 9-22-1955," MT Papers.

86. "Odakyū SE-sha kōsoku shiken kara 20-nen," 22.

87. Ibid., 15; and Ubukata, *Odakyū monogatari*, 40–41.

88. Yamamoto, "Kōsoku kansetsu densha SE ni tsuite," 312.

89. Odakyū dentetsu kabushiki kaisha, *Super Express 3000* (1957), 8–10.

90. "Shōwa 30-7-12 Nihon sharyō seizō kabushiki kaisha Tokyo shiten SE-sha M-jyūryō gaisan," MT Papers.

91. This style fixed the units with little flexibility during the high-speed run. The resulting vibrations were more complex in nature and more frequent than the common articulation in which each car had two trucks. Matsui Nobuo, "SE-sha ni yoru kōsoku shaken," *Tetsudōgijutsu kenkyūjo Shōwa 32-nendo kenkyū seika gaiyō shisetsukyoku kankei* (March 1958): 72–73, RTRI archive.

92. Yamamoto, "Kōsoku kansetsu densha SE ni tsuite," 312.

93. Miki Tadanao and Takabayashi Morihisa, "Keiryō recycling-seat shisaku hōkoku, dai 1-pō," *Shōwa 30-nen kamihanki kenkyū seika gaiyō, kōsakukyoku kankei, Shōwa 31-nen 2-gatsu, Tetsudō gijutsu kenkyūjo*, RTRI archive.

94. Ubukata, *Odakyū monogatari*, 53; and Takabayashi Morihisa, letter to author, 4 August 2003.

95. Miki, "Kōzō kyōdo kara mi ta densha no dōkō," 7.

96. Ibid., 9–10.

97. Yamamoto, "Kōsoku kansetsu densha SE ni tsuite," 316.

98. Miki Tadanao, "Odakyū 3000-kei SE-sha sekkei no tuioku," *Tetsudō Fan* 32, no. 375 (1992): 94; Miki et al., "Kōsoku sharyō no kūki rikigakuteki kenkyū," *Nihon kikai gakkaishi* 61, no. 478 (November 1958): 34–43; Miki, "Kōsoku tetsudō sharyō no kūki rikigakuteki shomondai, Part 1," *Kikai no kenkyū* 12, no. 7 (1960): 17–24; Miki, "Kōsoku tetsudō sharyō no kūki rikigakuteki shomondai, Part 2," *Kikai no kenkyū* 12, no. 8 (1960): 13–18; and Miki, "Kōsoku tetsudō sharyō no kūki rikigakuteki shomondai, Part 3," *Kikai no kenkyū* 12, no. 9 (1960): 25–30.

99. Yamamoto, "Kōsoku kansetsu densha SE ni tsuite," 316.

100. "SE-sha (kyōki sekai kiroku) 20-shūnen kinen zadankai," *Denkisha no kagaku* 31, no. 1 (July 1978): 25.

101. Miki Tadanao, conversation with author, 22 December 2002.

102. Nihon kokuyū tetsudō tetsudō gijutsu kenkyjo, "SE-sha ni yoru kōsokudo shiken hōkoku," *Tetsudō gijutsu kenkyūjo sokuhō* No. 58-17 (January 1958): 13, RTRI archive.

103. Miki Tadanao and Takabayashi Morihisa, "Keiryō recycling-seat shisaku hōkoku, dai 1-pō," 16–17, RTRI archive; and Miki Tadanao, conversation with author, 22 December 2002.

104. Ubukata and Morokawa, *Odakyū Romance Car monogatari*, 75.

105. Odakyū dentetsu kabushiki kaisha, *Super Express 3000*, 52.

106. Ubukata, *Odakyū monogatari*, 45.

107. "SE-sha (kyōki sekai kiroku) 20-shūnen kinen zadankai," *Denkisha no kagaku* 31, no. 1 (July 1978): 29.

108. "Shōwa 30.3.16 Nihon sharyō seizō kabushiki kaisha Tokyo shiten SE-sha gisō jūryō oyobi kōhi no hikaku," MT Papers.

109. Kōtsūshinbun henshūkyoku, *Atarashii tetsudō no tankyū: Tetsudō gijutsu kenkyū no kadai* (Tokyo: Kōtsū kyōryokukai, 1959), 114.

110. Ibid., 78; and Kanō Yasuo, interview by author, Tokyo, 22 March 2004.

111. Kōtsūshinbun henshūkyoku, ed., *Atarashii tetsudō no tankyū*, 82–83.

112. Ibid., 121.

113. Kanō Yasuo, interview by author, Tokyo, 22 March 2004.

114. Nomura Yoshio and Kajikawa Atsuhiko, "Soren kokutetsu ni okeru saikin no brake ni kansuru kenkyū (3)," *Tetsudō gijutsu kenkyūjo sokuhō* (April 1960), RTRI archive.

115. *Kodama-gō kōsokudo shiken kiroku, shōwa 34-nen 9-gatsu 10-ka, Nihon kokuyū tetsudō gishichōshitu* (1959), 87–89, RTRI archive.

116. Miyasaka Masanao, "Rinji sharyō sekkei jimusho," *Kokuyū tetsudū* 94 (1957): 20.

117. Ibid.

118. Nihon kokuyū tetsudō, *Business tokkyū densha* (1958), 6–7; and Odakyū dentetsu kabushiki kaisha, *Super Express 3000*, 7.

119. Nihon kokuyū tetsudō, *Business tokkyū densha* (1958), 3.

120. Fukuhara, *Business tokkyū o hashiraseta otokotachi*, 66–68.

121. Ibid., 75.

Chapter 7 · Former Military Engineers and the Development of the Shinkansen, 1957–1964

1. *Mainichi shinbun*, Evening Edition, 1 October 1964; and *Asahi shinbun*, Evening Edition, 1 October 1964; and *Saga shinbun*, 2 October 1964.

2. Shinohara Takeshi, *Omoide no ki* (Tokyo: Kenyūsha, 1994), 48; and Harada Yutaka, *Omoide* (Tokyo: Harada Yutaka, 1991), 39.

3. Takei Akemichi "Hongoku tetsudō ressha sokudo no hattatsu," *Kikai gakkaishi* 41, no. 251 (1938): 113–119.

4. Nishitani Tetsu, "Tōkaidō-sen ressha sokudo no hensen," *Kōtsū gijutsu* 126 (1956): 17–18.

5. Maema Takanori, *Dangan ressha: Maboroshi no Tokyo-hatsu Beijing-yuki chōtokkyū* (Tokyo: Jitsugyō-no-nihonsha, 1994).

6. Nihon kokuyū tetsudō, *Kokutetsu rekishi jiten* (Tokyo: Nihon kokuyū tetsudō, 1973), 71.

7. *Shinkansen 10-nen shi* (Tokyo: Nihon kokuyū tetsudō shinkansen sōkyoku, 1975), 4.

8. "Shinkansen kakū densha senro kenkyū hōkoku," *Tetsudō gyōmu kenkyū shiryō* 2, no. 11 (1943): 2–3; and Nihon kokuyū tetsudō, *Tetsudō gijutsu hattatsushi*, 1: 151.

9. Nihon kokuyū tetsudō, *Tetsudō gijutsu hattatsushi*, 1: 154.

10. Ibid., 134.

11. Fujishima Shigeru, "Shinkansen jisoku 200-kiro no missitsu," in *Bungei shunjū ni miru Shōwa-shi* (Tokyo: Bungei shunjū, 1988), 568.

12. "Ryūsenkei sharyō mokei no fūdō shiken seiseki ni tsuite," *Gyōmu kenkyū shiryō* 25, no. 2 (1937): 1–34.

13. Mainichi shinbunsha, ed., *Speed 100-nen* (Tokyo: Mainichi shinbusha, 1969), 86–87.

14. Nihon kokuyū tetsudō, *Tetsudō gijutsu hattatsushi, dai-1hen, sōsetsu*, 72–75; and Shima Hideo, *D-51 Kara shinkansen made: Gijutsusha no mita kokutetsu* (Tokyo: Nihon keizai shinbunsha, 1977), 41.

15. Shima Yasujirō, "Tetsudō kikan no sunpō to sharyō no kidō ni taisuru atsuryoku no kankei," *Kikai gakkaishi* 28, no. 95 (1925): 129–136.

16. Nihon kokuyū tetsudō, *Tetsudō gijutsu hattatsushi, Tetsudō gijutsu hattatsushi, dai-1hen, sōsetsu*, 86.

17. Shima Hideo ikōshū henshū iinkai-hen, *Shima Hideo ikōshū: 20-seiki tetsudōshi no shōgen* (Tokyo: Nihon tetsudō gijutsu kyōkai, 2000), 114.

18. *Shinkansen 10-nen shi*, 5.

19. Ariga Sōkichi, *Sogō Shinji* (Tokyo: Sogō Shinji-denki kankōkai, 1988), 499–501.

20. Shima Hideo, *D-51 kara Shinkansen made*,113.

21. Nihon kokuyū tetsudō, *Kokutetsu rekishi jiten*, 71.

22. *Asahi shinbun*, 14 February 1956.

23. Shima Hideo ikōshū henshū iinkai-hen, *Shima Hideo ikōshū*, 148.

24. Tetsudō gijutsu kenkyūjo, *Tetsudō gijutsu kenkyūjo sōritsu 70-shūnen*, 281, 84.

25. Nakamura Hiroshi, interview by author, 29 April 2004; Nakamura Kazuo, interview by author, 7 November 2002. Personal experiences with the director are inherently subjective, ranging from pleasant to unpleasant. But anonymous informants who had opposed to him in various ways supported my characterization.

26. Tetsudō gijutsu kenkyūjo, *10-nen no ayumi: Sōritsu 60-shūnen* (Tokyo: Tetsudō gijutsu kenkyūjo, 1967), 212.

27. Ibid., 212; Shinohara Takeshi, *3-gatsukai News*, 20 January 1985, 1, MT Papers; Shinohara Takeshi and Takaguchi Hideshige, *Shinkansen hatsuansha no hitorigoto: Moto Nihon tetsudō kensetsu kōdan sōsai Shinohara Takeshi no network-gata Shinkansen no kōsō* (Tokyo: Pan research shuppan, 1992), quotes from 82–83; and Shinohara Takeshi, *Omoide no ki*, 33–34.

28. Shinohara Takeshi, *Omoide no ki*, 36–38.

29. Hoshino Yōichi, "Kidō no kōzō," *Kōtsū gijutsu* 135 (1957), 9–11; Matsudaira Tadashi, "Anzen to norigokochi," *Kōtsū gijutsu* 135 (1957), 5–8; Miki Tadanao, "Sharyō kōzō," *Kōtsū gijutsu* 135 (1957), 2–5; Shinohara and Takaguchi, *Shinkansen hatsuansha no hitorigoto*, 85; and Mainichi Shinbunsha, ed., *Speed no 100-nen*, 40.

30. Shinohara and Takaguchi, *Shinkansen hatsuansha no hitorigoto*, 88, 94; Shinohara, *Omoide no ki*, 37; and Tetsudō gijutsu kenkyūjo, *10-nen no ayumi*, 219.

31. *Asahi shinbun*, 4 May 1957.

32. Aoki Kaizō, *Kokutetsu* (Tokyo: Shinchōsha, 1964), 284.

33. Shinohara and Takaguchi, *Shinkansen hatsuansha no hitorigoto*, 96.

34. Uchihashi Katsuto, *Zoku, zoku takumi no jidai* (Tokyo: Sankei shuppan, 1979), 46.

35. Ibid., 44–45; and Miki Tadanao, interview by author, 1 August 2003; Kōtsū shinbun henshūkyoku, *Atarashii tetsudō no tankyū: Tetsudō gijutsu kenkyū no kadai* (Tokyo: Kōtsū kyōryokukai, 1959), 1; and Miki, "Kōsoku tetsudō sharyō no kūki rikigakuteki shomondai, 1," 17–18.

36. Aoki, *Kokutetsu*, 312–313.

37. Shin'ichi Tanaka, "Shinkansen sharyō: sono kaihatsu no zengo," *Denkisha gakkai kenkyūkai shiryō* (2002): 20; and Matsudaira Tadashi, *3-gatsukai News*, 20 January 1985, 3, MT papers.

38. Tetsudō tetsudō gijutsu kenkyūjo, *10-nen no ayumi: Sōritsu 60-shūnen*, 213.

39. Kōtsū shinbun henshūkyoku, *Atarashii tetsudō no tankyū*, 3; and *Tetsudō gijutsu kenkyjo gaiyō 1959*, 6, 8–9, RTRI archive.

40. Kōtsū shinbun henshūkyoku, *Atarashii tetsudō no tankyū*, i.

41. *Sharyō shikendai: Shōwa 34-nen 10 gatsu Tetsudō gijutsu kenkyūjo*, RTRI archive.

42. *Tetsudō gijutsu kenkyūjo kotei shisan ichiranhyō, Shōwa 46-nen 3-gatsu 31-nichi genzai*, RTRI archive.

43. Shinohara and Takaguchi, *Shinkansen hatsuansha no hitorigoto*, 4, 98.

44. Hōmu daijin kanbō shihō hōsei chōsabu shihō hōsei-ka, *Saikō saibansho hanreishū*, vol. 12 (Tokyo: n.p., 1958), 123–39.

45. Aoki, *Kokutetsu*, 292.

46. Kokudo kōtsūshō, Kokutetsu kaikaku ni tsuite, accessed 7 November 2011, http://www.mlit.go.jp/tetudo/kaikaku/01.htm.

47. Japanese National Railways, *General Statement Supporting the Loan from International Bank for Reconstruction and Development, December 1959*, 36, RTRI archive; and Nihon kokuyū tetsudō, *Kokutetsu arakaruto* (Tokyo: Nihon kokuyū tetsudo, 1965), 30.

48. Nihon kokuyū tetsudō, *Kokutetsu arakaruto*, 33; and Kubota, "Sengo nihon tetsudōshi no ronten," *Tetsudō shigaku* 6 (1988): 43.

49. Japanese National Railways, *General Statement Supporting the Loan from International Bank for Reconstruction and Development*, 33, RTRI archive; and Kubota, "Sengo nihon tetsudōshi no ronten," 43.

50. Kubota, "Sengo nihon tetsudōshi no ronten," 44.

51. Sangiin kessan iinkai, "Dai 41-kai kokkai heikaigo kaigiryoku 10-gō," 4 December 1962, NDL.

52. *Nihon kokuyū tetsudō kansen chōsakai secchi ni tsuite, Shōwa 32-nen 8-gatsu 30-nichi kakugi kettei*, NDL.

53. Kansen chōsakai, *Nihon kokuyū tetsudō kansen chōsakai, Dai 1-bunkakai giji gaiyō*, RTRI arvhive.

54. *Shinkansen 10-nenshi*, 6; and Japanese National Railways, *General Statement Supporting the Loan from International Bank for Reconstruction and Development*, 41, RTRI archive.

55. *Nihon kokuyū tetsudō kansen chōsakai secchi ni tsuite, Shōwa 32-nen 8-gatsu 30-nichi kakugi kettei*, NDL.

56. Japanese National Railways, *General Statement Supporting the Loan from International Bank for Reconstruction and Development*, 41, RTRI archive.

57. Nihon kokuyū tetsudō, *Kokutesu rekishi jiten*, 70.

58. *Shinkansen 10-nenshi*, 8; and Mainichi shinbunsha, ed., *Speed no 100-nen*, 49.

59. Ariga, *Sogō Shinji*, 580–81, Nihon kokuyū tetsudō Shinkansen sōkyoku, *Shinkansen: Sono 20-nen no kiseki* (Tokyo: Nihon kokuyū tetsudō Shinkansen sōkyoku, 1984), 43.

60. Shima Hideo ikōshū henshū iinkai-hen, *Shima Hideo ikōshū*, 146.

61. Tetsudō tetsudō gijutsu kenkyūjo, *10-nen no ayumi: Sōritsu 60-shūnen*, 217; and Ariga, *Sogō Shinji*, 602.

62. Tsumura Takumi, *3-gatsukai News*, 20 January 1985, 4, MT Papers; Yamanouchi Shūichirō, *Shinkansen ga nakattara* (Tokyo: Asahi bunko, 2004), 159–160; and Japanese National Railways, *Data on Engineering Submitted to International Bank for Reconstruction and Development*, 30 May 1960, RTRI archive.

63. Satō Yoshihiko, "Sekai ginkō ni yoru Tōkaidō Shinkansen Project no hyōka," *Tetsudō shigaku* 19 (2001): 70.

64. Nihon kokuyū tetsudō tetsudō gijutsu kenkyūjo, *Tokaido Shinkansen ni kansuru kenkyū*, (Tokyo: Nihon kokuyū tetsudō tetsudō gijutsu kenkyūjo, 1964), 4: 1.

65. Nihon kokuyū tetsudō tetsudō gijutsu kenkyūjo, *Tokaido Shinkansen ni kansuru kenkyū* (Tokyo: Nihon kokuyū tetsudō tetsudō gijutsu kenkyūjo, 1962), 3: 9–69; *Kokutetsu shokuin meibo (gijutsu gakushi) shōwa 25-nen 8-gatsu 10-ka genzai* (1950); Kumezawa Yukiko, letter to author, 22 March 2004; Kanō Yasuo, interview by author, Tokyo, 22 March 2004; Shinohara Osamu, letter to author, 11 March 2004; Matsubara Keiji, *Shūsenji teikoku rikugun zen gen'eki shōkō shokumu meikai* (Tokyo: Senshi kankō iinkai, 1985); Kaigunshō,

Gen'eki kaigun shikan meibo vol. 3 (Tokyo: Kaigunshō, 1944); and Denkikei dōsōkai, *Sepiairo no 3-gōkan: Rekishi archive*, accessed 26 October 2011, http://todaidenki.jp/hist/?cat=16. Some engineers moved from one research institute to another, and thus, their wartime affiliations were not always fixed. For instance, Shinohara worked at the NTRI at one point in time. Kumezawa briefly served in the army. Hirakawa inherited the laboratory from Hoshino Yōichi, a specialist in track structure and one of the four presenters in the RTRI public forum in 1957. Hoshino entered the JNR in 1931 and stayed with the organization as the laboratory leader until he retired in 1959.

66. Miki Tadanao, "Kōsoku ressha no kūki rikigakuteki shomondai," *Kōtsū gijutsu* 113 (1950): 30.

67. "Kōsoku sharyō mokei no fūdō shiken," *Tetsudō gijutsu kenkyūjo itaku kenkyū hōkoku, Showa 36-nendo* (1961), 10, RTRI archive.

68. Tanaka Shin'ichi, interview by author, 28 November 2002.

69. Japanese National Railways, *Data on Engineering: Submitted to International Bank for Reconstruction and Development* (May 1960), 3, RTRI archive.

70. Japanese National Railways, *Technical Aspects on the Tokaido Line*, 82, RTRI archive.

71. Sawano Shūichi, "Shinkansen no sharyō," *JREA* 2, no. 1 (1959): 14.

72. Ishizawa Nobuhiko, "Sekkei to shūzen (1) Pantograph," *Tetsudō kōjō* 9, no. 1 (1958): 16–17.

73. Kumezawa Ikurō, "Kasen to shūden," *Denki gakkai zasshi* 84, no. 10 (1964): 35.

74. Arimoto Hiroshi and Kunieda Masaharu, "Pantograph fūdōnai shiken," *Tetsudō gijutsu kenkyū shiryō* 125 (1960): 1–3, RTRI archive; and Kumezawa Ikurō, "Shinakansen no shūden ni tsuite," *Kōtsū gijutsu* 159 (1963): 11.

75. Japanese National Railways, *Data on Engineering: Submitted to International Bank for Reconstruction and Development*, 41, RTRI archive.

76. Yasuda Akio and others, "Shinkansen no kensetsu kijun," *JREA* 2, no. 1 (1959): 8.

77. Kakumoto Ryōhei, *Shinkansen kaihatsu monogatari* (Tokyo: Chuō kōronsha, 2001), 36–37; and Shima Hideo, "Shinkansen no kōsō," 149.

78. Nakayama Yasuki, "Tunnel nai ni okeru ressha no kuki teikō," *Tetsudō gijutsu kenkyū shiryō* 16, no. 6 (1959): 38, RTRI archive.

79. Satō Yutaka, "Kidō ni kuwawaru suichoku shōgeki atsuryoku," *Tetsudō gijutsu kenkyū shiryō* 16 (1958): 1–3, RTRI archive; Satō Yutaka, "Kurikaeshi kajū ni yoru dōshō chinka no jikken," *Tetsudō gijutsu kenkyū shiryō* 65 (1959): 1–13, RTRI archive; Satō Yutaka, "Rail no kyokubu ōryoku," *Tetsudō gijutsu kenkyū shiryō* 27 (1958): 1–3, RTRI archive; Satō Yutaka, "Yokoatsu ni taisuru kidō kyōdo no kenkyū," *Tetsudō gijutsu kenkyū shiryō* 110 (1960): 1–7, RTRI archive; and Satō Yutaka and Toyoda Masayoshi, "Kakushu dōryokusha no kyokusen yokoatsu no rail yokomage ni yoru sokutei kekka," *Tetsudō gijutsu kenkyū shiryō* 57 (1959): 1–3, RTRI archive.

80. Ikari Yoshirō, *Chō kōsoku ni idomu: Shinkansen kaihatsu ni kaketa otokotachi* (Tokyo: Bungei shunjū, 1993), 244–45.

81. Ikari, *Kaigun Gijutsushatachi no taiheiyō sensō*, 256–58.

82. Tetsudō tetsudō gijutsu kenkyūjo, *10-Nen No Ayumi: Sōritsu 60-shūnen*, 64.

83. "Ki makuragi," in *Kōsoku tetsudō no kenkyū: Shu to shite Tōkaidō Shinkansen ni tsuite* (Tokyo: Kenyūsha, 1967), 134; and Nihon kokuyū tetsudō tetsudōdan gijutsu kenkyūjo, *Tokaido Shinkansen ni kansuru kenkyū* (Tokyo: Nihon kokuyū tetsudō tetsudō gijutsu kenkyūjo, 1964), 5: 175–207.

84. Enomoto Shinsuke, letter to author, 5 April 2004; and Nakamura Hiroshi, interview by author, 29 April 2004. The train's axles were heat treated for the same reason.

85. Japanese National Railways, *Technical Aspects on the Tokaido Line* (1963), 76, RTRI archive.

86. Nihon kokuyū tetsudō Shinkansen sōkyoku, *Shashin to illusto de miru Shinkansen: sono 20-nen no kiseki* (Tokyo: Shinkansen sōkyoku, 1984), 46.

87. Ibid.

88. Sawano Shūichi, "Shinkansen no sharyō," *JREA* 2, no. 1 (1959): 15.

89. "Kōsoku resshayō fūatsu brake no kenkyū," in *Tetsudō gijutsu kenkyūjo itaku kenkyū hōkoku: Showa 35-nendo* (Tokyo: n.p., 1960); Kōtsū shinbun henshūkyoku, *Atarashii tetsudō no tankyū*, 61; and Miki Tadanao, "Kōsoku tetsudō sharyō no kūki rikigakuteki shomondai, 3," *Kikai no kenkyū* 12 (1960): 26–27.

90. Sogō Shinji, "Shingijutsu to speed up," *JREA* 2, no. 1 (1959) 1.

91. "Sangiin yosan iinkai kaigiroku dai 5-gō," 18 November 1959, NDL.

92. "Shūgiin un'yu iinkai kaigiroku dai 32-gō," 26 May 1961, NDL.

93. Shima Hideo, "Shinkansen no kōsō," in *Sekai no tetsudō* (Tokyo: Asahi shinbunsha, 1964), 145.

94. Tanaka Shin'ichi, "Sharyō," *Tetsudō gijutsu* 42, no. 1 (1985): 19.

95. *Shinkansen-yō shisaku ryokyaku densha* (Tokyo: Nihon kokuyū tetsudō, 1962), 4, RTRI archive.

96. Tanaka Shin'ichi, interview by author, 28 November 2002.

97. Kōtsū shinbun henshūkyoku, *Atarashii tetsudō no tankyū*, 64; and Matsudaira, "Kōsoku tetsudō gijutsu no raimei 2," 31.

98. Matsudaira Tadashi, "Tōkaidō Shinkansen ni kansuru kenkyū kaihatsu no kaiko: Shu to shite sharyō no shindō mondai ni kanren shite," *Nihon kikai gakkaishi* 75, no. 646 (1972): 105.

99. "Kōryū kiden," in *Kōsoku tetsudō no kenkyū: Shu to shite Tōkaidō Shinkansen ni tsuite* (Tokyo: Ken'yūsha, 1967), 463; and Hayashi Masami, "Sekai ni hokoru kōsoku tetsudōyō daideiryoku kyōkyū hōshiki: AT kiden hōshiki," *Hatsumei* 76, no. 8 (1979): 52.

100. *Shinkansen-yō shisaku ryokyaku densha*, 1, RTRI archive.

101. Hayashi Masami, "Kūgishō no omoide to sengo no watashi," in *Kaigun kōkū gijutsushō denkibu*, 100.

102. Mark Aldrich, "Combating the Collision Horror: The Interstate Commerce Commission and Automatic Train Control, 1900–1939," *Technology and Culture* 34, no. 1 (1993): 49–77; and Hobara Mitsuo, "Ressha shūchū seigyo," *Denki gakkai zasshi* 84, no. 10 (1964): 51.

103. Ministry of Land, Infrastructure, Transport and Tourism, *Shinkansen Japanese High-speed Rail*, accessed 25 October 2011, http://www.mlit.go.jp/en/tetudo/tetudo_fr2 _000000.html.

104. Kawanabe Hajime, "Atarashii shanai shingō," *JREA* 2, no. 4 (1959): 2.

105. Amamiya Yoshifumi and Kurita Nobuo, "Tetsudōyō hyoumenha radar no kōsō to kiso jikken," *Tetsudō gijutsu kenkyūjo sokuhō* 62 (1963): 1, RTRI archive.

106. Ibid.; and Miki Tadanao, "Ressha no speed-up o habamu mono," *Shindenki* 12, no. 1 (1958): 24.

107. Kawanabe Hajime, "Jidō ressha seigyo," *Denki gakkai zasshi* 84, no. 10 (1964): 43.

108. Japanese National Railways, *Technical Aspects on the Tokaido Line*, 84, RTRI archive.

109. *Yomiuri shinbun*, 27 February 2004.

110. Ministry of Land, Infrastructure, Transport and Tourism, *Shinkansen Japanese High-Speed Rail*, accessed 25 October 2011, http://www.mlit.go.jp/en/tetudo/tetudo_fr2_000000.html.

111. Amamiya Yoshifumi and Kurita Nobuo, "Tetsudōyō hyōmenha radar no kōsō to kiso jikken," 1, 23–24, RTRI archive.

112. Nihon kokuyū tetsudō tetsudō gijutsu kenkyūjo, *Tōkaidō Shinkansen ni kansuru kenkyū* (Tokyo: Nihon kokuyū tetsudō tetsudō gijutsu kenkyūjo, 1962), 3: 492–506.

113. Ikari, *Chō kōsoku ni idomu*, 268.

114. Shima Takashi and Tani Masao, "Shinkansen sharyō no haishō sōchi," *JREA* 7, no. 4 (1964): 45.

115. *Shinkansen-yō shisaku ryokyaku densha*, 7; and Shima Hideo, *D-51 Kara Shinkansen Made*,111.

116. Nihon kokuyū tetsudō-hen, *Nihon kokuyū tetsudō 100-nenshi* (Tokyo: Nihon kokuyū tetsudō, 1973), 14: 611.

117. Fujishima, "Shinkansen jisoku 200-kiro no missitsu," 570–571.

118. Tani Masao, "Shinkansen sharyō no kimitsu," *JREA* 7, no. 7 (1964): 5; "Japon: Un An d'exploitation de la nouvelle ligne du Tokaïdo," *La Vie du rail*, 26 December 1965, 11; and quote from "Le Japon inaugure son 'chimen de fer de demain'!: Naissance d'un super train . . . ," *La Vie du rail*, 7 March 1965, 12.

119. Fujishima, "Shinkansen jisoku 200-kiro no missitsu," 568–569.

120. Miki, "Kōsoku tetsudō sharyō no kūki rikigakuteki shomondai 3," 30.

121. Nihon kokuyū tetsudō-hen, *Nihon kokuyū tetsudō 100-nenshi*, 547.

122. Japanese National Railways, *Technical Aspects on the Tokaido Line*, 79, RTRI archive.

123. Hoshino Yōichi, "Shin kōzō kidō," *JREA* 1, no. 7 (1958): 10.

124. Ikari, *Chō kōsoku ni idomu*, 242–43.

125. Shima Hideo, "Shinkansen no kōsō," 149.

126. Sumida, *Sekai no kōsoku tetsudō to speed up*, 71.

127. Shima Hideo ikōshū henshū iinkai-hen, *Shima Hideo ikōshū*, 163.

128. Hara Tomoshige, "Ressha ga kōsoku de suidō ni totsunyū suru baai no ryūtai rikigakuteki shomondai," *Tetsudō gijutsu kenkyū hōkoku* 153 (1960): 1–3.

129. Ariga, *Sogō Shinji*, 546.

130. Sumida, *Sekai no kōsoku tetsudō to speed up*, 73; and Shima Hideo, *D-51 kara Shinkansen made*, 135.

131. Hoshikawa Takeshi, ed., *Shinkansen zenshi* (Tokyo: Gakken kenkyūsha, 2003), 168.

132. Aoki Eiichi and others, *A History of Japanese Railways*, 144.

133. Nihon Olympic Iinkai, "Tokyo Olympic 1964," accessed 8 November 2011, http://www.joc.or.jp/past_games/tokyo1964.

134. Nihon Olympic Iinkai, "Tokyo Olympic 1964," accessed 8 November 2011, http://www.joc.or.jp/past_games/tokyo1964/story/vol01_01.html.

135. "Japon: Un An d'exploitation de la nouvelle ligne du Tokaïdo," *La Vie du rail*, 26 December 1965, 8.

136. *Ferro Carriles: Catátalogo De Sellas Temáticos* (Barcelona: DOMFIL, 2001).

137. Sumida, *Sekai no kōsoku tetsudō to speed up*, 147.

138. Tetsudō gijutsu kenkyūjo, *Tetsudō gijutsu kenkyūjo sōritsu 70-shūnen: 10-nen no ayumi*, 4.

139. Yamanouchi Shūichirō, *Shinkansen ga nakattara* (Tokyo: Asahi bunko, 2004), 145.

140. "Le Japon inaugure son 'chimen de fer de demain'!: Naissance d'un super train . . . ," *La Vie du rail*, 7 March 1965, 12.

141. "Quelques particularités techniques de la nouvelle ligne Japonaise du Tokaïdo," *La Vie du rail*, 7 March 1965, 29.

142. Jacob Meunier, *On the Fast Track* (Westport, CT: Praeger, 2002), 4, 76.

143. Ibid., 76, 89; Yamanouchi, *Shinkansen ga nakattara*, 248; and Sumida, *Sekai no kōsoku tetsudō to speed up*, 148.

144. Yamanouchi, *Shinkansen ga nakattara*, 248.

145. Meunier, *On the Fast Track*, 5, 91.

146. Stefan Zeilinger, *Wettfahrt auf der Schiene: Die Entwicklung von Hochgeschwindig-keitszügen im europΩischen Vergleich* (Franfurt: Campus Verlag, 2003), 105.

147. *Railway Gazette International* 124 (5 July 1968); and Mainichi shinbunsha, *Speed 100-nen*, 164–165.

148. Sumida, *Sekai no kōsoku tetsudō to speed up*, 59–60, 199.

Conclusion • Legacy of War and Defeat

1. John W. Dower, "Useful War," in *Japan in War & Peace: Selected Essay* (New York: New Press, 1991), 9–32.

2. Chalmers Johnson, *MITI and the Japanese Miracle: The Growth of Industrial Policy, 1925–1975* (Stanford, CA: Stanford Univ. Press, 1982), 309; Mizuno Hiromi, *Science for the Empire: Scientific Nationalism in Modern Japan* (Stanford, CA: Stanford Univ. Press, 2008), 184; Daqing Yang, *Technology of Empire: Telecommunications and Japanese Expansion in Asia, 1883–1945* (Cambridge, MA: Harvard Univ. Press, 2010), 388–390; and Dower, "Useful War," 10–11.

3. Koizumi Kenkichirō, "In Search of Wakon: The Cultural Dynamics of the Rise of Manufacturing Technology in Postwar Japan," *Technology and Culture* 43 (2002): 29–49; and Morris Low, *Science and the Building of a New Japan* (New York: Palgrave, 2005).

4. Aoki Eiichi et al., *A History of Japanese Railways*, 198.

5. Sumida, *Sekai no kōsoku tetsudō to speed up*, 58–59.

6. Tetsudō gijutsu kenkyūjo, *Senpai genshokusha meibo, Shouwa 52-nen 1-gatsu 1-nichi genzai; Kokutetsu shokuin meibo (gijutsu gakushi) shōwa 25-nen 8-gatsu 10-ka genzai* (1950); and Zasshi maru henshūbu, ed. *Suisei/99-kanbaku*, 23.

7. Tetsudō gijutsu kenkyūjo, *10-nen no ayumi: Sōritsu 60-shūnen*, 175.

8. Kyōtani Yoshihiro, interview by Furuse Yukihiro, "Tsunawatari no Shinkansen," *Furuse Yukihiro no off side 2001*, no. 19, accessed 17 August 2001, http://www.honya.co.jp/contents/offside/index.cgi?20010817.

9. Tetsudō gijutsu kenkyūjo, *10-nen no ayumi: Sōritsu 60-shūnen*, 58

10. Tanaka Hisashi, "Sōritsuki no fujōshiki tetsudō 1: Fujōshiki tetsudō no reimeiki," *RRR* 58, no. 1 (2001): 26–27; and Aoki Eiichi et al., *A History of Japanese Railways*, 198.

11. "Shūgiin un'yu iinkai 9-gō," 6 June 1978, NDL.

12. Kokudo kōtsushō, "Kokutetsu kaikaku ni tsuite," accessed 7 November 2011, http://www.milt.go.jp/tetudo/kaikaku/01.htm.

13. Fujii Shigeki, "Kokutetsu chōki sekimu no shorimondan to sono keizaiteki fukui ni kansuru ichikōsatsu," *Kaikei kensa kenkyū* 17, accessed 7 November 2011, http://jbaudit.go.jp/effort/study/mag/17-5.html; Aoki Eiichi et al., *A History of Japanese Railways*, 182; and Kakumoto Ryōhei, *Shinkansen kaihatsu monogatari* (Tokyo: Chūō bunko, 2001), 200.

14. Kakumoto, *Shinkansen kaihatsu monogatari*, 200.

15. Kyōtani Yoshihiro, Oku Takeshi, and Sanuki Toshio, *Chō kōsoku Shinkansen* (Tokyo: Chūō kōronsha, 1971) (quoted from pages 173, 179).

16. Sasagawa Yōhei, "Linear Shinkansen to Kyōtani Yoshihiro," *Nihon zaidan kaichō Sasagawa Yōhei blog* (blog), 27 February 2009, http://blog.canpan.info/sasakawa/archive/1807; and Louis Frédéric, *Japan: Encyclopedia* (Cambridge, MA: Harvard Univ. Press, 2002), 398, 825.

17. "Inauguration TGV + discours Mitterrand," accessed 16 November 2011, http://www.ina.fr/economie-et-societe/vie-sociale/video/CAB8100656501/inauguration-tgv-discours-mitterrand.fr.html.

18. Yamanouchi, *Shinkansen ga nakattara*, 255.

19. "Sangiin tochi mondai ni kansuru tokubetsu iinkai, Kaigiroku dai 2-gō (11th kokkai)," 7 December 1987, NDL.

20. "Dai 1-rui dai 15-gō Shūgiin yosan iinkai gijiroku dai 4-gō," 3 February 1988, NDL.

21. "Dai 112-kai kokkai shūgiin un'yu iinkai giroku 2-gō," 2 March 1988, NDL.

22. This phenomenon is not unique to postwar Japan. As excellent studies by Mizuno Hiromi and Daqing Yang amply show, Japanese technocrats have used nationalistic discourse to promote science and technology for World War II.

Archival Sources: Japan

Bōeishō bōeikenkyūjo senshi kenkyū center, Tokyo (NIDS, National Institute for Defense Studies)

Kaiyō anzen gijutsu kenkyūjo, Tokyo (NMRI, National Maritime Research Institute)

Kōbunshokan Asia rekishi shiryō center, Tokyo (NAJ, Japan Center for Asian Historical Records, National Archives of Japan)

Kokkai toshokan, Tokyo (NDL, National Diet Library)

Kōtsū hakubutsukan, Tokyo (TM, Modern Transportation Museum)

Miki Tadanao shuki, Zushi (MT Papers)

Mitsubishi jūkōgyō minami komaki kōjō toshoshitsu, Nagoya (MHI, Mitsubishi Heavy Industries, Minami Komaki Plant)

JR higashi nihon shiryōshitsu, Tokyo (JRE, JR-East: East Japan Railway Company)

Shōwakan, Tokyo (Shōwakan, National Shōwa Memorial Museum)

Tetsudō gijutsu kenkyūjo toshoshitsu, Tokyo (RTRI archive, Railway Technical Research Institute)

Tokyo daigaku kōgakubu toshoshitu, Tokyo (UTA, Tokyo University Department of Aeronautics and Astronautics Library)

Tokyo daigaku senpaku kōgaku toshoshitsu, Tokyo (UTN, Tokyo University Department of Systems Innovation Library)

Tokyo daigaku sentan kagaku gijutsu kenkyū center tosho shiryōshitsu, Tokyo (RCAST archive, Research Center for Advanced Science and Technology)

Tokyo daigaku shiryō henshanjo, Tokyo (UTEO, Tokyo University Historiographical Institute)

Tokyo daigakushi shiryōshitsu, Tokyo (UT archive, Tokyo University Archive)

Archival Sources: United States

National Air & Space Museum, Washington, D.C. (NASM)

National Archives, Washington, D.C. (NA)

Maxwell United States Air Force Base Library, Montgomery, Alabama (MUSAFB)

Wright-Patterson United States Air Force Base Library, Dayton, Ohio (WPAFB)

Archival Source: Great Britain

Imperial War Museum, London (BWM)

Published Sources: Chinese

Zhongguo zhongri guanxi shixuehui, ed. *Youyizhuchunqiu: Weixin zhongguo zuochu gongxian de ribenren.* Beijing: Xinhua chuban, 2002.

Published Sources: English

Abramitzky, Ran, Leah Boustan, and Katherine Eriksson. "Productivity and Migration: New Insights from the 19th Century." *Stanford News,* 1 May 2010. http://news.stanford.edu/news/2010/may/siepr-productivity-migration-050310.html.

Adams, Philip. *Historical Sociology.* Ithaca, NY: Cornell Univ. Press, 1982.

Akera, Atsushi. *Calculating a Natural World: Scientists, Engineers, and Computers during the Rise of U.S. Cold War Research.* Cambridge, MA: MIT Press, 2006.

Aldrich, Mark. "Combating the Collision Horror: The Interstate Commerce Commission and Automatic Train Control, 1900–1939." *Technology and Culture* 34, no. 1 (1993): 49–77.

Anderson, John, Jr. *A History of Aerodynamics and Its Impact on Flying Machines.* Cambridge: Cambridge Univ. Press, 2000.

Aoki, Eiichi, et al. *A History of Japanese Railways, 1872–1999.* Tokyo: East Japan Railway Culture Foundation, 2000.

Bartholomew, James R. "Japanese Modernization and the Imperial Universities, 1876–1920."*Journal of Asian Studies* 37, no. 2 (1978): 251–271.

———. *The Formation of Science in Japan: Building a Research Tradition.* New Haven, CT: Yale Univ. Press, 1989.

Bennett, Marion. "The Immigration and Nationality (McCarran-Walter) Act of 1952, as Amended to 1965." *Annals of the American Academy of Political and Social Science* 367 (1966): 127–136.

Bilstein, Roger E. *Orders of Magnitude: A History of the NACA and NASA, 1915–1990.* Washington, D.C.: National Aeronautics and Space Administration, 1989.

Clancey, Gregory. *Earthquake Nation: The Cultural Politics of Japanese Seismicity, 1868–1930.* Berkeley: Univ. of California Press, 2006.

Cogdell, Christina. *Eugenics Design: Streamlining America in the 1930s.* Philadelphia: Univ. of Pennsylvania Press, 2004.

Constant, Edward W., II, "Science in Society: Petroleum Engineers and the Oil Fraternity in Texas, 1925–65." *Social Studies of Science* 9, no. 3 (1989): 439–472.

Coombs, Gary. "Opportunities, Information Networks and the Migration-Distance Relationship." *Social Networks* 1 (1979): 257–276.

Dees, Bowen C. *The Allied Occupation and Japan's Economic Miracle: Building the Foundations of Japanese Science and Technology.* Surrey, England: Japan Library, 1997.

Dower, John W. "Useful War." In *Japan in War and Peace: Selected Essays,* 9–32. New York: New Press, 1991.

———. *War without Mercy: Race and Power in the Pacific War.* New York: Pantheon Books, 1986.

Eckerson, Helen. "Immigration and National Origins." *Annals of the American Academy of Political and Social Science* 367 (1966): 4–14.

Eckert, Michael. "Strategic Internationalism and the Transfer of Technical Knowledge: The United States, Germany, and Aerodynamics after World War I." *Technology and Culture* 46, no. 1 (2005): 104–131.

Edwards, Paul N. *The Closed World: Computers and the Politics of Discourse in Cold War America.* Cambridge, MA: MIT Press, 1996.

Entrikin, Nicholas J. *The Betweenness of Place: Towards a Geography of Modernity.* Baltimore: Johns Hopkins Univ. Press, 1991.

Ericson, Eric. *Identity, Youth, and Crisis.* New York: W. W. Norton, 1968.

Francillon, Rene J. *Japanese Aircraft of the Pacific War.* Annapolis, MD: Naval Institute Press, 1970.

Frédéric, Louis. *Japan: Encyclopedia.* Cambridge, MA: Harvard Univ. Press, 2002.

Fruin, Mark. "Peasant Migrants in the Economic Development of Nineteenth-Century Japan." *Agricultural History,* 54 no. 2 (1980): 261–277.

Fujitani, Takashi. *Splendid Monarchy: Power and Pageantry in Modern Japan.* Berkeley: Univ. of California Press, 1998.

Gimbel, John. *Science, Technology, and Reparations: Exploitation and Plunder in Postwar Germany.* Stanford, CA: Stanford Univ. Press, 1990.

Gooday, Graeme, and Morris Low. "Technology Transfer and Cultural Exchange: Western Scientists and Engineering Encounter Late Tokugawa and Meiji Japan." *OSIRIS: Beyond Joseph Needham: Science, Technology, and Medicine in East and Southeast Asia* 13 (1998): 99–128.

Gusterson, Hugh. *Nuclear Rites: A Weapons Laboratory at the End of the Cold War.* Berkeley: Univ. of California Press, 1996.

Hanle, Paul. *Bringing Aerodynamics to America.* Cambridge, MA: MIT Press, 1982.

Harris, Sheldon. *Factories of Death: Japanese Biological Warfare, 1932–1945, and the American Cover-up.* New York: Routledge, 2002.

Hasegawa, Akio. "My Father Tatsuo Hasegawa (1916–2008)." Accessed 22 September 2010. http://www.geocities.jp/pinealguy/tatsuo.tatsuo.htm.

Hashimoto, Takehiko. "Theory, Experiment, and Design Practice: The Formation of Aeronautical Research, 1909–1930." PhD diss., Johns Hopkins Univ., 1990.

Hatton, Timothy, and Jeffrey Williamson. *Global Migration and the World Economy: Two Centuries of Policy and Performance.* Cambridge, MA: MIT Press, 2005.

Havens, Thomas R. *Architects of Affluence: The Tsutsumi Family and the Seibu-Saison Enterprise in Twentieth-Century Japan.* Cambridge, MA: Harvard Univ. Press, 1994.

Hecht, Gabrielle. *The Radiance of France: Nuclear Power and National Identity after World War II.* Cambridge, MA: MIT Press, 1998.

Hein, Carola. "Shaping Tokyo: Land Development and Planning Practice in the Early Modern Japanese Metropolis." *Journal of Urban History* 36, no. 4 (2010): 447–484.

Hill, Peter. "Kamikaze, 1943–5." In *Making Sense of Suicide Missions,* 1–42. New York: Oxford Univ. Press, 2005.

Horikoshi, Jirō. *Eagles of Mitsubishi: The Story of the Zero Fighter.* Seattle: Univ. of Washington Press, 1981.

Houstoun, Marion, Roger Kramer, and Joan Barrett. "Female Predominance in Immigration to the United States since 1930: A First Look." *International Migration Review* 18, no. 4 (1984): 908–963.

Hughes, Thomas P. *Networks of Power: Electrification in Western Society, 1880–1930.* Baltimore: Johns Hopkins Univ. Press, 1983.

Itō, Kenji. "Gender and Physics in Early 20th Century Japan: Yuasa Toshiko's Case." *Historia Scientiarum* 14 (2004): 118–135.

Japan Air Raids Org. "Timeline." Accessed 5 October 2011. http://www.japanairraids.org /timeline.

Jedlicka, Davor. "Opportunities, Information Networks and International Migration Streams." *Social Networks*, 1 (1979): 277–284.

Johnson, Chalmers. *MITI and the Japanese Miracle: The Growth of Industrial Policy, 1925–1975.* Stanford, CA: Stanford Univ. Press, 1982.

Kane, Robert. *All Transportation.* 14th ed. Dubuque, IA: Kendall Hunt, 2003.

Kármán, Theodore von. *The Wind and Beyond: Pioneer in Aviation and Pathfinder in Space.* Boston: Little, Brown, 1967.

Koizumi, Kenkichirō. "In Search of Wakon: The Cultural Dynamics of the Rise of Manufacturing Technology in Postwar Japan." *Technology and Culture* 43 (2002): 29–49.

Layton, Edwin T., Jr. *The Revolt of the Engineers: Social Responsibility and the American Engineering Profession.* Baltimore: Johns Hopkins Univ. Press, 1971.

Lepage, Jean-Denis G. G. *Aircraft of the Luftwaffe, 1935–1945: An Illustrated Guide.* Jefferson, NC: McFarland, 2009.

Leslie, Stuart. *The Cold War and American Science: The Military-Industrial-Academic Complex at MIT and Stanford.* New York: Columbia University Press, 1994.

Low, Morris. *Science and the Buidling of a New Japan.* New York: Palgrave, 2005.

Lutz, Catherine. *Homefront: The Military City and the American Twentiety Century.* Boston: Beacon Press, 2001.

Mackenzie, Donald. *Inventing Accuracy: A Historical Sociology of Nuclear Missile Guidance.* Cambridge, MA: MIT Press, 1990.

Madeley, Christopher. "Britain and the World Engineering Congress: Tokyo 1929." In *Britain and Japan in the Twentieth Century: One Hundred Years of Trade and Prejudice,* 46–61. London: I. B. Tauris, 2007.

Mannheim, Karl. *Essays on the Sociology of Culture.* New York: Oxford Univ. Press, 1956.

Marshall, Byron. *Academic Freedom and the Japanese Imperial University, 1868–1939.* Berkeley: Univ. of California Press, 1992.

Masai, Yasuo. "Tokyo: From a National Centre to a Global Supercity." *Asian Journal of Communication* 2, no. 3 (1992): 68–80.

McCormick, Kevin. *Engineers in Japan and Britain: Education, Training, and Employment.* London: Routledge, 2000.

McGraw-Hill Encyclopedia of Science and Engineering. New York: McGraw Hill, 1997.

McKeown, Adam. *Melancholy Order: Aisan Migration and the Globalization of Borders.* New York: Columbia Univ. Press, 2008.

McQuillan, Kevin. "Family Composition, Birth Order and Marriage Patterns: Evidence from Rural Alsace, 1750–1885." *Annales de démographie historique* 1 (2008): 57–71.

Meunier, Jacob. *On the Fast Track French Railway Modernization and the Origins of the TGV, 1944–1983.* Westport, CT: Praeger, 2002.

Mindell, David. *Between Human and Machine: Feedback, Control, and Computing before Cybernetics*. Baltimore: Johns Hopkins Univ. Press, 2000.

Ministry of Land, Infrastructure, Transport and Tourism. "Shinkansen Japanese High-Speed Rail." Accessed 25 October 2011. http://www.mlit.go.jp/en/tetudo/tetudo_fr2_000000.html.

Mizuno, Hiromi. *Science for the Empire: Scientific Nationalism in Modern Japan*. Stanford, CA: Stanford Univ. Press, 2008.

Naitō, Hatsuho. *Thunder Gods: The Kamikaze Pilots Tell Their Story*. New York: Dell Book, 1982.

Nathan, John. *Sony: The Private Life*. New York: Houghton Mifflin, 1996.

National Physical Laboratory. "NPL's History Highlights." Accessed 18 April 2011. http://www.npl.co.uk/content/ConMediaFile/4279.

Ngai, Mae M. "The Architecture of Race in American Immigration Law: A Reexamination of the Immigration Act of 1924." *Journal of American History* 86, no. 1 (1999): 67–92.

Nye, David. *Technology Matter: Questions to Live With*. Cambridge, MA: MIT Press, 2006.

Partner, Simon. *Assembled in Japan: Electrical Good and the Making of the Japanese Consumer*. Berkeley: Univ. of California Press, 1999.

Peattie, Mark. *Sunburst: The Rise of Japanese Naval Air Power, 1909–1941*. Annapolis, MD: Naval Institute Press, 2002.

Pekkanen, Saadia, and Paul Kallender-Umezu. *In Defense of Japan: From the Market to the Military in Space Policy*. Stanford, CA: Stanford Univ. Press, 2010.

Pilcher, Jane. "Mannheim's Sociology of Generations: An Undervalued Legacy." *British Journal of Sociology* 45, no. 3 (1994): 481–495.

Railway Gazette International 124 (5 July 1968).

Roland, Alex. "Technology and War: The Historiographical Revolution of the 1980s." *Technology and Culture* 34 (1993): 117–134.

Rubin, Ernest. "The Demography of Immigration to the United States." *Annals of the American Academy of Political and Social Science* 367 (1966): 15–22.

Russell, Edmund. *War and Nature: Fighting Humans and Insects with Chemicals from World War I to Silent Spring*. Cambridge: Cambridge Univ. Press, 2001.

Samuels, Richard J. *Rich Nation Strong Army: National Security and the Technological Transformation of Japan*. Ithaca, NY: Cornell Univ. Press, 1994.

Satō, Yasushi. "Systems Engineering and Contractual Individualism: Linking Engineering Processes to Macro Social Values." *Social Studies of Science* 37, no. 6 (2007): 909–934.

Schatzberg, Eric. *Wings of Wood, Wings of Metal*. Princeton, NJ: Princeton Univ. Press, 1998.

———. "Wooden Airplanes in World War II: National Comparisons and Symbolic Culture." In *Archimedes: New Studies in History and Philosophy of Science and Technology*, Vol. 3 of *Atmospheric Flight in the Twentieth Century*, edited by Alex Roland and Peter Galison. Dordrecht, Netherlands: Springer, 2000.

Schuman, Howard, and Cheryl Rieger. "Historical Analogies, Generational Effects, and Attitudes Toward War." *American Sociological Review* 57, no. 3 (1992): 315–326.

Sherry, Michael S. *The Rise of American Air Power: The Creation of Armageddon*. New Haven, CT: Yale Univ. Press, 1987.

Smith, Merritt Roe, ed. *Military Enterprise and Technological Change: Perspectives on the American Experience*. Cambridge, MA: MIT Press, 1985.

Smith, Richard. "The Intercontinental Airliner and the Essence of Airplane Performance, 1929–1939." *Technology and Culture* 24, no. 33 (1983): 428–449.

Sulloway, Frank. *Born to Rebel: Birth Order, Family Dynamics, and Creative Lives.* New York: Pantheon Books, 1996.

Suzuki, Tessa-Morris. *The Technological Transformation of Japan from the Seventeenth Century to the Twenty-First Century.* Cambridge: Cambridge Univ. Press, 1994.

Tsutsui, William M. *Manufacturing Ideology: Scientific Management in Twentieth-Century Japan.* Princeton, NJ: Princeton Univ. Press, 1998.

Turkle, Sherry. *Alone Together: Why We Expect More from Technology and Less from Each Other.* New York: Basic Books, 2011.

Vincenti, Walter. *What Engineers Know and How They Know It: Analytical Studies from Aeronautical History.* Baltimore: Johns Hopkins Univ. Press, 1993.

Walter, Grunden. *Secret Weapons and World War II: Japan in the Shadow of Big Science.* St. Lawrence: Univ. of Kansas Press, 2005.

White, James. "Internal Migration in Prewar Japan." *Journal of Japanese Studies* 4, no. 1 (1978): 81–123.

Yamamura, Kōzō. "Success Illgotten? The Role of Meiji Militarism in Japan's Technological Progress." *Journal of Economic History* 37, no. 1 (1977): 113–135.

Yang, Daqing. *Technology of Empire: Telecommunications and Japanese Expansion in Asia, 1883–1945.* Cambridge, MA: Harvard Univ. Press, 2010.

Young, Louise. *Japan's Total Empire: Manchuria and the Cultural Wartime Imperialism.* Berkeley: Univ. of California Press, 1998.

Published Sources: French

"Le Japon inaugure son 'chimen de fer de demain'!: Naissance d'un super train . . . " *La Vie du Rail.* 7 March 1965: 12–17.

"Quelques particularités techniques de la nouvelle ligne Japonaise du Tokaïdo" *La Vie du rail.* 7 March 1965: 17–29.

"Japon: Un An d'exploitation de la nouvelle ligne du Tokaïdo." *La Vie du rail.* 26 December 1965: 8–13.

"Inauguration TGV + discours Mitterrand." Accessed 16 November 2011. http://www.ina .fr/economie-et-societe/vie-sociale/video/CAB8100656501/inauguration-tgv-discours -mitterrand.fr.html.

Published Sources: German

Ciesla, Burghard. "Das 'Project Paperclip'—deutsche Naturwissenschaftler und Techniker in den USA (1946 bis 1952)." In *Historische DDR-Forschung: Aufsätze und Studien.* Vol. 1. Berlin: Academie-Verlag, 1994.

Koch, Matthias. *Rüstungskonversion in Japan nach dem Zweiten Weltkrieg: Von der Kriegswirtschaft zu einer Weltwirtschaftsmacht.* Tokyo: Deutsches Institut für Japanstudien, 1998.

Vogt, Richard. *Weltumspannende Memoiren eines Flugzeug-Konstrukterurs.* Steineback/ Woerthsee: Flieger-Verlag, 1976.

Zeilinger, Stefan. *Wettfahrt auf der Schiene: Die Entwicklung von Hochgeschwindigkeitszügen im europäischen Vergleich.* Frankfurt: Campus Verlag, 2003.

Published Sources: Japanese
Newspapers

Asahi shinbun
Chūgai shōgyō shinpō
Kagaku bunka shinbun
Mainichi shinbun
Nihon keizai shinbun
Osaka Asahi shinbun
Saga shinbun
Yomiuri shinbun

Books, Journals, and Web Sources

Akio Morita Library. "Morita Asset Management." Accessed 23 September 2010. http://www.akiomorita.net/profile/life.html.

Amamiya, Yoshifumi, and Yoshikata Maki. "Zatsuon denryoku ni chakumoku shita zatsuongen tanchiki." *Tetsudō gijutsu kenkyū hōkoku* 132, no. 23 (1960): 1–11.

Amamiya, Yoshifumi. "Kokutetsu densha yori hassei suru chūtanpa musen zatsuon ni kansuru kenkyū." *Tetsudō gijutsu kenkyū hōkoku* 58, no.9 (1959): 1–10.

Aoki, Kaizō. *Kokutetsu.* Tokyo: Shinchōsha, 1964.

Ariga, Sōkichi. *Sogō Shinji.* Tokyo: Sogō Shinji-den kankōkai, 1988.

Asahi shinbunsha. *Sekai no tetsudō.* Tokyo: Asahi shinbunsha, 1964.

Bōeichō bōei kenshūjo seishishitsu. *Senshi sōsho.* 102 vols. Tokyo: Asagumo shinbusha, 1966–1980.

Bungei shunjū-hen. *Ningen bakudan to yobarete: Shōgen ōka tokkō.* Tokyo: Bungei shunjū, 2005.

Dainihon teikoku gikaishi kankōkai. *Dainihon teikoku gikaishi.* Vol. 5. Tokyo: Dainihon teikoku gikaishi kankōkai, 1927.

Denkikei dosōkai. *Sepiairo no 3-gōkan: Rekishi archive.* Accessed 26 October 2011. http://todaidenki.jp/hist/?cat=16.

Doi, Takeo. *Hikōki sekkei 50-nen no kaisō.* Tokyo: Suitōsha, 1989.

Doi, Zenjirō. *Kessen heiki maruyu: Rikugun sensuikan.* Tokyo: Kōjinsha, 2003.

Enomoto, Shinsuke. "Kinzoku zairyo no hirō to naibu masatsu ni kansuru kenkyū." *Chūō kōkū kenkyūjo ihō* 2, no.7 (1943): 177–189.

———. "Kinzoku zairyo no hirō to naibu masatsu ni kansuru kenkyū," *Chūō kōkū kenkyūjo ihō* 2, no. 10 (1943): 305–324.

Fujii, Shigeki. "Kokutetsu chōki sekimu no shorimondan to sono keizaiteki fukui ni kansuru ichikōsatsu." *Kaikei kensa kenkyū* 17. Accessed 7 November 2011. http://jbaudit.go.jp/effort/study/mag/17-5.html.

Fujishima, Shigeru. "Shinkansen jisoku 200-kiro no missitsu." In *Bungei shunjū ni miru Shōwa-shi.* Vol. 2. Tokyo: Bungei shunjū, 1988.

Fukuhara, Shun'ichi. *Business tokkyū o hashiraseta otokotachi.* Tokyo: JTB, 2003.

Fukushima, Mutsuo. "Zero Inspired Today's Innovations." *Japan Times.* 14 January 2004.

Gunji gijutsu kara minsei gijutsu eno tenkan: Dainiji sekai taisen kara sengo eno wagakuni no keiken. 2 vols. Tokyo: Nihon gakujutsu shinkōkai, 1994–1996.

Gunyōki kaihatsu monogatari. Vol. 2. Tokyo: Kōinsha, 2002.

Hakone onsen ryokan kyōdō kumiai. "Hakone onsei kōshiki gaido: Hakopita." Accessed 28 October 2011. http://www.hakone-ryokan.or.jp/002_rekishi.html.

Hakone onsen ryokan kyōdō kumiai. *Hakone onsenshi.* Hakone: Hakone onsen ryokan kyōdō kumiai, 1986.

"Hakone Yumoto onsen, Hakone onsen no rekishi." Accessed 28 October 2011. http://www.hakone-yado.jp/hakone-rekishi.html.

Hara, Takeshi and Akio Yasuoka. *Nihon rikukaigun jiten.* Tokyo: Shinjinbutsu ōraisha, 1997.

Hara, Tomoshige. "Ressha ga kōsoku de suidō ni totsunyū suru baai no ryūtai rikigakuteki shomondai." *Tetsudō gijutsu kenkyū hōkoku* 153 (1960): 1–22.

Harada, Yutaka. *Omoide.* Tokyo: Harada Yutaka, 1991.

Hashimoto, Kōichi. "Kokutetsu ni okeru kyōryō kyōdo shindō shiken no genjō to shōrai." *Doboku gakkaishi* 33, no. 5–6 (1948): 31–34.

Hashimoto, Kōichi, and Fumihito Itō. "Rosen dōro Miyagino-bashi no kyōdo sokutei." *Doboku gakkaishi* 37, no. 4 (1952): 13–17.

Hashimoto, Takehiko. *Hikōki no tanjō to kūkirikigaku no keisei.* Tokyo: Tokyo University Press, 2012.

Hata, Ikuhiko, ed. *Nihon rikukaigun sōgō jiten.* Tokyo: Tokyo daigaku shuppankai, 1991.

Hata, Ikuhiko. *Shōwa no nazo o ou.* Vol. 2. Tokyo: Bungei shunjū, 1999.

Hatano, Isamu. *Kindai nihon no gunsangaku fukugōtai.* Tokyo: Sōbunsha, 2005.

Hattensuru tetsudō gijutsu: Saikin 10-nen no ayumi. Tokyo: Nihon tetsudō gijutsu kyōkai, 1965.

Hayashi, Masami. "Sekai ni hokoru kōsoku tetsudōyō daideiryoku kyōkyū hōshiki: AT kiden hōshiki." *Hatsumei* 76, no. 8 (1979): 51–57.

Hayashi, Shōzō. "Keiryō 3-tōsha naha 10-keishiki." *Kōtsū gijutsu* 108 (1955): 260–263.

Heisei 11-nendo sangyō gijutsu no rekishi ni kansuru chōsa kenkyū hōkokusho. Tokyo: Nihon kikai kōgyō rengōkai, 2000.

Heisei 5-nendo bun'yabetsu kagakugijututaikei no genjō to shōrai ni kansuru chōsa kenkyū hōkokusho. Tokyo: Nihon kikai kogyō rengōkai, 1994.

Hiroshige, Tetsu, *Kagaku no shakaishi: Sensō to kagaku.* Vol. 1. Tokyo: Iwanami shoten, 2002.

Hirota, Kōzō. *Mantetsu no shūen to sonogo: Aru chūō shikenjoin no hōkoku.* Tokyo: Sōgensha, 1990.

Hobara, Mitsuo. "Ressha shūchū seigyo." *Denki gakkai zasshi* 84, no. 10 (1964): 51–56.

Hokkaidō daigaku. *Hokudai 100-nen shi, bukyokushi 2.* Sapporo: Gyōsei, 1980–82. Hōmu daijin kanbō shihō hōsei chōsabu shihō hōsei-ka. *Saikō Saibansho Hanreishū.* Vol. 12. Tokyo, 1958.

Horikoshi, Jirō and Masatake Okumiya. *Zerosen.* Tokyo: Asahi sonorama, 1992.

———. *Zerosen: Sono eikō to kiroku.* Tokyo: Kōbunsha, 1995.

———. *Zerosen: Sono tanjō to eikō no kiroku.* Tokyo: Kōbunsha, 1970.

Hoshi, Akira. *Sharyō no keiryōka.* Tokyo: Nihontosho kankōkai, 1956.

Hoshikawa Takeshi, ed. *Shinkansen zenshi.* Tokyo: Gakken kenkyūsha, 2003.

Hoshino, Yōichi. "Kidō no kōzō." *Kōtsū gijutsu* 135 (1957): 9–11.

———. "Shin kōzō kidō." *JREA* 1, no. 7 (1958): 9–11.

Hosokawa, Hachirō. "Ningen bakudan ōka to tomoni." *Kōkūfan* 13, no. 14 (1964): 76–79.

Ikari, Yoshirō, et al. *Nihon no gunji technology.* Tokyo: Kōjinsha, 2001.

Ikari, Yoshirō. *Ikiteiru Zerosen.* Tokyo: Yomiuri shinbunsha, 1970.

————. *Kaigun gijutsushatachi no taiheiyō sensō.* Tokyo: Kōjinsha, 1989.

————. *Kaigun kūgishō.* 2 vols. Tokyo: Kōjinsha, 1985.

————. *Kōkū technology no tatakai.* Tokyo: Kōjinsha, 1996.

————. *Sentōki hayabusa: Shōwa no meiki sono eikō to higeki.* Tokyo: Kōjinsha, 2003.

————. *Chō kōsoku ni idomu: Shinkansen kaihatsu ni kaketa otokotachi.* Tokyo: Bungei shunjū, 1993.

Ikehara, Teruo. "Corolla kaihatsu no Hasegawa Tatsuo-shi ni okeru 'shusa 10-kajō." *Nikkei Business Online.* Accessed 22 September 2010. http://business.nikkeibp.co.jp/article /tech/20060825/108606.

Imamura, Yōichi. "Yokosuka, Kure, Sasebo, Maizuru ni okeru kyūgunyōchi no tenyō ni tsuite: 1950–1976 nendo no kyūgunkōshi kokuyū zaisan shori shingikai ni okeru kettei jikō no kōsatsu o tōshite." *Nhon toshi keikaku gakkai toshikeikaku ronbunshū* 43, no. 3 (2008): 193–198.

Ishii, Kanji. *Nihon no sangyō kakumei: Nisshin, nichiro sensō kara kangaeru.* Tokyo: Asahi shinbunsha, 1997.

Ishikawa, Junkichi. *Kokka sōdōinshi.* 13 vols. Tokyo: Kokka sōdōinshi kankōkai, 1975–1987.

Ishizawa, Nobuhiko. "Sekkei to shūzen (1) Pantograph." *Tetsudō kōjō* 9, no. 1 (1958): 16–17.

Itō, Hiromitsu. "Haikyo no entotsu." *Suikō* 343 (1982): 23–25.

Itokawa, Hideo. *Kyōi no jikan katsuyōjutsu: Naze koredake saga tsukunoka?* Tokyo: PHP kenkyūjo, 1985.

Izawa, Katsumi. "Kyakusha kōtai ka." *Kōtsū gijutsu* 40 (1949): 15–17.

Japan Automobile Hall of Fame, 2009. "Kūriki no tokusei to kihon jūshi no kōseinō o kaihatsu: Kubo Tomio." Accessed 22 September 2010. http://www.jahfa.jp/JAHFA _PR2_2009.pdf.

Kagaku gijutsu seisakushi kenkyūkai. *Nihon no kagaku gijutsu seisakushi.* Tokyo: Mitō kagaku gijutsu kyōkai, 1990.

Kawasaki kōkūki kogyō kabushiki kaisha. *Kōkūki seizō enkaku.* 1946.

Kaigun henshū iinkai-hen. *Kaigun.* 15 vols. Tokyo: Seibun tosho, 1981.

Kaigunshō. *Gen'eki kaigun shikan meibo.* Vol. 3. Tokyo: Kaigunshō, 1944.

Kaigun jinrai butai senyūkai henshūiinkai. *Kaigun jinrai butai.* Tokyo: Kaigun jinrai butai senyūkai, 1996.

Kaigun kōkū gijutsushō denkibu. Tokyo: Kūgishō denkibu no kai, 1987.

Kaigun kōkū gijutsushō zairyōbu no kai. *Kaigun kōkū gijutsushō zairyōbu shūsen 50-shūnen kinenshi.* Tokyo: Kaigun kōkū gijutsushō zairyōbu no kai, 1996.

Kaigun kōkū gijutsushō. Tokyo: Gakushū kenkyūsha, 2008.

Kaikūkai, ed. *Umiwashi no kōseki: Nihon kaigun kōkū gaishi.* Tokyo: Hara shobō, 1982.

Kakumoto, Ryōhei. *Shinkansen kaihatsu monogatari.* Tokyo: Chūō kōronsha, 2001.

Kanematsu, Manabu. *Shūsen zengo no ichi shōgen: Aru tetsudōjin no kaisō.* Tokyo: Kōtsū kyōkai, 1986.

Kawamura, Atsuo. *Kyaku kasha no kōzō oyobi riron.* Tokyo: Kōyūsha, 1952.

Kawamura, Yutaka. "Kyū kaigun gijutsu kenkyūjo ni miru kenkyū kaihatsu taisei no tokuchō." *Gijutsushi* 2 (2001): 18–30.

Kawanabe, Hajime. "Atarashii shanai shingō." *JREA* 2, no. 4 (1959): 2–7.

————. "Jidō ressha seigyo." *Denki gakkai zasshi* 84, no. 10 (1964): 43–50.

Kimura, Hidemasa. *Waga hikōki jinsei.* Tokyo: Nihon tosho center, 1997.

Kobashi, Yasujirō, ed. *Tani Ichirō sensei koki kinen kōenshū*. 1977.

Kofukuda, Terufumi. *Zerosen kaihatsu monogatari: Nihon kaigun sentōki zenkishu no shōgai*. Tokyo: Kōjinsha, 1985.

Kōgaku kogyōshi henshū kai. *Heiki o chūshin to shita nihon no kōgaku kogyōshi*. Tokyo: Kōgaku kogyōshi henshū kai, 1955.

Kohagura, Yasuyoshi. *Kiseki: Zōsen shikan Fukuda Tadashi no tatakai*. Tokyo: Kōjinsha, 1996. *Kōkū gijutsu no zenbō*. 2 vols. Tokyo: Kōyōsha, 1953.

Kōkū kōgyōshi henshū iinkai. *Minkan kōkūki kogyōshi*. 1948.

Kokudo kōtsushō. "Kokutetsu kaikaku ni tsuite." Accessed 7 Nov 2011. http://www.milt.go.jp/tetudo/kaikaku/01.htm.

Kōseishō, Jinkō mondai kenkyūjo. *Jinkō mondai kenkyū* 2, no. 12 (February 1941).

"Kōsoku resshayō fūatsu brake no kenkyū." *Tetsudō gijutsu kenkyūjo itaku kenkyū hōkoku: Showa 35-nendo*. Tokyo, 1960.

Kōsoku tetsudō no kenkyū: Shu to shite Tokaido Shinkansen ni tsuite. Tokyo: Kenyūsha, 1967.

Kōtsūshinbun henshūkyoku. *Atarashii tetsudō no tankyū: Tetsudō gijutsu kenkyū no kadai*. Tokyo: Kōtsū kyōryokukai, 1959.

Kubo, Masaki. "Tetsudō gijutsu kenkyūjo no genzai to shōrai eno michi." *Kōtsū gijutsu* 30 (1949): 10–15.

Kubota, Hiroshi. *Nihon no tetsudō sharyō shi*. Tokyo: Grandpri shuppan, 2001.

———. *Tetsudō jūdai jiko no rekishi*. Tokyo: Grandpri shuppan, 2000.

———. "Sengo nihon tetsudōshi no ronten." *Tetsudō shigaku* 6 (1988): 39–46.

Kumezawa, Ikurō. "Kakū denshasen no kyōdo oyobi sei oshiageryō keisanhō." *Tetsudō gijutsu kenkyū hōkoku* 120 (1960): 1–12.

———. "Kasen to shūden." *Denki gakkai zasshi* 84, no. 10 (1964): 35–42.

———. "Shinkansen no shūden ni tsuite." *Kōtsū gijutsu* 159 (1963): 10–14.

Kyōtani, Yoshihiro, Takeshi Oku, and Toshio Sanuki. *Chō kōsoku Shinkansen*. Tokyo: Chūō kōronsha, 1971.

Kyoto daigaku 70-nen shi henshū iinkai. *Kyoto daigaku 70-nen shi*. Kyoto: Kyoto daigaku, 1967.

Kyūshū daigaku sōritsu 50-shūnen kinenkai. *Kyūshū daigaku 50-nen shi, tsūshi 1*. Fukuoka: Kyūshū daigaku sōritsu 50-shūnen kinenkai, 1967.

Maema, Takanori. *Dangan ressha: Maboroshi no Tokyo-hatsu Beijing-yuki chōtokkyū*. Tokyo: Jitsugyō-no-nihonsha, 1994.

———. *Fugaku: Bei hondo o bakugeki seyo*. 2 vols. Tokyo: Kōdansha, 1995.

———. *Gijutsusha tachino haisen*. Tokyo: Sōshisha, 2004.

———. *Man machine no shōwa densetsu: Kōkūki kara jidōsha e*. 2 vols. Tokyo: Kōdansha, 1996.

———. *Senkan Yamato no iseki*. 2 vols. Tokyo: Kōdansha, 2005.

———. *YS-11 Kokusan ryokakki o tsukutta otokotachi*. Tokyo: Kōdansha, 1994.

Mainichi shinbunsha. *Speed 100-nen*. Tokyo: Mainichi shinbunsha, 1969.

Makita, Kōji, and Yōetsu Fujita. "Kyū Yokosuka kaigun kōshō kōin kishukusha o tenyō shita shiei jūtaku ni tsuite: Yokosukashi ni okeru sengo shiei jūtaku ni kansuru kenkyū sono 5." Accessed 7 September 2010. http://www.cit.nihon-u.ac.jp/kenkyu/kouennkai/reference/No_38/4_kenchiku/4-050.pdf.

Maruhama, Tetsurō. "Kokutetsu shuyō kansenkei SHF kaisen o kaerimite." *Kōtsū gijutsu* 176 (1960): 22–25.

Matsubara, Keiji. *Shūsenji teikoku rikugun zen gen'eki shōkō shokumu meikai*. Tokyo: Senshi kankō iinkai, 1985.

Matsudaira, Tadashi, et al. "Nijiku kasha no banetsuri sōchi kaizō ni yoru kōsokuka."
Tetsudō gyōmu kenkyū shiryō 10, no. 18 (1953): 5–9.

Matsudaira, Tadashi "Anzen to norigokochi." *Kōtsū gijutsu* 135 (1957): 5–8.

———. "Kōsoku tetsudō gijutsu no raimei I." *Railway Research Review* 50, no. 3 (1993): 25–39.

———. "Kōsoku tetsudō gijutsu no raimei II." *Railway Research Review* 50, no.4 (1993): 28–34.

———. "Kyakusha oyobi densha no koyū shindōsū." *Tetsudō gyōmu kenkyū shiryō* 6, no. 2 (1949): 3–14.

———. "Sharinjiku no dakōdō." *Tetsudō gyōmu kenkyū shiryō* 9, no. 1 (1952): 16–26.

———. "Tokaido Shinkansen ni kansuru kenkyū kaihatsu no kaiko: Shu to shite sharyō no shindō mondai ni kanren shite." *Nihon kikai gakkaishi* 75, no. 646 (1972): 100–108.

Matsuura, Yoshinari. *Tokushu kōgekiki ōka fukugen ni tsuite* (Unpublished manuscript, December 2001).

Meikū kōsakubu no senzen sengoshi: Moriya sōdanyaku, watashi to kōkūki seisan. Nagoya: Mitsubishi jūkōgyō kabushiki kaisha Nagoya kōkūki seisakujo, 1988.

Miki, Tadanao. "Chō tokkyū ressha (Tokyo-Osaka 4-jikan han) no ichi kōsō." *Kōtsū gijutsu* 89 (1954): 2–6.

———. *Jinrai tokubetsu kōgekitai.* Tokyo: Sannō shobō, 1968.

———. "Kōsoku ressha no kūki rikigakuteki shomondai." *Kōtsū gijutsu* 113 (1950): 30- 32.

———. "Kōsoku tetsudō sharyō no kūki rikigakuteki shomondai, 3." *Kikai no kenkyū* 12 (1960): 25–30.

———. "Ressha no speed-up o habamu mono." *Shindenki* 12, no. 1 (1958): 18–24.

———. "Sharyō kōzō." *Kōtsū gijutsu* 135 (1957): 2–5.

———. "Kōsoku tetsudō sharyō no kūki rikigakuteki shomondai, Part 1." *Kikai no kenkyū* 12, no. 7 (1960): 17–24.

———. "Kōsoku tetsudō sharyō no kūki rikigakuteki shomondai, Part 2." *Kikai no kenkyū* 12, no. 8 (1960): 13–18.

———. "Kōsoku tetsudō sharyō no kūki rikigakuteki shomondai, Part 3." *Kikai no kenkyū* 12, no.9 (1960): 25–30.

———. "Kōzō kyōdo kara mi ta densha no dōkō." *Denkisha no kagaku* 10, no. 4 (April 1957): 6–10.

———. "Monorail 45-nen no tsuioku." *Monorail* 82 (1994): 1–13.

———. "Odakyū 3000-kei SE-sha sekkei no tuioku." *Tetsudō Fan* 32, no. 375 (1992): 91–99.

Miki, Tadanao, et al. "Jiko kara mita kyakusha・densha no kōzō sekkei shiryō." *Tetsudō gyōmu kenkyū shiryō* 7, no. 4 (1950): 4–10.

———. "Kōsoku sharyō no kūki rikigakuteki kenkyū." *Nihon kikai gakkaishi* 61, no. 478 (November 1958): 34–43.

———. "Ōgata trailer bus shaken hōkoku." *Tetsudō gyōmu kenkyū shiryō* 6, no. 1(1949):5–12.

———. "Shōnan denshayō tsūfūki shaken." *Tetsudō gyōmu kenkyū shiryō* 7, no. 15 (1950): 4–10.

Miyasaka, Masanao. "Rinji sharyō sekkei jimusho," *Kokuyū tetsudū* 94 (1957): 20.

Mizusawa, Hikaru. "Asia taiheyō sensōki ni okeru kyūrikugun no kōkū kenkyū kikan eno kitai." *Kagakushi kenkyu* 43 (2004): 22–30.

———. "Rikugun ni okeru kōkūkenkyūjo no setsuritsu kōsō to gijutsuin no kōkū jūtenka." *Kagakushi kenkyū* 42 (2003): 31–39.

Monbu kagakushō. *Gakusei 100-nenshi.* Accessed 4 October 2011. http://www.mext.go.jp
    /b_menu/hakusho/html/hpbz198101/hpbz198101_2_121.html.

Monbushō. *Atarashii kenpō no hanashi* (1947).

Morita, Akio, and Masaru Ibuka. "Ibuka taidan, Innen ni michibikareruyō ni (2) Guest:
    Morita Akio." Aaccessed 23 September 2010. http://www.sony-ef.or.jp/library/ibuka
    /pdf/taidan_no62_2.pdf.

Nagoya daigakushi henshū iinkai. *Nagoya daigaku 50-nenshi, bukyokushi 2.* Nagoya: Nagoya
    daigaku shuppankai, 1989.

Naitō, Hatsuho. *Gunkan sōchō: Hiraga Yuzuru.* Tokyo: Bungei shunjū, 1987.

———. *Kaigun gijutsusenki.* Tokyo: Tosho shuppan, 1976.

———. *Ōka: Kyokugen no tokkōki.* Tokyo: Chūōbunko, 1999.

Nakagawa, Keiichirō, ed. *Kigyō keiei no rekishiteki kenkyū.* Tokyo: Iwanami shoten, 1990.

Nakagawa, Yasuzō. *Kaigun gijutsu kenkyūjo: Electronics ōkoku no senkusha tachi.* Tokyo: Ni-
    hon keizai shinbunsha, 1987.

Nakamura, Kazuo. "Hizumi gauge tanjō 50-nen." *Kyowa gihō* 370 (1988): 1–11.

Nakaoka, Tetsurō. *Nihon kindai gijutsu no keisei: Dentō to kindai no dynamics.* Tokyo: Asahi
    shinbunsha, 2006.

Nakata, Kin'ichi. "Tetsudō gijutsu kenkyūjonai ni okeru gas turgine no kenkyū." *Tetsudō
    gyōmu kenkyū shiryō* 7, no. 17 (1950): 4–9.

Nakayama, Shigeru, Kunio Gotō, and Hitoshi Yoshioka, eds. *A Social History of Science and
    Technology in Contemporary Japan.* 4 vols. Melbourne: Trans Pacific Press, 2001–2006.

NHK Project X seisakuhan. *Project X chōsenshatachi.* Tokyo: Nihonhōsō kyōkai, 2000.

NHK Special shuzaihan. *Nihon kaigun 400 jikan no shōgen: gunreibu · sanbōtachi ga katatta
    haisen.* Tokyo: Shinchōsha, 2011.

Nihon kagakushi gakkai-hen. *Nihon kagaku gijutsushi taikei, dai-2kan tsūshi.* Tokyo: Daiichi
    hōki shuppan, 1967.

Nihon kaigun kōkūshi hensan iinkai. *Kaigun kōkūshi.* 4 vols. Tokyo: Jiji tsūshinsha, 1969.

Nihon kōkū gakujutsushi henshū iinkai. *Nihon kōkū gakujutsushi (1910–1945).* Tokyo:
    Maruzen shuppan, 1990.

Nihon kōkū kyōkai. *Nihon kōkūshi: Showa zenki-hen.* Tokyo: Nihon kōkū kyōkai, 1975.

Nihon kokuyū tetsudō. *Kokutesu rekishi jiten.* Tokyo: Nihon kokuyū tetsudō, 1973.

———. *Kokutetsu á la carte.* Tokyo: Nihon kokuyū tetsudō, 1965.

———. *Tetsudō gijutsu hattatsushi.* 8 vols. Tokyo: Nihon kokuyū tetsudō, 1958–1959.

———. *Business tokkyū densha* (1958).

———. *Tetsudō 80-nen no Ayumi, 1872–1952.* Tokyo: Nihon kokuyū tetsudō, 1952.

———. *Tetsudō sengo shorishi.* Tokyo: Taishō shuppan, 1981.

Nihon kokuyū tetsudō Shinkansen sōkyoku. *Shinkansen: Sono 20-nen no kiseki.* Tokyo: Ni-
    hon kokuyū tetsudō Shinkansen sōkyoku, 1984.

Nihon kokuyū tetsudō tetsudō gijutsu kenkyūjo. *Tokaido Shinkansen ni kansuru kenkyū.*
    Tokyo: Nihon kokuyū tetsudō tetsudō gijutsu kenkyūjo, 1962.

———. *Tokaido Shinkansen ni kansuru kenkyū.* Tokyo: Nihon kokuyū tetsudō tetsudō gi-
    jutsu kenkyūjo, 1964.

Nihon kokuyū tetsudō tetsudō gijutsu kenkyūjo 50nen-shi kankō iinkai. *Nihon kokuyū
    tetsudō tetsudō gijutsu kenkyūjo 50nen-shi.* Tokyo: Kenyūsha, 1957.

Nihon kokuyū tetsudō-hen. *Nihon kokuyū tetsudō 100-nenshi.* 19 vols. Tokyo: Nihon kokuyū
    tetsudō, 1969–1974.

Nihon Olympic Iinkai. "Tokyo Olympic 1964." Accessed 8 November 2011. http://www.joc
.or.jp/past_games/tokyo1964.

Nihon sangyō gijutsushi gakkaihen. *Nihon sangyōshi jiten*. Kyoto: Shibunkaku shuppan,
2007.

Nishii, Kazuo, ed. *Shōwa-shi zenkiroku*. Tokyo: Mainichi shinbunsha, 1989.

Nishikawa Akiji no omoide henshū iinkai. *Nishikawa Akiji no omoide*. Nagoya, 1964.

Nishitani, Tetsu. "Tokaidō-sen ressha sokudo no hensen." *Kōtsū gijutsu* 126 (1956): 17–19.

Noma, Sawako, ed. *Shōwa 20,000 nichi no zenkiroku*. Vols. 7 and 8. Tokyo: Kōdahsha, 1989.

Odakyū dentetsu kabushiki kaisha. *Super Express 3000*. 1957.

Odakyū dentetsu. "Kaisha gaiyō." Accessed 24 November 2011. http://www.odakyu.jp
/company/about/outline.

———. "Odakyū 80-nenshi." Accessed 20 November 2011. http://www.odakyu.jp/com
pany/history80/01.html.

"Odakyū SE-sha kōsoku shiken kara 20-nen." *Denkisha no kagaku* 30, no. 11 (October 1977):
15–20.

Ōjibōbō: Mitsubishi jūkō Nagoya 50-nen no kaiko. Vol. 3. Nagoya: Ryōkōkai, 1970.

Okamura, Jun, ed. *Kōkūgijutsu no zenbō*. Vol. 1. Tokyo: Kōyōsha, 1953.

Ōkubo, Toshikane, et al., eds. *Kindaishi shiryō*. Tokyo: Kikkawa kōbōkan, 1965.

Onda, Shigetaka. *Tokkō*. Tokyo: Kōdansha, 1991.

Osaka daigaku 25-nen shi. Osaka: Osaka daigaku, 1956.

Ozaki, Toshio. "Ōka 43-otsugata sekkei no omoide." *Kōkūfan* 13, no. 14 (1964): 68–73.

Rokuda, Noboru and NHK Project X seisakuhan. *Project X chōsenshatachi: Shūnen ga unda
Shinkansen*. Tokyo: Nihonhōsō kyōkai, 2002.

"Ryūsenkei sharyō mokei no fūdō shiken seiseki ni tsuite." *Gyōmu kenkyū shiryō* 25, no.
2 (1937): 1–34.

Sagawa, Shun'ichi. "Ressha tono tsūshin I." *Japan Railway Engineer's Association* 5, no.
1(1962): 50–53.

Saitō, Michinori. *Chōhō sen: Rikugun noborito kenkyūjo*. Tokyo: Gakushū kenkyūsha, 2001.

Sasagawa, Yōhei. "Linear Shinkansen to Kyōtani Yoshihiro." *Nihon zaidan kaichō Sasagawa
Yōhei blog* (blog). 27 February 2009. http://blog.canpan.info/sasakawa/archive/1807.

Satō, Yasushi. "*Dainiji sekai taisen zengo no kokutetsu gijutsu bunka*." *Kagakushi Kenkyū* 46
(2007): 209–219.

Satō, Yoshihiko. "Sekai ginko ni yoru Tokaido Shinkansen Project no hyōka." *Tetsudō
shigaku* 19 (2001): 69–80.

Sawai, Minoru. "Daigaku (senzenki)." In *Nihon sangyō gijutsushi jiten*. Kyoto: Nihon shibunkaku shuppan, 2007.

———. "Senjiki nihon teikoku ni okeru gijutsusha kyōkyū." In *Kindai higashi Asia keizei
no shiteki kōzō: Higashi Asia shihon shugi keiseishi*. Tokyo: Nihon hyōronsha, 2007.

Sawano, Shūichi. "Shinkansen no sharyō." *JREA* 2, no. 1 (1959): 14–16.

"SE-sha (kyōki sekai kiroku) 20-shūnen kinen zadankai." *Denkisha no kagaku* 31, no. 1 (July
1978): 21–29.

Sekkeisha no shōgen. 2 vols. Tokyo: Suitōsha, 1994.

Sentanken tankendan. *Sentanken tankendan dai 3-kai hōkoku*. Tokyo: Sentanken tankendan, 1997.

Shiba, Chūsaburō. "Kōkūkenkyūjo no jigyō ni tsuite." *Kikai gakkaishi* 27, no. 89 (1924):
813–822.

Shima, Hideo. *D-51 kara Shinkansen made: Gijutsusha no mita kokutetsu.* Tokyo: Nihon keizai shinbunsha, 1977.

Shima Hideo ikōshū henshū iinkai-hen. *Shima Hideo ikōshū: 20-seiki tetsudōshi no shōgen.* Tokyo: Nihon tetsudō gijutsu kyōkai, 2000.

Shima, Takashi and Masao Tani. "Shinkansen sharyō no haishō sōchi." *JREA* 7, no. 4 (1964): 45–48.

Shima, Yasujirō. "Tetsudō kikan no sunpō to sharyō no kidō ni taisuru atsuryoku no kankei." *Kikai gakkaishi* 28, no. 95 (1925): 129–136.

"Shinkansen kakū densha senro kenkyū hōkoku." *Tetsudō gyōmu kenkyū shiryō* 2, no. 11 (1943): 2–3.

Shinkansen 10-nen shi. Tokyo: Nihon kokuyū tetsudō shinkansen sōkyoku, 1975. Shinohara, Takeshi. *Omoide no ki.* Tokyo: Kenyūsha, 1994.

Shinohara, Takeshi and Hideshige Takaguchi. *Shinkansen hatsuansha no hitorigoto: Moto Nihon tetsudō kensetsu kōdan sōsai Shinohara Takeshi no network-gata Shinkansen no kōsō.* Tokyo: Pan research shuppan, 1992.

Shinohara, Yasushi and Hiroyuki Kimoto. "Osaka-Himeji kan S.H.F. no sekkei to shaken kekka," *Kōtsū gijutsu* 99 (1954): 32–35.

Shōwa 16 · 17-nen Kōkūnenkan. Tokyo: Dainihon higō kyōkai, 1943.

Sogō, Shinji. "Shingijutsu to speed up." *JREA* 2, no. 1 (1959): 1.

Sōmushō. "Ippan sensai homepage." Accessed 14 September 2011. http://www.soumu .go.jp/main_sosiki/daijinkanbou/sensai/index.html.

Subaru Museum. "Saishō gen no tuning ni yoru chōsen: Subaru hakubutsukan." Accessed 18 February 2011. http://members.subaru.jp/know/museum/subaru360.

Suga, Miya. "Beikoku 1952-nen imin kikahō to nihon ni okeru 'imin mondai' kan no henyō." *Tokyo gakugei daigaku kiyō jinbun shakai kagaku-kei II,* 61 (2010): 127–141.

Sumida, Shunsuke. *Sekai no kōsoku tetsudō to speed up.* Tokyo: Nihon tetsudō tosho, 1994.

Takei, Akemichi. "Hongoku tetsudō ressha sokudo no hattatsu." *Kikai gakkaishi* 41, no. 251 (1938): 113–119.

Tanaka, Hisashi. "Sōritsuki no fujōshiki tetsudō 1: Fujōshiki tetsudō no reimeiki." *RRR* 58, no. 1 (2001): 26–27.

Tanaka, Shin'ichi. "Shinkansen sharyō: sono kaihatsu no zengo." *Denkisha gakkai kenkyūkai shiryō* (2002).

———. "Sharyō." *Tetsudō gijutsu* 42, no. 1 (1985): 18–21.

Tani, Masao. "Shinkansen sharyō no kimitsu." *JREA* 7, no. 7 (1964): 5–7.

Tani, Seiichirō. "Bōfu makuragi ni tsuite." *Kōtsū gijutsu* 64 (1951): 30–31.

Tetsudō gijutsu kenkyūjo. *10-nen no ayumi: Sōritsu 60-shūnen.* Tokyo: Tetsudō gijutsu kenkyūjo, 1967.

———. *Tetsudō gijutsu kenkyūjo sōritsu 70-shūnen: 10-nen no ayumi.* Tokyo: Tetsudō gijutsu kenkyūjo, 1977.

———. *Tōkaidō Shinkansen ni kansuru kenkyū: sōron.* Tokyo: Japan National Railways, 1960.

"Tokyo-Osaka 4-jikan 45-fun Tōkaidōsen dangan ressha kakū shijōki." *Popular Science* 2, no. 1 (January 1954): 69–74.

Tokyo daigaku 100-nenshi henshū iinkai. *Tokyo daigaku 100-nenshi.* 10 vols. Tokyo: Tokyo daigaku, 1987.

Tokyo daigaku kōgakubu senpaku kōgakka. *Senpaku kōgakka no 100-nen.* Tokyo: Tokyo daigaku, 1983.

Tokyo daigaku seisan gijutsu kenkyūjo-hen. *Tokyo daigaku dai-2 kōgakubushi.* Tokyo: Tokyo daigaku seisan gijutsu kenkyūjo, 1968.

Tokyo daigakushi shiryōshitu. *Tokyo daigaku no gakuto dōin, gakuto shutsujin.* Tokyo: Tokyo daigaku, 1997.

Tokyo Kogyōdaigaku 90-nen shi. Tokyo: Zaikai hyōronshinsha, 1975.

Tokyo teikoku daigaku gakujutsu taikan kankōkai. *Tokyo teikoku daigaku gakujutsu taikan kōgakubu kōkū kenkyūjo hen.* Tokyo: Tokyo teikoku daigaku, 1944.

Tōkyū sharyō seizō kabushiki kaisha. *Tōkyū sharyō 30-nen no ayumi.* Yokohama: Tōkyū sharyō seizō kabushiki kaisha, 1978.

Tomita, Tetsuo. "Gijutsu no juyō ni oyobosu shijōkōzō oyobi fūdo kankyō ni kansuru jisshōteki bunseki." Ph. D diss., Tokyo Institute of Technology, 1999.

Tomizuka, Kiyoshi. *80-nen no shōgai no kiroku.* Tokyo: Tomizuka Kiyoshi, 1975.

———. *Kōkenki.* Tokyo: Miki shobō, 1998.

———. *Meiji umare no waga oitachi.* Tokyo: Tomizuka Kiyoshi, 1977.

———. *Showa umare no waga oitachi.* Tokyo: Tomizuka Kiyoshi, 1977.

Toyota Motors. "*Corolla no tetsugaku.*" Accessed 22 September 2010. http://toyota.jp/infor jmation/philosophy/corolla/history/index.html.

"Tsunawatari no Shinkansen," *Furuse Yukihiro no off side 2001*, no. 19, 17 August 2001, http://www.honya.co.jp/contents/offside/index.cgi?20010817.

Ubukata, Yoshio and Hisashi Morokawa. *Odakyū Romance Car monogatari.* Osaka: Hoikusha, 1994.

Ubukata, Yoshio. *Odakyū monogatari.* Kawasaki: Tamagawa shinbunsha, 2000.

Uchihashi, Katsuto. *Zoku zoku takumi no jidai: Kokutetsu gijutsujin zero hyōshiki kara no nugai tabi.* Tokyo: Sankei shuppan, 1979.

Uchimaru, Saiichirō. "Jiron: Tokyo teikoku daigaku kikai kōgakuka ni okeru kyōiku no genjō." *Kikai gakkaishi* 40, no. 237 (January 1937): 1–2.

Ujike, Yasuhiro. "Kyū nihongun ni okeru bunkan nado no ninyō ni tsuite: hannin bunkan o chūshin ni." *Bōei kenkyūjo kiyō* 8, no. 2 (2006): 69–85.

Un'yu gijutsu kenkyūjo. *10-nen shi.* Tokyo: Transportation Technology Research Center, 1960.

Un'yushō. *Kokuyū tetsudōno genjō: Kokuyū tetsudō jissō hōkokusho.* Tokyo, 1947.

———. *Kokuyū tetsudō no fukkō: Tetsudō 75-nen kinen shuppan daiisshū.* 1948.

Uno, Hiroshi, ed., *Asahi shinbun ni miru Nihon no ayumi: Shōdo ni kizuku minshu shugi.* Vols. 2 and 3. Tokyo: Asahi shinbunsha, 1973.

Unoki, Jūzō. "Zoku keiryō kyakusha sonogo." *Kōtsū gijutsu* 136 (1957): 306–310.

Waseda daigaku daigakushi henshūjo. *Waseda daigaku 100-nen shi.* Tokyo: Waseda daigaku, 1987.

Yamamoto, Risaburō. "Kōsoku kansetsu densha SE ni tsuite." *Sharyō gijutsu* 40 (1958): 311–317.

Yamana, Masao. "Suisei ga dekirumade." *Kōkfan*13, no. 15 (1964): 74–78.

Yamanouchi, Shūichirō. *Shinkansen ga nakattara.* Tokyo: Asahi bunko, 2004.

Yamaoka, Shigeki. "Mitsubishi ZC707 Chijō ni orita engine." *Tetsudōshigaku* 11 (December 1992): 7–13.

Yanagida, Kunio. *Zerosen moyu.* Vol. 5. Tokyo: Bungei shunjū, 1990.

———. *Reishiki sentōki.* Tokyo: Bungei shunjū, 1977.

Yang, Daqing. "Chū goku ni todomaru nihon gijutsusha: Seiji to gijutsu no aida." In *1945-*

nen no rekishi ninshiki: Shūsen o meguru nicchū taiwa no kokoromi. Tokyo: Tokyo Univ. Press, 2009.

Yasuda, Akio, et al. "Shinkansen no kensetsu kijun." *JREA* 2, no. 1 (1959): 5–14.

Yasutomi, Shigeyoshi. "'Sensō hanayome' to Nikkei community (III) stereotypes ni moto-zuku haiseki kara juyō e." *Kaetsu daigaku kenkyū ronshū* 44, no. 2 (2002): 55–82.

Yokobori, Shōichi. "Kyaku densha no gijutsu teki mondai ten." *Kōtsū gijutsu* 76 (1952): 438–441.

Yokosukashi, ed. *Senryōka no Yokosuka: Rengō kokugun no jōriku to sono jidai.* Yokosuka: Yokosukashi, 2005.

Yokosukashi, ed. *Yokosukashishi.* Vol. 1. Yokosuka: Yokosukashi, 1988.

Yokosukashishi hensan iinkai. *Yokosukashishi.* Yokosuka: Yokosuka shiyakusho, 1957.

Zasshi maru henshūbu, ed. *Ginga/Ichishiki rikkō.* Tokyo: Kōjinsha, 1994.

———. *Suisei/99-kanbaku.* Tokyo: Kōjinsha, 2000.

Published Source: Spanish

Ferro Carriles: Catátalogo De Sellas Temáticos. Barcelona: DOMFIL, 2001.